Agroecological Innovations

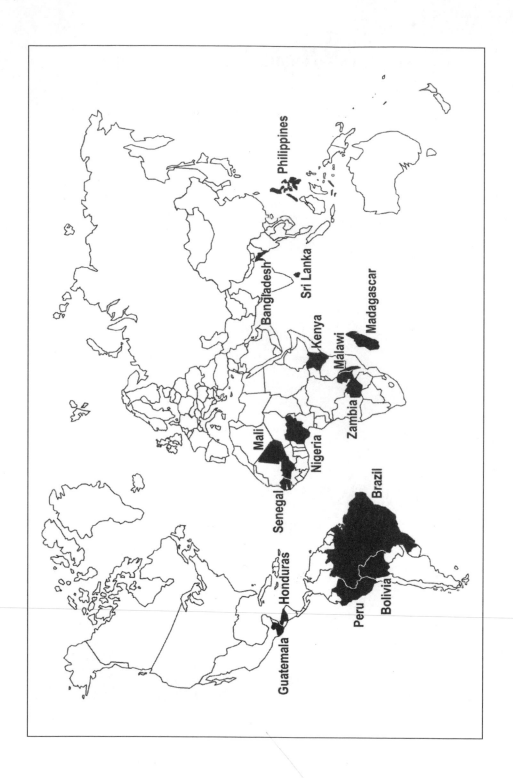

Philippines

Bangladesh
Sri Lanka

Madagascar

Kenya
Malawi

Zambia

Mali

Nigeria

Senegal

Brazil

Honduras

Guatemala

Peru
Bolivia

Agroecological Innovations

Increasing Food Production with Participatory Development

Edited by

Norman Uphoff

EARTHSCAN

Earthscan Publications Ltd
London • Sterling, VA

First published in the UK and USA in 2002 by
Earthscan Publications Ltd

Copyright © Norman Uphoff, 2002

ISBN: 1 85383 857 8 paperback
       1 85383 856 X hardback

Typesetting by PCS Mapping & DTP, Newcastle upon Tyne
Printed and bound by Creative Print and Design (Wales), Ebbw Vale
Cover design by Danny Gillespie

For a full list of publications please contact:
Earthscan Publications Ltd
120 Pentonville Road
London, N1 9JN, UK
Tel: +44 (0)20 7278 0433
Fax: +44 (0)20 7278 1142
Email: earthinfo@earthscan.co.uk
http://www.earthscan.co.uk

22883 Quicksilver Drive, Sterling, VA 20166–2012, USA

A catalogue record for this book is available from the British Library

Library of Congress Cataloging-in-Publication Data

Agroecological innovations : increasing food production with participatory
development / edited by Norman Uphoff.
       p. cm.
   Includes bibliographical references (p.).
   ISBN 1-85383-857-8 (pbk.) – ISBN 1-85383-856-X (hardback)
    1. Agricultural innovations. 2. Agricultural ecology. 3. Agricultural productivity. I.
Uphoff, Norman Thomas.

   S494.5.I5 A329 2002
   338.1'6--dc21

                                                                    2001007058

Earthscan is an editorially independent subsidiary of Kogan Page Ltd and publishes in
association with WWF-UK and the International Institute for Environment and
Development

This book is printed on elemental chlorine-free paper

Contents

PART 3 ADVANCING AGROECOLOGICAL AGRICULTURE WITH PARTICIPATORY PRACTICES

List of Tables

List of Figures

List of Contributors

Miguel Altieri. Professor of Environmental Science, Policy and Management, University of California, Berkeley; general coordinator of Sustainable Agriculture Networking and Extension (SANE), UNDP, and technical advisor, Latin American Consortium on Ecology and Development (CLADES).

Marco Barzman. Research associate, Sustainable Agriculture Research and Education programme, University of California, Davis; former project coordinator of the New Options for Pest Management (NO-PEST) Project, CARE/Bangladesh, Dhaka.

Randall Brummett. Regional research coordinator for Africa, for the International Center for Living Aquatic Resource Management (ICLARM), presently in Cameroon.

Roland Bunch. Coordinator for COSECHA (Consultants in People-centred Eco-agriculture), a non-governmental organization (NGO) in Tegucigalpa, Honduras; former Central American regional representative for the international NGO World Neighbours.

Ademir Calegari. Senior soil researcher, Agronomy Institute, Londrina, Paraná, Brazil.

Sylvie Desilles. Former project coordinator for the New Options for Pest Management (NO-PEST) Project, CARE/Bangladesh, Dhaka.

Amadou Makhtar Diop. Technical director of the Sustainable Agriculture Program, Rodale Institute, Kutztown, Pennsylvania; former director of Rodale programme in Senegal.

Erick Fernandes. Assistant professor of Crop and Soil Sciences, co-leader of the African Food Systems and Natural Resource Management Initiative, Cornell University, and former coordinator of the Alternatives to Slash-and-Burn Network, International Centre for Research in Agroforestry (ICRAF), Nairobi, Kenya.

Mamby Fofana. Project director for the Unitarian Service Committee of Canada in Bamako, Mali.

Dennis Garrity. Director-general of the International Centre for Research in Agroforestry (ICRAF); former coordinator of ICRAF's Southeast Asia regional programme, and previously head of the Farming Systems Programme, International Rice Research Institute (IRRI), Los Baños, Philippines.

Keith Jones. Natural Resources Institute, University of Greenwich, UK, former consultant for CARE integrated pest management (IPM) programme in Sri Lanka.

Peter Kenmore. Coordinator of the Global Integrated Pest Management (IPM) Faculty, UN Food and Agriculture Organization (FAO), Rome.

Arie Kuyvenhoven. Head of Department of Development Economics, and co-coordinator, Sustainable Agriculture Group, Wageningen University, Wageningen, The Netherlands.

Alice Pell. Professor of Animal Science, and co-leader of the African Food Systems and Natural Resource Management Initiative, Cornell University.

Alison Power. Associate professor of Science and Technology Studies, and former director of Agricultural Ecosystems Program, Cornell University; member of National Academy of Science's Committee on Sustainable Agriculture in the Humid Tropics.

Jules Pretty. Director of the Centre for Environment and Society, University of Essex, UK, and former director of the Sustainable Agriculture Programme, International Institute for Environment and Development (IIED), London, UK.

Ruerd Ruben. Associate professor of Development Economics, Department of Social Sciences, Wageningen University, Wageningen, The Netherlands.

Edward Ruddell. Formerly Andean regional representative for World Neighbours.

Pedro Sanchez. Former director-general of the International Centre for Research in Agroforestry (ICRAF), Nairobi, Kenya; professor emeritus of Soil Science and Forestry, North Carolina State University.

Lori Ann Thrupp. Director for Sustainable Agriculture, World Resources Institute (WRI), Washington, DC.

Mary Tiffen. Consultant, Drylands Research, Crewkerne, Somerset, UK; former senior research fellow, Overseas Development Institute (ODI), London.

Norman Uphoff. Director, Cornell International Institute for Food, Agriculture and Development (CIIFAD); professor of Government, Cornell University.

* * * * *

Marco Barzman, Ademir Calegari and Ruerd Ruben were not able to participate in the conference at Bellagio because of limitations on the number of people that the conference centre could accommodate, however, they joined in the collaborative writing task afterwards.

Other participants in the Bellagio Conference on Sustainable Agriculture were: Pierre Crosson, senior fellow, Energy and Natural Resources Division, Resources for the Future, Washington, DC; Doug Forno, senior advisor, Department of Rural Development, World Bank; Per Pinstrup-Andersen, director-general of the International Food Policy Research Institute (IFPRI); Vernon W Ruttan, Regents professor emeritus of Applied Economics, University of Minnesota; and Jean Marc van der Weid, executive director, Assessoria e Servicos a Projetos em Agricultura Alternative (AS-PTA), Rio de Janiero, Brazil. Their papers are cited in the references, and their contributions to the group discussions are gratefully acknowledged.

Acronyms and Abbreviations

BNF	biological nitrogen fixation
CARE	an international NGO that has been promoting IPM, along with other development initiatives
CBA	cost–benefit analysis
CEC	cation exchange capacity
CGIAR	Consultative Group for International Agricultural Research
CIIFAD	Cornell International Institute for Food, Agriculture and Development
CIMMYT	International Centre for Improvement of Maize and Wheat
COSECHA	Associaciòn de Consejeros una Agricultura Sostenible, Ecològia y Humana, an NGO in Honduras
DANIDA	Danish International Development Agency
DFID	Department for International Development (formerly ODA), UK
EC	European Community
EMBRAPA	Brazilian National Agency for Agricultural Research
ESSA	Faculty of Agriculture, University of Antananarivo, Madagascar
FAO	Food and Agriculture Organization of the United Nations
FFS	farmer field-schools, established in FAO-supported IPM programmes
FHM	farm household modelling
FSRP	Farmer–Scientist Research Partnership, developed by ICLARM
GMCC	green manures and cover crops
GNP	gross national product
GTZ	Gesellschaft für technische Zusammenarbeit (German Agency for Development Cooperation)
ha	hectare (2.54 acres)
IAPAR	Agronomic Research Institute of Paraná, Brazil
ICM	integrated crop management (broader than, but including, IPM)
ICLARM	International Center for Living Aquatic Resource Management
ICRAF	International Centre for Research in Agroforestry
IFPRI	International Food Policy Research Institute
IIED	International Institute for Environment and Development, London
IMF	International Monetary Fund
IPM	integrated pest management
IRRI	International Rice Research Institute
ISRA	Senegalese Institute for Agricultural Research
LTTE	Liberation Tigers for Tamil Eelam, Sri Lanka

MCA	multi-criteria analysis
N	nitrogen
NGO	non-governmental organization
NO_3	nitrate
NPK	chemical fertilizer containing nitrogen, phosphorous and potassium
NRM	natural resource management
NVS	natural vegetative strips
ODA	Overseas Development Administration (now DFID), UK
OFC	other field crops
OXFAM	an international NGO supporting grassroots development efforts
P_2O_5	phosphate
P	phosphorous
PFA	production function analysis
R	rupees (Sri Lanka)
RARC	Regenerative Agriculture Research Centre, Senegal
SAI	sustainable agricultural intensification
SALT	sloping agricultural land technology, developed in the Philippines
SRI	a system of rice intensification (developed in Madagascar)
SWC	soil and water conservation
T&V	Training and Visit System (Malawi)
TAC	Technical Advisory Committee (CGIAR)
t/ha	tons per hectare (metric)
USAID	United States Agency for International Development
USCC	Unitarian Service Committee of Canada

Introduction

By the middle of the 21st century, world food production will need to be at least twice what it is now if we are to meet both economic demand and human needs. Failure to achieve this increase will slow economic growth and add to the presently unacceptable levels of poverty, hunger and disease. Thus both the rich and the poor, and everyone in between, have a stake in the continued expansion of food production around the world – in ways that do not (further) degrade our natural resource base. While having adequate food supply is not a sufficient condition to ensure food security and economic prosperity, it is a necessary one.

Doubling food production will be a difficult task, with at least one-third less land available per capita by 2050, even with reduced rates of population growth. The supplies of water available for agriculture will probably be reduced even more, and neither crops nor livestock can survive without adequate water. Moreover, present methods of agricultural production are contributing to environmental pollution through toxic agrochemicals and inorganic fertilizer runoff and infiltration. These methods are very dependent on fossil fuels and other forms of energy whose prices and supplies are likely to be less favourable several decades from now.

Certain technological changes could make agrochemicals more benign, and other forms of energy more widely available. There are currently high hopes that biotechnology can raise yields substantially through genetic modification. Agriculture has been one of the most progressive sectors of the world economy in technological terms. However, many millions of farmers, indeed the majority worldwide, have not been able to take advantage of these new opportunities, because of cost and other constraints. In some areas, indeed, new technologies have led to displacement and increased poverty for rural households. So technological change will not necessarily bring greater food security.

At a conference on the future of the world's food supply organized by the Keystone Center and held at Airlie House in rural Virginia in March 1997, some participants asked whether we should be trying to meet food needs entirely by projecting the present strategies for agricultural research and development indefinitely into the future. Professionals from half a dozen disciplines were not convinced that expanding production along the present technological trajectory – 'doing more of the same' – would ensure food security in ways that are environmentally acceptable and socially desirable, or maybe even economically sustainable.

Proponents of agroecological approaches argued that these could contribute significantly to meeting world food needs in ways that support rather than degrade the environment, though no quantification is possible at present. They suggested, further, that these approaches could enhance human resources by improving people's management capacities at the same time as redressing disparities in distribution, because agroecological methods are well suited for use by less-favoured households.

Participants willing to rely on present approaches cited the Green Revolution's success in doubling world grain output over a 30-year period as a precedent for expansion of production using mainstream technologies. But it was acknowledged that the rates of agricultural growth and technological advance with this high-input strategy have slowed during the 1990s. The gap between farmers' best production attainments and what scientists can achieve on their experiment stations has been narrowing as farmers catch up with researchers. Moreover, better human nutrition, a more important goal for agriculture than food production alone, will not be achieved simply by greater output of grains. While total caloric consumption is the main determinant of nutritional status, there is increasing concern about essential micronutrients, less available in grains.

Critics of alternative approaches maintain that agriculture without modern inputs must necessarily produce low outputs, contributing to food shortages and creating pressures to expand the area under cultivation. They credited the Green Revolution's advances in land productivity with having saved many millions of hectares of forest, and consequently with having preserved more biodiversity than alternative production approaches could. Supporters of alternative agricultural approaches, on the other hand, pointed to the environmental costs resulting from today's high-external-input agriculture. At the same time, they rejected the assumption that agroecologically-based systems must be less productive than 'modern' technologies are, citing some impressive cases where alternative methods were doubling or tripling yields, including those of staple crops.

A discussion, involving Robert Herdt, at the time the agricultural sciences director of the Rockefeller Foundation; Miguel Altieri, a leading contributor to the agroecological literature from the University of California, Berkeley; and Norman Uphoff, director of the Cornell International Institute for Food, Agriculture and Development (CIIFAD), led to ideas for a follow-up conference. This would bring together people who had experience with agroecological approaches to consider what kind of case could be made for devoting more attention and investment to these alternatives, together with others who could help evaluate these ideas critically.

It was not assumed that so-called 'alternative agriculture' should or could replace present modes of 'modern' agriculture, or that it can meet all world food needs by itself. Indeed, it is misleading to talk about 'alternative agriculture' because the concept encompasses a great variety of practices and techniques. Some of these are new while others are based on age-old principles and practices that have been little studied. It was agreed that this is a subject about which more should be known empirically, and for which there is a need to begin establishing some theoretical bases.

The Rockefeller Foundation accepted a proposal from Altieri and Uphoff to hold an international conference on this subject at its conference centre in

Bellagio, Italy. CIIFAD provided administrative and some financial support for the conference, held on 26–30 April 1999, and the World Bank's Rural Development Department made a grant for travel support available to some of the participants from developing countries.[1]

The Bellagio centre offers an incomparably fine and congenial setting for focused and fruitful discussion. While there were a few heated exchanges, most of the sharing of ideas and experience was amicable and productive. The basic and shared concern was: what will benefit people? – especially the poor and marginalized, and urban consumers as well as rural producers – and at the same time: what will sustain the natural resource base on which agriculture and indeed all human and other life depends? Being able to achieve higher levels of production biophysically is very important but not sufficient, since alternative practices need to be economically efficient and profitable for households, as well as socially and nutritionally beneficial.

There was too little systematic data to draw firm conclusions or make confident generalizations. Only a tiny fraction of the resources put into mainstream agricultural development have thus far been invested in agroecological approaches. But the case studies provided evidence of impressive possibilities for increasing production using mostly local resources and knowledge that would probably stand the test of rigorous economic cost–benefit analysis. The adoption rate among farmers was taken as a practical test of the economics of new approaches, since this reflected their net benefits from innovation.

The case studies focused disproportionately on African experiences, partly because some of the most innovative work is going on there, but mostly because this is the global region in which food shortages are most likely to be severe in the decades ahead. If agroecological approaches can raise food production under such adverse soil and water conditions, they will accomplish gains where conventional modern agricultural methods have largely failed over the past 40 years. The cases from Latin America and Asia were different from, but consistent with, what is being learned from Africa. We were not looking for specific technologies to be extended to other countries, because local conditions always require adaptation and often different solutions. Rather, we sought principles and practices that can be applied, with appropriate adjustments, to a wide variety of circumstances. A good number of such principles and practices were identified.

By the end of the conference, there was enough agreement among participants that a book was planned to share more widely the learning gained from the papers and discussions. A number of the papers presenting some of the more technical aspects of the subject were edited by Altieri for a special issue

---

1 In addition, the International Centre for Research in Agroforestry (ICRAF), the International Center for Living Aquatic Resources Management (ICLARM), the International Food Policy Research Institute (IFPRI), the Plant Protection Service of the Food and Agriculture Organization (FAO), Resources for the Future in Washington DC, and the University of Essex in the UK, covered the travel costs of participants from their institutions.

of *Environment, Development and Sustainability* (vol 1, nos 3–4, 1999). Uphoff undertook to work with contributors to prepare an integrated volume that assesses both the biophysical and socioeconomic dimensions of this large subject.

Because the 24 participants formed a diverse set of professionals, representing more than a dozen nationalities and coming from non-governmental organizations (NGOs), universities, research institutions and other international organizations in 13 countries, this book speaks with many different voices. It tells many different stories. But all contribute to a better understanding of the twin themes of agroecology and participation. At the conference, people with biological training often found themselves dwelling on sociological learning processes, while social scientists engaged themselves with subjects like plant physiology and soil dynamics. The interdisciplinary nature of the group helped members to gain many new and broader insights into their own work.

The editing process has sought to integrate knowledge across cases and topics, but without homogenizing language and style. To have created a single voice would have done an injustice to the diversity of disciplinary and personal perspectives brought together. The varied voices also make for more interesting reading.

The contributors to this volume are and will continue to be engaged with various aspects of agroecological and participatory development. We hope that readers will find enough merit and challenge in the following chapters that the number of people working on the empirical and theoretical dimensions of this subject will grow greatly in the future. Whether or not analysts and critics become more actively and systematically engaged with this subject, it is already taking root in many ways in many countries. In particular, many NGOs are getting more involved with these new/old approaches.

The question that cannot be answered at present is whether enough people in the scientific establishment, donor agencies and government circles will conclude that this subject area now warrants serious investigation, and will support the move of so-called 'alternative' approaches from the margin of agricultural research and development towards the centre of future strategies to deal with food security and economic development in the 21st century.

Norman Uphoff
Cornell International Institute for
Food, Agriculture and Development, Ithaca, NY

Part 1

Issues for Analysis and Evaluation

Chapter 1

The Agricultural Development Challenges We Face

Norman Uphoff

The agricultural technologies that were developed and extended over the past four decades have contributed to unprecedented growth in world food production. The doubling of grain output globally between 1965 and 1990 was a remarkable achievement that drew on the skills and innovations of thousands of scientists and extensionists and millions of farmers, backed by the supportive decisions of policy-makers. Without what is now referred to as 'the Green Revolution', there would be large food deficits in the world today, with adverse environmental impacts from having to bring extensive areas of less suitable land under cultivation (Crosson and Anderson, 1999).

But there are many and growing concerns that this strategy of agricultural development may not be the best or the only one to promote in the future, since it has costs as well as benefits. Conway (1997) has framed the issue in terms of needing in the future 'a doubly green revolution', one that reverses environmental deterioration at the same time that it augments the supply of food. It should also ensure that the food produced meets the nutritional, economic and social needs of the millions of people who are hungry and malnourished, presently numbering about 800 million (Pinstrup-Andersen and Cohen, 1999). The aim of the agricultural enterprise should be to produce secure and healthy people, not just food.

Can future world food needs be met by more of the same kinds of investments supported in past decades through research, extension, infrastructure and policies? Should we be continuing mostly or exclusively with what has become the predominant approach to agriculture? Or should policy-makers, scientists and producers be looking also for other approaches that could increase food supplies in ways that are more environmentally sustainable, more economically efficient in terms of total factor productivity and more socially just? The dominant approach has relied heavily on the use of agrochemicals to increase available soil nutrients and to protect crops and animals from insects, pathogens and weeds, as well as on genetic changes and on mechanization and inanimate energy sources. Alternative approaches, as discussed in Chapter 2, rely more on the combination and interaction of biological processes that can be explained and utilized in agroecological terms.

That food production will need to be increased substantially in the future is not in dispute. Greater public and private investments in agricultural research and extension are clearly justified and urgently needed, considering the lengthy lag times before new practices are widely adopted and fully exploited. The experiences from Africa, Latin America and Asia reported here suggest that a greater share of research and extension efforts – and a growing share to the extent that it is justified by results – should focus on approaches grounded in *agroecological* concepts and concerns. In this process, farmers should be involved actively as *partners* with scientists and extensionists for devising, testing and evaluating new practices, not just adopting them.[1]

Alternative kinds of agriculture are often not new. Frequently they draw on traditional knowledge and practices, although they are increasingly supported by scientific explanations. With appropriate development and application, we find that they can offer opportunities to increase food production not just by increments but sometimes by multiples. The case studies presented in Part 2 show how new and better combinations of plant, soil, water and nutrient management practices, combined with livestock and/or fish in intensified farming systems and protected by integrated pest management (IPM), are achieving production increases of 50 to 100 per cent, and sometimes 200 or 300 per cent – under a wide variety of conditions, and even in environments that are quite adverse. The crops grown in the cases reported include main staples such as rice, maize, beans and potatoes, and reports from researchers at international agricultural research centres show similar increases in production of wheat and cassava utilizing practices like those considered here. The universe of experience presented here is not one of particular technologies for selected crops but rather one applying various *principles* that can capitalize more fully on existing genetic potentials.

A good example of agroecological approaches was reported recently from Yunnan province in China by Zhu et al (2000). There, crop losses were reduced and yields were raised by intercropping rice varieties that are susceptible to blast disease with non-susceptible varieties. By varying management practices to capitalize on natural disease resistance – at first on all the rice fields in five townships in 1998 and then in ten townships in 1999 – blast disease was reduced by 94 per cent compared with rice grown in monoculture. The yield from otherwise-susceptible rice varieties was raised by 89 per cent. Reduction in disease was so successful that after two years, farmers no longer used fungicidal sprays, and in 2000, the method was being used on 40,000ha.

Another recent example comes from Kenya, where serious food crop losses are caused by stem-boring insects and by the parasitic witchweed *striga*, which strangles other plants' roots. A push–pull strategy has been devised in which maize and sorghum crops are interplanted with sudan grass, napier grass and molasses grass, plus two legume species (silverleaf, *Desmodium uncinatum*, and greenleaf, *D. intortum*). The first two plants function as trap crops, drawing insect pests away from the maize and sorghum, while the latter three repel borers and keep them from laying eggs on the crops. For reasons not fully understood (possible allelopathic affects), this combination of plants sharply reduces the incidence of striga (Khan et al, 2000). In controlled trials, statistically significant increases in yield of about 2 tons of maize per hectare are reported. Well-managed farms using this system in 1999 got 7 to 8 tons per hectare (t/ha), compared with 4 to 5 tons on control plots. This strategy has the extra value of producing needed forage for livestock, something that more costly chemical means of control do not provide.

A growing number of such advances in knowledge and practice should satisfy governments, researchers and donor agencies, as well as farmers, that there are many promising agroecological opportunities worth investigating and supporting. Indeed, taking these alternatives seriously – refining, adapting and disseminating them – may determine how successfully the people of this world can meet their needs for food in the future, and at the same time have livable natural and social environments in the century ahead.

THE SITUATION

Projections differ as to exactly when in this century food producers around the world will need to be producing twice the present level of agricultural output to meet the requirements of a larger and, everyone hopes, more prosperous population. Such calculations should always factor in the currently large unmet food needs that a humane world will not continue tolerating indefinitely. Achieving major increases in production should be a worldwide objective, although these by themselves will not assure food security for all. It is less important to know whether food supply needs to be doubled by 2030 or by 2050 than to have sound ideas of how to accomplish the immense task of eliminating food insecurity.

Technological Contributions

Concern with the future world food situation is sometimes dismissed by pointing to the doubling of cereal grain production achieved over recent decades. This remarkable acceleration in food production combined higher-yielding varieties with increased use of irrigation water, fertilizer and other agrochemicals. It is questionable, however, to what extent this strategy for agricultural growth will continue to suffice, particularly in less favoured areas and for poorly endowed farmers. Over the past decade, yield increases from the Green Revolution technologies have been decelerating, and in some cases even stagnating (Pingali et al, 1995; Hobbes and Morris, 1996). Production levels have been affected, as always, by shifts in weather and national policies. More grain could have been produced worldwide during the 1990s if producers had been offered higher market prices. But eliciting more production by raising food prices is not a satisfactory way to achieve food security, certainly not for the poor. Gains in productivity will be more beneficial.

The question of yield potential – how to keep moving up the maximum attainable yield – has begun to trouble many scientists and some policy-makers. The present strategy is to obtain higher yields by using ever-greater inputs of fertilizer and irrigation water, even though these inputs now have diminishing returns in many places, making their increased use less and less productive at the margin. Moreover, at high input levels, we are seeing adverse environmental impacts from production that is very chemical- and fossil-fuel-intensive.

By the middle of this century, there will be about one-third less *arable land* available per capita, and probably an even greater reduction in the availability of *water* for agricultural purposes (Smil, 2000, pp31–46).[2] The productivity of land and water and labour will need to be more than doubled if food supply is to meet demand when these essential resources become less available. Also, unless greater conservation efforts are made and are successful, there will be continuing reductions in *biodiversity*, which is an important source of genetic material for further advances in plant and animal breeding. Agricultural systems in the future will probably have to cope also with greater climatic uncertainty. Evidence of global warming continues to accumulate, but we may need to be more concerned about the effects of *weather variability*. Extreme events such as droughts, floods, storms, heatwaves and frosts are more devastating to crop production than is gradual temperature change. This would make diversity and robustness in agricultural systems more valuable traits.

Biotechnology is regarded by some as a means for achieving large future increases in agricultural production, 'to feed the world'. But major benefits from biotechnology remain still largely over the horizon, and given the incentives and the predominance of the private sector in this domain, increasing food crop yields has not been the main focus of investment in biotechnology (Ruttan, 1999). It is welcome that some attention is being given to the nutritional quality of food, such as enhancement of the vitamin A content in what is being called 'golden rice'. Its increased iron content may be even more important for human nutrition, especially of women, than are increased vitamins. Some advanced technological breakthroughs may indeed transform production possibilities in agriculture. But given the critical importance of food to human wellbeing and to maintaining dynamic economies, it is hardly advisable to put all of our agricultural eggs in the biotechnology basket. Neither the agrochemical paradigm nor the biotechnological paradigm therefore appear to be sufficient models for determining development policy and investment.

Changes in Population Growth and Demand

Some good news is that the rate of population growth is beginning to slow globally, in some places quite dramatically. For example, the average number of children born to women in Bangladesh declined from 6.2 to 3.4 in just a decade, and population growth rates are now dropping in most developing countries. But previous rapid population expansion has given the world a very young age structure, with billions of men and women now in, or entering, their most fertile years.

Demographers have scaled back their estimates of expected maximum human population from an earlier anticipated peak of 15 billion or more to between 9 and 10 billion. But even this reduced additional growth means that there will be half to two-thirds more people than are now living on Earth. Almost all of this new population will be in developing countries, a large share of whose people are poor and likely to be undernourished. Ensuring food security for them and for their descendants is a huge challenge that will be much more difficult to meet to the extent that overall food supply is not growing sufficiently (Pinstrup-Andersen and Cohen, 1999).

With population growth slowing, the strongest force increasing demand for more agricultural production will be *rising incomes*, which are desired by practically all governments and individuals. Although richer people spend smaller proportions of their income on food, in total they consume more food – and richer food, which contributes to various kinds of disease and debilitation. The changes in diet that usually accompany higher incomes will require relatively greater increases in the production of feed grains, rather than food grains, as foods of animal origin partly displace plant-based foods in people's diets. It takes two to six times more grain to produce food value through animals than to get equivalent value directly from plants.[3] It is thus quite credible to estimate that in order to meet economic and social needs within the next three to five decades, the world should be producing more than twice as much grain and agricultural products as at present, but in ways that these are accessible to the food-insecure.

Economic and Distributional Considerations

Simply increasing food supply will, of course, not assure food security to all households, communities and nations. This will depend on more equitable distributions of income and of food. Access to food is mediated by purchasing power, however this income is obtained. Thus, it is poverty, *not inadequate supply*, that is the main and most immediate cause of hunger. However, adequate food supply remains a necessary if not sufficient condition for eliminating hunger and poverty. To the extent that the poor are poorly fed, they are too weak and too prone to disease to make the most of whatever resources they do have. And being poor, they have little bargaining power, which means they get poorly compensated for their labour, regardless of their skill levels.[4]

Whenever there are food shortages, it is the consumption of the poorest that is curtailed. Figuratively, and sometimes quite literally, they stand at the end of the queue when food is distributed from the front of the line, ie, from the top of the socioeconomic hierarchy. When all available food has been given out, those people who remain in line must go hungry. So supply limitations adversely affect the poorest. Moreover, whenever demand exceeds supply, food prices go up, sometimes drastically. This reduces people's real incomes, and particularly the meagre incomes of the poor, while prices pinch to a lesser extent the incomes of the middle class.

Clearly, many complex socioeconomic and policy issues need to be resolved to reduce poverty and hunger. We do not consider increased food supply to be a cure for these societal ills. But efforts to ensure adequate food supply are justifiable in both practical and ethical terms. Those who are underfed need adequate food and good nutrition to achieve their full productive and human potentials. This will benefit not only themselves but also others in their societies.

Whenever agricultural productivity lags and food shortages ensue, those who are presently well fed will experience some deceleration in the growth of the economies on which their wellbeing depends. Resources that could be devoted to non-agricultural investments and consumption have to be diverted to meet basic food needs. Capital investments in the expansion of manufacturing and services diminish whenever the world is less able to feed all of its inhabitants because of slow growth in agricultural productivity.

THE CHALLENGE AND OPPORTUNITY

Conventional thinking holds that a doubling of food supply is possible – and in the opinions of some, will only be possible – through redoubled efforts to 'modernize' agriculture worldwide (Avery, 1995). The success of such agriculture in temperate zone areas with its mechanization of production (generally not very suitable in the tropics), its reliance on fossil fuels and agrochemicals, and its large investments of capital per worker and per hectare have created a presumption within many governments, research institutions and donor agencies that such agriculture is the best, and probably the only promising way to increase world food production. There is no single alternative approach to mainstream practices and technologies, but rather a variety of approaches. So we are not proposing any 'alternative' that could and should replace present forms of agriculture; for example, there are some important potential complementarities between different kinds of agricultural practice.[5] The designation 'sustainable agriculture', though widely used and often used by most of the contributors to this volume, is also a contestable designation because sustainability is multiply-contingent on many factors. It is not inherent in any particular practice or farming system, nor is it something that can be assessed for more than a short period of time.

Nobody can know which if any practices, let alone systems, will remain robust and profitable under any and all possible future conditions. It is easier to suggest what is likely to be *un*sustainable than to know with any certainty what will retain its productivity some or many years from now. It is thus more useful to consider practices and technologies along a *continuum* between likely-to-be-sustainable and unlikely-to-be-sustainable, rather than to categorize practices and technologies – and their proponents – into separate and opposing camps. We use the designation 'agroecological' in a descriptive way that avoids making categorical statements about 'alternative agriculture' or 'sustainable agriculture'.

Proponents of Green Revolution technologies can point to many benefits resulting from those innovations. The declining real price of cereal grains over the past three decades has contributed in a major way to increased food security around the world (Crosson and Anderson, 1999). However, these innovations have not benefited millions of households to whom these innovations are not well suited because of environmental, infrastructural, economic, social or other limitations. Moreover, these technologies – when utilized in large-scale, industrialized systems of agriculture – have led to certain environmental problems that have adverse effects for ecosystems and human health. For a number of reasons, it should not be assumed that these technologies can meet all future needs.

About 1 billion people – one-sixth of the world's population, and a much greater percentage of the poor – live and work in situations where their farming, herding or fishing operations benefit little from the agricultural technologies currently favoured by policy-makers and the research community. Factors such as landholding size, inadequate rainfall or groundwater, poor soil fertility, unfavourable topography and remoteness from markets, infrastructure and institutions make these technologies either unavailable or not appropriate. This should not surprise anyone since most modern technologies have been developed and evaluated under more favourable (rather than less favourable) conditions, using research to achieve maximum yield increases rather than to find solutions assuring secure production.

Even in some of the better-endowed areas, the sustainability of mainstream technologies is now problematic. Water depletion and soil erosion have emerged as serious problems for industrialized modes of agriculture. Falling water tables in the Indian Punjab, the North China Plains and the Great Plains of the United States, for example, could shut down 'thirsty' production practices in the decades ahead (Postel, 1997). Controls are being placed on modern agriculture to reduce chemical runoffs, residues and toxic nutrient build-ups from use of agrochemicals and chemical fertilizers, especially application of large quantities of inorganic nitrogen (Pretty, 1999).

But this book was not intended to evaluate the future potentials and limitations of Green Revolution technologies. This has been done by Conway (1999), Smil (2000) and others, and it will continue to be a subject deserving attention. Our purpose here is to consider the potentials and problems associated with various *alternatives* or *complements* to these more capital- and chemical-intensive approaches to raising agricultural production. We need systems of agriculture that utilize whatever methods are most beneficial in human and ecological terms. As in agriculture, the most productive systems are likely to be hybrids: optimizing strategies are usually mixed ones.

What we are proposing, supported by field experience presented in Part 2, is that agricultural development strategies that are less dependent on external inputs and with a skilled rather than deskilled labour force have much to offer, especially for the more marginal and vulnerable sections of the population, and should not be dismissed as inefficient or backward. Productivity should be assessed in terms of available factors of production and the multiple objectives to be served, not being assumed a priori on the basis of practices that serve best the interests of a minority of producers or on the basis of an ambiguous but popular concept like modernity.

AGROECOLOGICAL APPROACHES TO
AGRICULTURAL INNOVATION

A common feature of most approaches considered here as alternatives to mainstream production strategies is that they are based on agroecological principles and thinking, either explicitly or tacitly. They capitalize on a more systematic understanding of the processes of microbiology and nature rather than rely primarily on chemical and engineering innovations. Agroecological approaches, discussed in Chapter 3, seek to create optimum growing conditions for plants and animals not as individual specimens but as parts of larger ecosystems in which ecological services are provided and nutrients are recycled in mutually supportive ways (Carrol et al, 1989; Altieri, 1995). In particular, the soil is regarded not as a repository for production inputs or as a terrain to be exploited and mined, but rather as a living system in which micro- and macro-organisms interact with organic and mineral materials to produce environments below and above ground in which plants, animals and humans thrive.

Such alternative approaches have previously been described as low-input technologies, eg Sanchez and Benites (1987). But this designation refers mainly to the external inputs required, so these are better characterized as *low-external-input* technologies. The amount of labour, knowledge, skills and management required to make land and other factors of production most productive agroecologically is actually quite substantial. Rather than focus on what is *not* being utilized, it is better to focus on what is most essential for raising food production – labour, knowledge, skill and management.

The combination of factors of production that will be most beneficial and sustainable economically in a given situation will, of course, depend on its productivity and on market opportunities and forces. There has been a long-term trend in agriculture to use more external inputs, particularly ones based on fossil energy sources. But this trend has been driven in part by subsidies that are no longer fiscally tenable. Moreover, the level of external inputs that produces net benefits for society is starting to look different now that we begin reckoning the associated environmental and other costs. If full costs are considered, the trend is less likely to proceed to an extreme, with optimization preferred over maximization.

Agroecological approaches do not reject the use of external inputs. Considerations of productivity and availability invariably shape farmers' decisions. The latter consideration is often neglected in economic analyses, as farmers cannot benefit from technologies that are not accessible, affordable and appropriate to their conditions. Purchased inputs present special problems and risks for small and poor farmers, especially where the supply of inputs and the credit to facilitate their purchase are problematic.

Agroecological approaches are better understood as *production systems* than simply as *technologies*. In such systems, a considerable range of inputs and outputs is managed with multiple objectives in mind, and with attention paid to the status and evolution of the system as a whole, not just to the ratios between certain inputs and outputs. For either more conventional or alternative approaches, appropriate policies and institutional arrangements are usually more critical for large-scale success than are the technologies themselves, and training and other opportunities for upgrading human capacities are usually involved. As discussed in Chapters 20 and 21, the advancement of agriculture of any sort requires much more than 'getting the technology right'.

Agroecological systems are not limited to producing low outputs, as critics such as Avery (1995) have asserted. Increases in production of 50 to 100 per cent have been fairly common with alternative production methods, as seen in Part 2. In some of these systems, yields of staple crops that the poor rely on most – rice, beans, maize, cassava, potatoes, barley – are being increased under a variety of conditions by several hundred per cent, relying more on labour and know-how than on expensive purchased inputs, by capitalizing on intensification and synergy in production strategies (Pretty and Hine, 2001).

Crop yields are important for households that must get the greatest output possible from their limited resources. But most important is the net production from all activities, reflecting total factor productivity (Chapter 5). Conventional agriculture, responding to the factor endowments prevailing in the United States, Europe and other temperate zones, has pursued *monoculture*. But there are major productive opportunities to be exploited in the *diversification* of farming systems, such as raising fish in rice paddies or growing crops on paddy bunds, or adding goats or poultry or agroforestry to household operations.

Agroecological approaches can increase overall output by moving away from focusing on single crops, thereby also enhancing the *stability* of production, which can be seen in lower coefficients of variation (Francis, 1988). It is difficult to quantify all of the potentials of diversified and intensified systems, however, because there is too little research and experience to establish what their limits are.

How sustainable such production systems will be cannot be determined at present because most are fairly recent in use, and data from older systems have seldom been gathered and analysed systematically. However, agroecological practices can be evaluated in terms of their ability to replenish nutrient supplies and maintain soil health and biodiversity, as discussed in Chapter 3. There is certainly no reason to think that these alternative systems will be any *less* sustainable than those that rely heavily on chemicals, mechanization and other external inputs; in fact, they should be more durable. A number of the systems reported below have sustained doublings of yield over a decade or more without signs of deterioration.

Agroecological approaches are already increasing production under environmental conditions that are far from ideal: on eroded hillsides in Central America, on high barren plateaus in the Andes, in semi-arid areas in the West African Sahel, on exhausted lands in Eastern and Southern Africa, in and around rain forests in Madagascar and Indonesia, in heavily populated areas of Malawi, on crowded flood plains in Bangladesh, on sloping areas in The Philippines, and within the war zone in Sri Lanka. Data on these systems and their productivity are presented in Part 2.

It may be objected that doubling yields is not very difficult when farmers are starting from such low levels of production. However, if such doubling is easy, one can ask why 'modern' technologies have fared so poorly when they have been introduced under these adverse conditions. Actually, some of the yield levels reported below are quite high in absolute terms; see, for example, the Madagascar case (Chapter 12) with yields reaching 10 to 15 tons of paddy rice per hectare or more, and the Andean case (Chapter 14) with potato yields up to 40 tons per hectare, in both cases without requiring purchased inputs.

A more important consideration is that these methods attain increased production in areas where the need for food is greatest. Considering the poor resource endowments and the urgent human needs in these areas, the augmentation of food production, whether judged in absolute or relative terms, is quite significant, providing food directly to households that are most vulnerable to food insecurity. Raising output in such regions and for such producers goes directly to the heart of the problem of meeting food needs.

Not all agricultural innovations will work under all sets of conditions – for example, where the soil lacks certain nutrients, or where rainfall is too little or unreliable. However, agroecological practices directly address such environmental constraints, seeking to remedy these while reaching reasonably high levels of production. They can enhance soils' nutrient status and water-retaining capacity, and some even make soil restoration or reclamation possible. Hardy leguminous species such as canavalia and tephrosia, for example, can be grown – and enrich the soil – in areas where it seems impossible that any plants will grow.

Where the labour supply is limited, some of these innovations will not be practical because they require more labour than presently invested. However, some agroecological practices are labour-saving. For example, intercropping the leguminous velvet bean (*mucuna*) with maize, using it as a slashed-and-then-mulched cover crop, reduces labour requirements at the same time that it protects the soil from erosion and enriches it through the fixation of nitrogen, raising yields by 35 to 40 per cent (Thurston et al, 1994).

EXPANDED ROLES FOR FARMERS

All of the technologies considered here are management- and knowledge-intensive, and most take considerable time to develop and diversify to users' satisfaction. Success depends in large part on the enhancement of human abilities to make decisions, manage resources, acquire information and evaluate results. Although such activities are conventionally regarded simply as a *cost* of production, farmers increase their levels of skill, knowledge and decision-making by engaging in them. This process makes farmers able to be more productive in the future. Activities that enhance human resources should be regarded as *benefits* for farmers and for society, not only as costs.

Agriculture that engages farmers in experimentation and evaluation is a more progressive kind of agriculture than where technological advance is a matter of farmers following instructions for new practices. It has the desirable effect of augmenting farmers' human capital, giving them greater confidence and skills to solve problems and advance their interests in other domains, as seen particularly in the Latin American case studies. Agroecological approaches thus are not simply grounded in biology and ecosystem interactions; they are connected to human resource development as well, as discussed in Chapter 20.

Emphasis on Process and Not Just on Products

The Bellagio conference on sustainable agriculture was planned to examine innovative technologies and production systems, but much of the discussion focused on the *processes* by which new agricultural practices are developed, improved, evaluated and extended. The new approaches have emerged largely from experience and experimentation, much of it by farmers themselves, though stimulation and support have come from a variety of sources – nongovernmental organizations (NGOs), international research institutions and universities. In some cases, government agencies have begun working in more collaborative relationships with farmers.

To some extent, practice is ahead of theory in this area, although agroecology provides a theoretical foundation for comprehending and assisting these changes in production practice. What are now considered innovations are often not really new, at least not to farmers. Realizing that such systems often go back many years prompted us to juxtapose 'new paradigms' with 'old practices' in the subtitle of our conference. Agroforestry, for example, discussed in Chapter 8, is practically as old as agriculture, occurring wherever perennials have been utilized in conjunction with annual crops and/or animals. It has now become an applied science in the field of natural resource management, though this does not mean it is a recent innovation (Izac and Sanchez, 2001).

As discussed in the following chapters, especially in Chapters 4 and 20, there is an emerging methodology for agricultural innovation that is as important as the technologies that result from it. This strategy is based on active farmer involvement – indeed, often on farmer leadership. The process starts with identification of problems and needs as well as opportunities. This is best done by or with farmers who participate in delineating and choosing among possible solutions; who test, monitor and evaluate the results of new practices, helping to disseminate those results that are considered beneficial. This process can be characterized as participatory technology development, as farmer-centred research and extension, or as farmer-to-farmer agricultural innovation.

This methodology is more important than any single technology it produces and promotes, because *sustainable agriculture requires continuing adaptation and change* in practices and strategies, as seen particularly in Chapter 6. Sustainability is not an intrinsic quality of any technology in itself, but of the 'fit' between that technology and the multifaceted context in which it is used. Continuous alterations are necessary to function beneficially under the changing environmental, economic and other conditions that affect the productivity and profitability of specific activities and crops. Participatory processes are more likely to produce a range of flexible options, rather than a single technological 'solution' to be promoted.

Local knowledge is essential for this process, but seldom sufficient. It is usually best complemented and elaborated by knowledge that scientists and researchers can bring to a collaborative process of advancing technological possibilities. This is particularly important for those households that have been by-passed by Green Revolution options. To solve their food security problems, the means for raising production must be within the comprehension as well as reach of these farmers themselves.[6]

Transfer vs Diffusion of Technology

What will be appropriate investments? Only in special cases will technologies that were developed for favoured areas, having different conditions, be equally productive in poorer and marginal areas. As a rule, new technologies for such areas will have to be considerably modified, or evolved de novo from existing knowledge and practices. 'Transfer of technology' from favoured to marginal areas will usually be inappropriate as a strategy because one has to deal with much greater variety and variability in the latter areas, doing more fitting and adapting than where good soils and reliable climate can support monoculture and large capital investments.[7]

Whether a technology will be sustainably productive or not depends in large part on local conditions, which vary widely and also change. Extra-local conditions, of course, also impinge, sometimes drastically, on sustainability, eg, international trade agreements or technological gains in competing countries. Agriculture can be developed more durably and productively under diverse and changing circumstances when rural people are actively involved in structuring and managing the process. Engagement in such a process will increase their knowledge, skills and confidence, making them better able to deal with future problems and challenges, whether these are in the domain of agriculture or outside it. Especially with the growing forces of globalization in national economies and cultures, it is crucial that farmers have the capacity for continuous change and adaptation, given that there are no permanent technological solutions for agriculture.

Some of the agricultural reorientations reported here are already operating on fairly large scales and are growing, such as integrated pest management (IPM) in Indonesia as reported by Oka (1997), with over 1 million farmers trained, and the spread of no-till agriculture in Brazil, now covering over 13 million hectares (Chapter 15). Other programmes are starting to scale up in a major way; tens of thousands of households are already adopting agroforestry practices in Central and Southern Africa (Chapter 8), and 1 million households are expected to come into an expanded programme for integrated farming systems in Bangladesh that starts in 2001 (Chapter 16). Pretty and Hine (2001) have calculated, based on analyses of 208 cases across Africa, Asia and Latin America, that about 9 million households are already using different combinations of agroecologically innovative practices on about 24 million hectares of land. As such large numbers of farmers become involved in and benefit from agroecological approaches to agriculture, there will be more political support emerging at all levels for these changes to continue.

Transitions in Rural Areas

It may be objected that farming systems that do not employ significant amounts of capital and chemicals will lock rural households into small-scale, low-productivity agriculture for generations to come, which critics consider unfortunate. They incorrectly equate scale with productivity, however, for advocates of agricultural modernization think it a mark of progress for most households to leave the rural areas and make way for a process of land consolidation to occur, where agriculture becomes larger-scale, more mechanized and, they believe, more productive.

This conception of agriculture, however, ignores the fact that although larger farms may be more *profitable* for their owners, they are seldom more *productive* in terms of the returns to labour and to land, which will become increasingly important in the future. Profitability and productivity are not the same things and are not both equally dependent on efficiency. Where good land is the most scarce factor of production, increasing its productivity is a primary concern. Larger holdings are almost always farmed less intensively than smaller ones, so returns per unit of land are lower in larger operations. Substituting capital for labour through mechanization in larger holdings does not necessarily raise yields, though it can raise profits for owners of capital, especially if subsidized. Larger, more extensive operations seldom surpass smaller, more intensively managed ones in terms of output per unit of land.[8]

Will the incomes from smallholdings be enough to satisfy people's aspirations as well as their needs? This is a crucial question, to be considered with regard to what are people's *real* alternatives. These are shaped by demographic and other macro-level changes as discussed in Chapter 6, as well as by specific local conditions. Large-scale and long-term factors influence this evolution in agricultural sectors, but the end-points will not be the same in all countries. Few if any regions in Africa and Asia are going to end up looking like the American Midwest.

Small farms are usually more productive per hectare than are large farms; the exception is when units are so small that households do not devote much attention or labour to them (Berry and Cline, 1979; Johnson and Ruttan, 1994). Intensification based on agroecological principles offers possibilities for substantially higher incomes. The increases are not often going to be as great as the ten-fold increase reported from Kenya, where *tithonia* has proved to be a very effective green manure (Chapter 8) or the very large increases in rice production possible with the synergistic management techniques developed in Madagascar (Chapter 12). But there can be very large increases from capitalizing on biological potentials. When land is a limiting factor, smallholdings using labour-intensive technologies will usually produce larger returns to labour than will big farms. The latter use labour and other resources extensively; their profitability comes more from economies of *size* than from technical economies of *scale*, which represents greater factor efficiency.

A rural lifestyle is preferred by many people now living in rural areas, as long as basic services and opportunities for their children are available. Higher incomes in urban areas are usually matched by higher costs of living, and by different quality of life. The greater opportunities for public services, amenities and entertainment in urban areas are often associated with unpleasant crowding, crime and other undesirable conditions.

It is not for us to decide for others the balance of advantages and disadvantages of rural vs urban life, since people should be free to choose for themselves what their futures will be. Surely many will prefer the challenges and opportunities of urban life. But if productivity is raised in rural areas, people will have less reason or need to change their location or lifestyle unless the opportunities available elsewhere are considered better than those at hand in rural areas.

It is appropriate for governments and external agencies to try to *increase the options* that rural people have, for themselves and for their children. They should not be confined to lives of rural poverty because of low productivity and diminishing natural resource quality. Nor should they be pushed by economic circumstances to migrate to urban areas due more to desperation than desire. Along with agricultural intensification, there should be opportunities created by rural agroindustries that add value and income in rural areas, with beneficial spread effects from agriculture. Already, in many countries, much income for rural households is coming from non-farm activities (Reardon, 1997; Reardon et al, 2001).

National development will certainly include greater urban and non-agricultural expansion. Agriculture should not be expected to employ in the future the same share of the labour force that it does now. Agroecological approaches are not intended to keep rural residents 'down on the farm', but rather to enable them to improve their livelihoods, and especially their knowledge and skills, so that they can have and can make more desirable choices.

Failure to promote people-centred agricultural and rural development of the kind considered here will surely accelerate migration to urban areas, in excess of the productive opportunities there that can utilize and support the population well. It was commented from experience in Bolivia that supplying externally developed technologies and getting them adopted, while sometimes adding to agricultural production, contributes little to human development, which is needed for making advances in all sectors. Farmers who have developed their analytical skills and their confidence from agricultural experimentation will be better able to be productive in cities if or when they are displaced to an urban environment.

Agroecological approaches are not limited to using local resources, we should re-emphasize. As seen in the Nigerian experience presented in Chapter 7, when the population became more dense, it was no longer possible to keep enough cattle for manure to maintain field fertility or to grow enough biomass for adequate compost; this made turning to chemical fertilizers necessary. Rock phosphate is an essential element for soil replenishment in phosphorous-deficient soils of sub-Saharan Africa. While an increase in soil nitrogen can be accomplished by agroforestry, phosphorous has to come from mineral sources.

In Madagascar, where soils are very deficient in phosphorous, few farmers can presently afford fertilizer given their low yields and income from rice. But if the system of rice intensification reported in Chapter 12 can continue to boost yields several-fold with existing resources, farmers should be able in the future to afford to purchase fertilizers to enhance their soil fertility. Thus, the process of agricultural development envisioned, while it emphasizes use of local resources, will evolve from and integrate elements from presently 'modern' agriculture and still newer innovations to come.

CONSIDERATIONS THAT SHAPE THE FUTURE

Small farmers in most parts of the developing world do not need to be food-deficient and as poor as they are today. Those who have not benefited from capital-intensive or chemical-based technologies because these were not appropriate for their ecological or economic conditions can profit from the knowledge-, skill- and management-intensive methods of production of agro-ecologically informed agriculture. Indeed, larger-scale farming units around the world can also benefit from understanding and adapting the principles and practices of such systems, as they are increasingly doing in the United States and Europe (Thrupp, 1998; Pretty, 1999). Agriculture in most parts of the world can become more productive and efficient by giving more consideration to biodiversity, synergy and other aspects of well functioning ecosystems.

Whether this potential will be realized is uncertain, however, because this will depend on appropriate and greater investments, both public and private; on supportive and consistent policies; on research to develop the scientific underpinnings of agroecological practices; and indeed, on rethinking what constitutes expertise, as discussed in Part 3. The investments made thus far in alternative agricultural approaches have been minimal – indeed, a tiny fraction of the resources that have gone into developing mainstream agriculture.

Economic and Social Considerations

Economic analysis, especially assessing the costs and benefits of labour inputs, is important because financial considerations guide and accelerate (or constrain) the process of agricultural change. For farmers, agronomic success is not enough. Net increases in income and food security are crucial criteria, and labour in rural communities almost never has zero opportunity cost. The slow spread of practices that are agroecologically sound has often been due to their substantial labour requirements. Returns to labour in particular need to be assessed when evaluating possibilities for the adoption and spread of agro-ecological systems, as considered in Chapter 5.

This acknowledged, economic profitability is not the only criterion affecting farmers' decisions. While income is important, especially for the poor, it is not an exclusive concern. For one thing, risk is ever-present in rural environments, and it is always a reason for discounting prospective returns. Moreover, where markets are unreliable or difficult or expensive to access, households will continue to regard self-sufficiency as a reasonable strategy for food security, no matter what advantages may be attributable in principle to market participation.

Households also have cultural values that need to be respected, and most parents attach great importance to opportunities for the next generation, being willing to make sacrifices in the present for their children's future. Maintaining intact, attractive rural communities is a value that gets considered alongside individual increases in income. So while economics need to be evaluated, because farmers must consider how innovations would affect their net income, it is not a sole determinant. It is one of several tests applied by farmers when assessing alternative agricultural practices.

Institutional and Knowledge Considerations

For small and marginal farmers to contribute significantly to future world food production, there need to be institutional changes and investments to realize this potential. These become more important as the processes of globalization in the economy and culture spread more widely. For farmers to be able to compete in larger markets and arenas, they must become more knowledgeable and 'agile'. The mutability of global opportunities and forces means that farmers have to be able to entertain many options and make quick adaptations. Economic specialization becomes more appropriate as market access increases, but the logic of specialization should not necessarily be taken to its extreme because market forces are rapidly changing. Being locked into a single mode of production or output through specialization, even if previously successful, can be economically fatal.

The process of transforming present agricultural practices towards more agroecologically suitable ones remains challenging in part because of our insufficient knowledge, as discussed in Chapter 19. However, the cases presented here justify some optimism. This approach to agricultural development, which draws on the accumulated knowledge and experience from the past that is possessed within farming communities, is forward-looking. The synergistic principles of agroecology, increasingly understood in formal scientific terms, should help to circumvent some of the constraints and undesired effects that result from production heavily dependent on capital, chemicals and machinery. Formal education and literacy are important, but are not in themselves sufficient. Most people with direct experience in farming can learn to practise and improve upon knowledge-intensive forms of agriculture that will transform rural people from their historical subordinated roles as 'hewers of wood and drawers of water'.

New Partnerships

Three decades ago, when the Green Revolution was being launched, this vision of rural people as being more than 'adopters' was held by few people outside of rural areas. It was considered that progressive change would be initiated outside rural communities, not by or with farmers themselves. The case studies presented here, however, give considerable evidence that the human capabilities available to be enlisted in a new kind of agricultural modernization have been underestimated and too narrowly conceived.

The technologies for the next era in agriculture will require many contributions from scientists, but technological development will proceed more effectively and broadly by their engaging with many people in other roles as partners. The approach presented here can be understood as walking on two legs: one of *agroecology*, which encompasses all resources, aspects and interactions of living systems, and the other of *participation*, which embraces a multiplicity of roles and talents but emphasizes those of farmers as co-generators as well as users of new technology.

NOTES

1   The term 'farmers' as used in this book refers to both female and male agriculturalists, who are likely to be producing more than just plant crops and to be involved in some or many off-farm economic activities.

2   Smil (2000) offers a thorough, detailed analysis of global food production prospects and requirements, which provides excellent background for the discussion in this volume. We will not attempt to replicate or repeat Smil's analysis.

3   This is a complex issue, arising wherever intensified production of poultry, pigs, cattle, fish or other animals is achieved through use of feed grains, fish meal or other protein sources that accelerate weight gain. Where animals are raised on non-competitive sources of nutrition, ie, on nutrients that could or would not be consumed by humans such as with foraging or grazing, the opportunity cost suggested in the text is avoided. On a global basis, human nutrition would be improved by more of this kind of (extensive) animal production because it is more likely to benefit the poorly nourished. In fact, however, at the margin now, most increases in the output of animal foods are being achieved by intensive methods. For a detailed and sophisticated discussion of these issues, see Chapter 5 in Smil (2000).

4   This was documented in an analysis of earnings for rural households in north central Java by Hart (1986). When controlling for educational and skill levels, adults in households owning no land received lower hourly remuneration than those from households with at least some land, as this permitted them to produce at least part of their own subsistence. Households with some land were not forced by desperate circumstances to accept employment on whatever terms were offered. Also, they were more valuable politically and socially as clients to the richer households that could provide employment and gleaning rights. The landless had to travel long distances for work that paid very little. Thus, the returns to labour were not simply a matter of 'returns to human capital'. Purely economic factors were less determinant of wage levels than was the political economy of land and village social and power relationships.

5   For example, chemical fertilizers and inputs of organic matter (composts and green manures), though often regarded as competing alternatives, can *each* be made more productive by adding appropriate amounts of the other kind of nutrient (eg, Palm et al, 1997; Schlather, 1998).

6   There is, however, no warrant for idealizing local knowledge, or for assuming that it is always complete or correct. The case in Chapter 12 shows that long-held

farmer (and scientist) beliefs about irrigated rice production practices can suppress production potential.

7   Two examples of technology transfer that has been clearly beneficial are vaccinations against communicable disease, and inexpensive, durable hand pumps for village water supply. The examples of successful 'technology transfer' cited in our discussions at Bellagio were more from the areas of health and infrastructure than from agriculture.

8   'Many studies of farming systems around the world have shown that there are few economies of scale in agriculture that might contribute advantages to farms larger than what a family could operate using its own labour. The lack of economies of scale in agriculture, coupled with the high cost of supervising wage labour, implies that a farm cultivated by an owner-operator without reliance on permanent outside labour – the family farm – is the most efficient unit of production. The few exceptions occur with plantation crops, or where large farms are able to overcome imperfections in other markets, such as those for outputs, inputs or credit' (Binswanger and Deininger, 1996, p11).

Chapter 2

Rethinking Agriculture for New Opportunities

Erick Fernandes, Alice Pell and Norman Uphoff

Over the last 30 years, the creation and exploitation of new genetic potentials of cereal crops, leading to what is called the Green Revolution, has saved hundreds of millions of people around the world from extreme hunger and malnutrition, and tens of millions from starvation. However, these technologies for improving crop yields have not been maintaining their momentum. The rate of yield increase for cereals worldwide – around 2.4 per cent in the 1970s and 2 per cent in the 1980s – was only about 1 per cent in the 1990s. Although the global food production system has performed well in recent decades, will further support of conventional agricultural research and extension programmes increase yields sufficiently to meet anticipated demand?

The next doubling of food production will have to be accomplished with less land per capita and with less water than is available now (Postel, 1996). The gains needed in the productive use of land and water are so great that both genetic improvements and changes in management will be required. The world needs continuing advances on the genetic front, especially of the sort proposed by Tanksley and McCouch (1997). However, food production is more often limited by environmental conditions and resource constraints than by genetic potential. Preoccupation with the methods that brought us the Green Revolution can divert attention from opportunities that can increase food supply without adversely affecting the environment, which are considered in this book.

Given appropriate research, policies, institutions and support, food production could be doubled with the existing genetic bases. Many of the needed advances in food production could be achieved by developing agricultural systems that capitalize more systematically on biological and agroecological dynamics rather than by relying so much on agrochemicals, mechanical and petrochemical energy and genetic modification.[1] This will require, however, some rethinking of what constitutes agriculture.

Although it has been argued that agricultural output will decline if 'modern' agriculture is not promoted to the maximum (eg Avery, 1995), 'low-tech' methods can be very productive with now-better-understood scientific bases. Where economically justifiable, these methods use available resources more efficiently than do high-input approaches. Farming systems reported in Part 2 such as those for rice in Madagascar (Chapter 12), maize and beans in Central America (Chapter 13) and potatoes and barley in the Andes (Chapter 14) demonstrate that output can be raised substantially, sometimes several-fold, with limited dependence on external resources. These crops are staples that are essential for meeting world food needs.

The potential of non-mainstream methods cannot be known until agro-ecological approaches are taken more seriously and evaluated systematically. Gains made through genetic improvement and use of external capital and chemical inputs over the last four decades have been substantial, and the first Green Revolution, despite the shortcomings some critics have pointed to, was one of the major accomplishments of the century.[2] But what will agricultural science do for an encore? While biotechnology holds out many promises, most of its benefits continue to be anticipated more than realized. Access to and widespread distribution of biotechnology's prospective benefits remain uncertain. The widely publicized 'golden rice' is still years from production in farmers' fields.

The challenge facing agriculture worldwide involves more than just achieving higher production, justifiable as that goal has been for previous scientific innovation when serious food deficits were an ominous possibility. Valid ecological and social considerations now make it imperative that further advances be environmentally friendly as well as economically sustainable and socially equitable. Also, more than increased food supply is needed; we should aim to ensure balanced and adequate supplies of nutrients that people can afford. In particular, adverse environmental and health externalities that result from modern agricultural methods – soil erosion, chemical hazards, soil and water pollution – are things that nobody would like to see increased, let alone doubled, as we seek to double the production of food.

Should resources for agricultural research be devoted, for example, to developing genetically engineered rice with high levels of vitamin A, assuming that cereal grain monoculture will continue to predominate? Or should we strive to incorporate nitrogen-fixing and nutrient-rich legumes and livestock into farming systems to better meet people's nutritional requirements with diversified diets – while simultaneously maintaining soil fertility? Such questions need to be addressed.

The next Green Revolution will depend at least in part on enlarging upon and diversifying the ideas that have guided past development efforts. The paradigms that presently organize and direct agricultural research and extension have been helpful for planning activities and producing theoretical explanations. But they have also created certain blind spots. The task of meeting world food needs will be more difficult if our vision of what is possible is limited by constraining conceptions of how best to raise agricultural output in effective, efficient and sustainable ways.

AGRICULTURE AS FIELD-CULTURE:
AN ETYMOLOGICAL PERSPECTIVE

The very concept of agriculture as it has been understood and practised in the West has been shaped by its semantic origins, coming from the Latin word *ager*, 'field'. Agriculture is mostly understood as the growing of plants in fields. (Similarly, in South Asia, most words for agriculture derive from the Sanskrit word for plough, *krsi*, so that agriculture in that region is characterized as 'plough work'.) Such a conceptualization, however tacit, makes the raising of livestock, fish, trees and other activities less central to the agricultural enterprise, except where cattle or oxen are necessary for ploughing, or where monocrop tree plantations substitute for fields. The full range and richness of the agricultural enterprise has not been well captured in the word that we use to refer to it.

Etymologically, it is not clear where livestock, fish, insects, microbes and trees fit in. Few sustainable farming systems exist that do not include several of these groups in addition to plants. But most often, those who work on other flora or on fauna have been accorded marginal status within agricultural ministries, or been assigned to separate ministries, leaving crop and soil specialists in charge of the agricultural sector.

Fishery departments are invariably marginal if located within an agricultural ministry, even though *aquaculture* integrated within farming systems has great potential, as shown in Chapter 9. Indeed, until 'agroforestry' was discovered (King, 1968; Bene et al, 1977) and the International Centre for Research in Agroforestry (ICRAF) was established, there was little concern with trees as part of agriculture, except in large-scale plantations where tree crops were commercially profitable. Otherwise, trees got respect and attention only if looked after by a separate ministry that was more concerned with forests or plantations than with farms.

Although *agroforestry* may sound like a kind of forest management, it is a comprehensive landuse management strategy that includes a range of woody perennials (particularly trees but also shrubs) in spatial and temporal associations with non-woody perennials, grasses and annual crops, together with a variety of animals, including cattle, sheep, goats, pigs, chickens, guinea pigs, fish and even bees (Lundgren and Raintree, 1982). While some agroforestry practices are extensive – for example, most agrosilvopastoral systems – these practices generally contribute to intensified production that is agroecologically sound and maintains soil fertility (Fernandes and Matos, 1995). Fortunately, the integration of perennial plants into otherwise annual farming systems is increasingly recognized as a mainstream opportunity to increase per-hectare output in future decades (Chapter 8).

A bias in favour of fields means that *horticulture* gets somewhat marginalized in most institutions dealing with agriculture, including universities. Gardens and orchards, being smaller, have lower status than fields, even if they produce several times more value per unit of land when intensively managed. Horticulture is devalued in part also because its produce is mostly perishable and hard to denominate. Heads of cabbage and baskets of apples are hard to compare with bags of rice or tons of wheat, their nutritional value notwithstanding. Historically, governments have gained more wealth and security from grains because these could be stored (or seized) more easily than fruits and vegetables.

Farming systems of most rural households around the world depend crucially upon *livestock and poultry*, large and/or small, together with home gardens and orchards and often with fish ponds and hedgerows. Efforts to improve single components of farming systems are likely to produce limited results unless the interdependence of land use, labour supply and seasonal activities for all of these farm enterprises is acknowledged.[3] In many areas of Asia, acceptance of the short-stalked, high-yielding cereal varieties that made the Green Revolution was low, for example, because the quantity and quality of the fodder produced by the new varieties was insufficient to meet livestock requirements. The goal of plant breeders had been to increase grain yield without considering forage needs. Farmers were willing to accept lower yields of grain in order to be able to feed their animals, which provided them with the manure they needed to maintain soil fertility and the traction required for tilling their land.

An argument sometimes made against livestock production is that animals are inherently wasteful; more calories can be produced per hectare from plants than from animals. If animals are fed on forages and by-products, however, rather than competing with humans for edible grain, such 'wastefulness' can be beneficial. In extensive and semi-extensive systems, animals that range freely during the day harvest plant nutrients from non-arable areas; at night when they are penned, most of these nutrients are deposited in their enclosure, later to be distributed as manure onto cropland. In parts of West Africa, pastoralists often negotiate grazing contracts with crop-growing neighbours. Pastoralists are encouraged to graze their cattle on fields with crop residues because the cattle deposit manure: their owners may even receive additional compensation for this service. If animals were in fact highly efficient in their conversion of harvested nutrients, there would be less transfer of nutrients from rangelands to croplands.

When green and animal manures are judiciously used in combination, nutrient availability can be nicely synchronized to meet plant demands. Manure is an important product of livestock raising. In sub-Saharan Africa, 25 per cent of agricultural domestic product comes from livestock even without considering manure or traction; when these are considered, this figure rises to 35 per cent (Winrock International, 1992). The quality and quantity of manure produced depends on what the animal consumes; in Java where 'cut and carry' tree-based fodder systems are common, animals are given extra feed to improve the quality of their manure (Somda et al, 1995; Tanner et al, 1995). Thus, animal production can be beneficial in ecological as well as human nutritional terms, as shown particularly in Chapters 10 and 11.

An additional consideration obscured by a preoccupation with fields is that *common property resources* for grazing and for forest products are an essential part of many households' economic operations (Berkes, 1989; Jodha, 1992).

Common lands often are the sites from which grazing livestock harvest nutrients that are brought back to the farm at night. As these areas are not fields, however, and do not belong to any specific user, evaluating their contributions to production is admittedly difficult. This is not, however, sufficient reason to overlook their role and potential, leaving their productivity to languish.[4] Privatization of these commons, often advised, removes the flexibility people need to withstand drought in dry regions. Farming systems improvement should encompass all the area and resources available to farmers and pastoralists.

Developing an adequate knowledge base for more productive and sustainable agriculture should start with explicit acknowledgment that agriculture involves much more than fields and field crops. To be sure, fields are commonly the main component of most farm production strategies. *Staple foods* are, after all, what their name implies – essential for food security. The world in general needs more, rather than less, of them, especially for the 800 million people who are currently undernourished. But other sources of calories are also important – potatoes, cassava, yams, sorghum, millet, sweet potatoes, taro, fish, meat, milk and so on – and these have been given much less support than rice, wheat and maize.[5] Calories, while necessary for survival, are not sufficient for human health. To achieve balanced diets, including essential micronutrients, the whole complex of flora and fauna that rural households manage to achieve food security and maintain their living standards should be better understood and utilized.

Not only should fixation on individual crops be avoided, but a broader understanding of the biophysical unit for agriculture is needed. A narrow focus on fields is giving way to a broader focus on *landscapes* and/or *watersheds*, within which fields function as interdependent units, especially as we gain a better agroecological understanding of agriculture (Conway, 1987; Carrol et al, 1990; Altieri, 1995).

ASSUMPTIONS ASSOCIATED WITH FIELD-CENTRED AGRICULTURE

Several limitations arise from this long-standing concept of agriculture. In different ways, each works against strategies for intensified and sustainable agricultural development that use the full set of local resources most productively.

The Time Dimension of Agriculture: A Cyclical View

In lore and literature, agriculture is described and celebrated as 'the cycle of the seasons'. How is agriculture practised with its field-based definition? By ploughing, planting, weeding, protecting and finally harvesting. Farmers then wait until the next growing season to plough, plant, weed, protect and harvest again, and wait once more for the next planting time. Planting defines agriculture in our minds as does the activity of harvesting. Yet if one looks beyond this standardized seasonal conception of agriculture, one finds trees that keep their leaves year-round, sheep that lamb twice a year, and microbes that continuously decompose soil organic matter with generation intervals measured in hours or minutes. These different time-frames all affect agricultural performance.

Fixation on an annual cycle of agriculture has arisen from its practice in temperate climates, where most modern scientific advances have been made. There, summer and winter seasons are the central fact of agricultural life. The year-round agriculture of tropical zones seems somehow irregular, almost unnatural, since it lacks periodic cultivation. This view is reflected in reports from early colonial administrators in tropical countries who regarded indigenous populations as 'lazy' because they did not work hard to produce their sustenance. There was no annual cycle of ploughing, planting and so on, which counterparts in colder climates had to maintain. People who harvested what they had not planted, or had not planted recently, were not regarded as 'real agriculturalists' by people from temperate zones.

There is seasonality in tropical regions, to be sure. The contrast between wet and dry seasons can be as stark as that between summer and winter. But with agriculture seen primarily as a matter of *cultivation*, annual crops get more attention and status than perennials. The latter have very important roles to play, however, particularly if one is concerned with the sustainability of agriculture. Their growth usually does not disturb or tax the soil as much, or as often, as does annual cropping. The latter invests in myriad biological 'factories' that produce food or fibre and then demolishes them at the end of the season. On the other hand, trees, vines or crops that rattoon keep all or most of that biological factory intact from year to year.

Since, usually, very little biomass is discarded in the farming systems operated by poorer farm families – it is used for fodder, fuel, mulch or other purposes – our point here is directed to research and extension priorities rather than to farmers. The latter have long known that combining a variety of perennials with annuals, animals and horticultural crops creates opportunities for more total output from given areas of land during the year, and with less pressure on soil resources; energy and nutrient flows are more efficient, and adverse pest and environmental impacts can be reduced by growing perennials rather than annuals.[6] Especially if the sustainability of agricultural production is an objective, giving perennials a larger role in agriculture makes sense.

Within agriculture understood in annualist terms, fallows are periods of rest and recuperation for the soil, a kind of gap in the cropping calendar. Many farmers, however, have thought of fallows differently, managing them so that they are more productive than land that is simply left alone. 'Managed fallows' are not an oxymoron but rather a source of supplementary income, providing fodder, fruit or other benefits while enriching the soil when leguminous species or plants otherwise considered to be weeds are allowed or encouraged to grow.[7] Cropping cycles are best looked at in terms of how soil fertility can be continuously enhanced while utilizing a wide variety of plant and animal species – a strategy described as 'permaculture' by Mollison (1990) – looking beyond crops that are planted periodically.

Spatial Dimensions of Agriculture: Thinking in Terms of Soil Volume Instead of Surface

Agriculture been defined and limited by a mental construction of agricultural space in much the same way that it has been stereotyped in terms of annual cycles. While farmers have long appreciated that agriculture is an enterprise best conducted in three dimensions, most agronomic and economic assessments consider agriculture essentially in *two dimensions*, as an enterprise carried out on a plane. The practice of agriculture is epitomized by ploughing, which breaks the surface of the soil in order to plant seeds and grow crops. This strategy suffices so long as the soil is deep, fertile and well supplied with water. But agriculture can be made more productive by conceiving and treating soil in *three-dimensional* terms, as volume, doing more than just breaking its surface and working it two-dimensionally.

Indeed, working the soil is a better term for agriculture than ploughing it, since working encompasses many functions.[8] This concept includes incorporating organic matter of various sorts into the soil and altering soil topography to capture and hold water, or to drain it. Getting crop residues and animal manures into the soil can promote greater synchrony between nutrient release from those residues and crop nutrient demand; soil organic matter promotes better water infiltration and retention at the same time that it creates better habitats for soil microflora and for micro- and macrofauna. In many traditional farming systems around the world, one finds soil being mounded into raised beds and even raised fields; terraces are constructed to retain and improve the soil and to make watering it easier, and drains are often installed. Soil-working activities are intended not just to exploit the soil's fertility but to improve it.

Alternately, in some farming systems one finds no ploughing, just the planting of seeds in undisturbed soil. This might be considered one-dimensional agriculture with activities concentrated on points rather than a surface, leaving the volume of soil beneath intact to nurture macro- and microbiological communities. To be sure, two-dimensional thinking accomplishes some important activities such as weed control and breaking the soil crust, but disturbances of the soil contribute to major erosional losses. Weeds can be controlled by other means than ploughing, and 'no-till agriculture' is now widely accepted as a modern practice, as noted below. How this practice has contributed to improving Brazilian agriculture is reported in Chapter 15.

In the coming decades, efforts to raise yields per hectare should not take the quality and durability of soil for granted, as the health and fertility of the soil are critical for productive and sustainable agriculture. Soil should be understood and managed in terms of its *volume* rather than its *surface*. Raising output sustainably will require more than working chemical fertilizers into the top horizon. Thinking of soil three-dimensionally should be part of any strategy for sustainable agricultural intensification.

Monoculture as 'Real' Agriculture

The standard view of agriculture as limited in time and space favours monocropping for achieving control and efficiency in production. Applying inputs is made easier with monoculture, whether calculating fertilizer applications or using mechanical power for weeding. But the conclusion that this is always the most productive way to use land is mistaken. This production method can raise the economic returns to labour or to capital, but it does not necessarily increase the returns to land. The latter resource will become ever more important in coming decades as the availability of arable land per capita declines.

Polyculture systems employing a combination or even a multitude of plants commonly have higher total yields per hectare, absorbing and generally requiring higher inputs of labour and nutrients. Where labour is relatively abundant and land is relatively scarce, this can be an efficient and economic system of resource use. The advantage of monocropping is that it makes mechanization, substituting capital for labour, more effective.[9] Only where mechanical power can bring into cultivation land that manual power cannot is greater physical production likely to result from mechanization. This generally makes agriculture more extensive than intensive.

Even when population is high in relation to arable area, it can be difficult to attract or retain labour to work in farm operations. Much of the impetus for farm mechanization has come from labour scarcities in the more economically advanced countries. When tractors and other machines have been introduced into developing countries with the mistaken idea that this will raise production, they have done more to displace labour than to make land more productive. Tractorization can raise profits for those who have greater access to land and capital, but it seldom leads to higher output per unit of land than using hand labour and animal traction, other things being equal.[10] In contrast to tractors, animals used for traction reproduce themselves, pay returns on the farmer's investment, and provide food, fuel and fertilizer at the same time. Since capital is so often subsidized by government policies, one should not consider the private profitability of using tractors and other capital inputs as a sole or sufficient justification for their use without analysing the full range of social costs and benefits.[11]

Because polyculture is less amenable to mechanization, it requires an adequate and reasonably skilled supply of labour. Many of the practices we discuss here are relatively labour-demanding, using human energy and skill instead of capital and chemicals to get more production from limited land resources. To the extent that investments of labour are made more productive by agroecological innovations, they can be better remunerated and lead to improvements in the agricultural sector and the rest of the economy.

It is widely believed, with more emotion than calculation, that clean-ploughed fields, sown uniformly in a single crop, planted neatly in rows with all extraneous plants removed, is the best kind of agriculture. Mulch makes fields look messy, and crop mixtures look chaotic rather than productive. But this assessment is more a matter of aesthetics than of science. Yields, yield stability and nutritional quality per unit of land from polyculture, although harder to measure, are usually greater than with monoculture.[12] Furthermore, keeping soils covered protects them against erosion.

Polycropping supported by a strategy of managing and recycling organic inputs offers many advantages and can raise yields with equivalent inputs. When maize and soybeans are intercropped, for example, there is about a 15 per cent gain in production that cannot be explained simply by the inputs applied, an increase reflecting synergy within the crops' growing environments (Vandermeer, 1989). Plant–animal intercropping yields comparable benefits. There are many situations, determined more by economic than by agronomic considerations, where monoculture will be a preferable strategy. But its superiority should not be assumed without proof, as happens now.

Mechanical Conceptions of Agriculture

Monocropping implicitly regards agriculture as a mechanical process, with inputs being converted into outputs by some fixed formula, whereas polycropping recognizes the inherently biological nature of agriculture. The relation posited between inputs and outputs is different for mechanical and biological paradigms. In the first, the ratio of outputs to inputs is predictable and proportional, fixed and usually linear. In the realm of nature, on the other hand, relationships are less predictable and seldom proportional. Large investments of inputs can come to nought, while under favourable conditions and with good management, modest inputs have many-times-larger effects.

Until something like the perpetual motion machine is invented, such disproportionality is not possible with mechanical phenomena, which depend on continuous inputs for their operation. Biological processes, on the other hand, can be self-sustaining and can adapt and evolve unassisted. Moreover, biological inputs can reproduce themselves. How one regards and utilizes *inputs* thus differs in subtle but important ways according to whether they are understood within a mechanical framework or in a biological context.

One area where 'modern' agriculture has rediscovered the advantages of biology is with so-called minimum tillage or no-till systems, now given the positive appellation 'conservation tillage' (Avery and Avery, 1996). Twenty years ago this was considered atavistic agriculture, harking back to the dibble stick in a modern era when heavy tractors and field machinery should be used to plough, plant, weed and harvest 'clean' fields. Yet no-till agriculture has now become state-of-the-art in many areas of the United States. Mechanical corn harvesters are designed to chop up plant stalks, leaves, husks and cobs to return this biomass to the land in biodegradable form to preserve soil fertility. In addition to recycling nutrients, conservation tillage protects the soil's surface and reduces wind and water erosion. The main limitation with little or no tillage is that weeds can become more of a problem unless farmers can afford chemical herbicides or use hand labour. (This new/old technology has become popular with businesses that sell herbicides to control weeds when there is no ploughing).

Innovative practices like the use of mulches, cover crops and green and animal manures, which were until recently largely ignored in 'modern' agriculture, can solve the problem of weeds, as seen in Chapter 15. These techniques capitalize on the large dividends that nutrient recycling can pay because of the multiplicative dynamics of biological processes. Whereas mechanical advantage is a well-accepted principle in physics and engineering, agricultural science should capitalize on the analogous and even more powerful principle of *biological advantage*.

FOUR EQUATIONS IN NEED OF REVISION

Efforts to raise agricultural productivity have been guided for many decades by four presumptions. These have produced some impressive results, so our objection is not that they are wrong. Rather, they have become too dominant in our thinking, with too hegemonic an influence on policy and practice. It has been taken for granted that they represent superior ways to boost production. This thinking can be stated in four tacit equations that have shaped contemporary agricultural research, extension and investment.

1   Control of pests and diseases = application of pesticides or other agrochemicals.
2   Overcoming soil fertility constraints = application of chemical fertilizers.
3   Solving water problems = construction of irrigation systems.
4   Raising productivity beyond these three methods = genetic modification.

Equating certain kinds of solutions with broad categories of problems limits the search for other methods to solve those problems, even when alternative practices might have a lower cost and be more beneficial in environmental and social terms. More progress in agriculture will be made if the above propositions are broadened. Fortunately, there is a good precedent in the way that the first equation has been substantially modified over the past 15 years.

Crops and Animals Can be Protected by Non-chemical Means

The modern-input paradigm for raising production has been most directly challenged with regard to pest and disease control through what is called *integrated pest management* (IPM). Adverse effects on human health as well as on the environment caused some scientists to explore ways to produce crops and animals with little and even no use of chemicals. Biological controls as well as alternative crop management practices have often turned out to be more cost-effective, and sometimes simply more effective. The chemical-based strategy of 'zero tolerance' for pests and diseases, rather than being a solution, exacerbates the problem, killing beneficial insects that are predators of crop pests. The widespread use of agrochemicals, particularly broad-spectrum ones, has had the consequence of making pest attacks worse.[13] Routine use of antibiotics to treat diseases and promote the growth of livestock has, unfortunately, increased the antibiotic resistance of pathogens that can infect humans and/or animals.

An IPM strategy does not preclude the use of chemicals. But the first lines of defence against pests and diseases are biological, trying to utilize the defensive and recuperative powers of plants and animals as well as the activity of beneficial and predator insects to farmers' advantage.[14] The Indonesian IPM programme, for example, taught farmers that spiders, previously viewed with antagonism, should be protected and preserved. Demonstrations showed that rice beyond a certain stage can sustain extensive leaf damage from insects, as much as 25 per cent, without depressing effects on yield, and even possibly some gain. When sheep in Australia and South Africa were fed leguminous forages containing tannins as part of their diets, their internal parasite loads were reduced, reducing expenditures on antihelmintic medicines and providing an alternative treatment when antihelmintic resistance is a problem (Kahn and Diaz-Hernandez, 2000). The presumptions of modern agricultural science regarding chemical means for pest and disease control have been broadly challenged, with such means being increasingly reduced and avoided where possible.

Soil Fertility can be Enhanced, Often More Effectively, by Non-chemical Means

The most broadly successful component of modern agriculture has been the introduction and use of inorganic fertilizers to supply soil nutrients, particularly nitrogen, phosphorous and potassium, where these were lacking. But this success has led many policy-makers and some scientists to equate soil fertility improvement with the application of fertilizers when, in fact, fertility depends on many additional factors. Indeed, the misuse or overuse of chemical fertilizer results in adverse effects on yield by negatively affecting the physical and biological properties of soil. The advantage of inorganic fertilizers is that they are easier to apply, often cheap (if subsidized) and have more predictable nutrient content. Also, organic nutrients are sometimes simply not available in sufficient supply.

When inorganic fertilizers are added to soils that possess good physical structure, with adequate soil organic matter and sufficient cation-exchange capacity, they can produce impressive improvements in yield. Where soils are acidic (low pH) and the nutrients needed for plants are in short supply, the application of appropriate amounts of lime (calcium carbonate) along with inorganic fertilizers can result in spectacular crop yield increases and can greatly improve farmer income. But in many circumstances, especially in the tropics, soils are not so well structured or well endowed. Then, inorganic fertilizers, especially if used in conjunction with tractors that compact the soil, can lead to changes in soil physics and biology that are counterproductive and diminish, sometimes sharply, the returns from adding chemical nutrients.

We have suggested to dozens of soil scientists in the United States and overseas that probably 60 to 70 per cent of soil research over the past 50 years worldwide has focused on soil chemistry and about 20 to 30 per cent on soil physics. This means that less than 10 per cent of soil research has been devoted to improving our understanding of its biology. This estimate has not been challenged by agronomists to date. Why such preoccupation with soil chemistry? It is the easiest kind of soil deficiency to study, giving quick, precise and replicable results, which point to simple remedies. The results of soil chemistry analyses are easy to interpret; by adding certain amounts and combinations of fertilizer nutrients, one can expect predictable increments to production. Moreover, such research gets funding easily, given the interests of fertilizer producers in such knowledge.

Yet even brief consideration of these three domains affecting soil fertility suggests that the amount of effort going into each, even if not necessarily equal, should be closer to parity. Any national research programme that deliberately allocated its scientific resources in the above disproportions would be considered misguided. Microbial activity is essential for nutrient availability and uptake. When one walks on ground that has been converted by leguminous species, compost, mulch or manures from something resembling concrete into absorbent, friable soil underfoot with good tilth, the contribution of soil microbiology is self-evident. But studying biological processes is more difficult than assessing differences in soil structure, and many times more difficult than measuring the chemical composition of soil samples.

Similarly, plant scientists with whom we have spoken have agreed that 90 per cent or more of their research effort over the past 50 years has been devoted to those parts of plants that are above ground, and less than 10 per cent to what is below ground. Indeed, plant scientists usually suggest that less than 5 per cent of their research has investigated sub-surface processes and dynamics. Yet any assessment of how plants grow and thrive suggests that a more balanced distribution of effort is desirable, with much more attention paid to the growth and functions of roots than in the past. However, just as it has been easier to study the chemistry of soil, it has been easier to analyse leaves and stalks than to probe the underground mechanisms of roots for uptake and transport of nutrients and water. Changing the soil's temperature by just a few degrees can alter significantly the microbial populations underground, for example, which makes such research difficult to replicate and validate.

Modern agricultural research's focus on soil chemistry and above-ground portions of plants has led to solutions that favour chemical and mechanical means. The belief that chemical fertilizers are the best way to deal with soil fertility limitations has arisen from – and has reinforced – the image of agriculture as a kind of industrial enterprise, where producing desired outputs is mostly a matter of investing certain kinds and amounts of inputs. Consequently, viewing agriculture more as a biological than as a mechanical process attaches greater value to the use of organic inputs. In recent years there has been a major increase in the application of biologically based technologies, such as vermiculture (raising worms) to enhance soil fertility and ameliorate the negative effects of industrial and agricultural wastes on soil (Appelhof et al, 1996; Acharya, 1997).

As in most things, combinations of factors are more likely to approach the optimum than one factor by itself. It is well known that for plants to utilize chemical fertilizer effectively, the soil in their root zone must have substantial capacity to retain and exchange nutrient cations, and that exchange capacity is considerably enhanced as soil organic matter content increases. Research shows the benefits of utilizing organic means to maintain soil fertility and also of adding some inorganic nutrients in combination with organic inputs to get the best results.[15]

Adding appropriate amounts and combinations of chemical nutrients can increase both plant productivity and the amount of crop residues (shoots and roots) that become available to increase and maintain soil organic matter. Augmenting organic matter is especially necessary in tropical soils, which, due to climatic and edaphic conditions, are more likely to need maintenance and restoration of organic material and nutrients. The bottom line is that chemical fertilizers by themselves are no substitute for incorporation of soil organic matter. Ideally both will be used in synergistic ways.[16]

Irrigation is Not the Only Way to Deal with Water Limitations

A mechanistic conception of agriculture reinforces the millennia-old fixation on irrigation as the best if not the only means of providing water for plants in water-scarce environments. In many places, given hydrological cycles and opportunities, irrigation is certainly necessary for the practice of agriculture. But its success over several thousand years has led people to look to this technology as the universal solution to water scarcity problems. When crops need water, the first thought is how to provide irrigation from surface or groundwater sources.

But there are other ways to meet crop requirements besides capturing water in a reservoir, by river diversion or by pumping it from some body of water above or below ground, and then conveying it through canals and other structures to deliver it to particular fields, in amounts and at times when it is needed.[17] In much the same way that assuming soil fertility problems are best solved by fertilizer applications, seeing water shortages as best handled by irrigation has made water harvesting and conservation almost lost arts. When farmers in semi-arid Burkina Faso, assisted by OXFAM, demonstrated that they could grow much better millet crops simply by placing rows of stones across their fields, to slow water runoff and store it in the soil, this was seen as a remarkable technology (Harrison, 1987, pp165–170). Chapters 10 and 11 report on the use of similar water retention methods in Senegal and Mali; numerous case studies with similar results have been documented in Reij et al (1996). Such practices should become part of the repertoire of soil and water management practices that farmers can adopt to utilize available rainfall most advantageously. Using mulch to capture water and slow evaporation is another simple method.

Measures to conserve and utilize water, like planting crops in certain rotations or seeding a new crop in a standing one to capitalize on residual moisture, should not be seen as something novel but rather as something normal, making the best use of water in combination with soil. Methods including collecting and storing water in small catchment dams, large clay jars or simply in porous soils should be experimented with to determine what designs can provide enough water to crops and animals (and for human uses) to justify the expenditure of labour and capital and sometimes land. Small catchment ponds are becoming more attractive and feasible options, as discussed in Chapter 6, providing water supplies in situ. We should also understand better how land preparation practices affect water retention and utilization.[18]

Irrigation will surely remain a major means for solving water problems, and we should be learning how to use scarce irrigation water more efficiently and effectively through means of social organization (Uphoff, 1986; 1996). But irrigation is not the only means to ensure that growing plants and animals have the water they need. Water scarcity will surely increase for agriculture around the world, so all possible means to acquire and conserve water need to be considered.

Genetic Manipulation is Not Always Necessary to Raise Production Significantly

The modern approach to agricultural improvement has stressed better plant and animal breeding, especially since the advent and success of the Green Revolution. Without denying the value of such efforts, or that there will be some future benefits from biotechnology, we think more attention should be paid to cultural practices, to soil preparation and management, to use of organic inputs, to more productive cropping patterns and systems, and to species that have previously been overlooked or underutilized.

A good example is the system of rice intensification (SRI) developed in Madagascar which can boost yields from any variety of rice by 100 to 200 per cent or more by changing management practices and without requiring any use of purchased inputs, as reported in Chapter 12. There are other examples of major yield increase potentials with staple crops. In the 1970s, a programme in Guatemala was able to help farmers raise their maize and bean yields from 400–600kg/ha to about 2400kg in just seven years, at a cost of about US$50 per household. Farmers who had become acquainted with experimentation and evaluation methods proceeded to double yields once more on their own after external assistance was withdrawn (see Chapter 13; also Krishna and Bunch, 1997). Very poor farmers working with an NGO in the high Andean regions have found that they could double or triple their yields of potatoes and barley by using lupine, a leguminous plant, as a green manure to add nitrogen to the very poor mountain soils and increase soil organic matter (Chapter 14). This method, like SRI in Madagascar, works with whatever varieties farmers are already planting and uses organic rather than chemical inputs from outside the community. Leguminous fallows, as reported in Chapter 8, can raise maize yields in southern Africa by two to four times.

The Mukibat technique, named after the farmer who devised it in Indonesia almost 50 years ago, can increase the yield of cassava by five times or more. It involves grafting cassava tubers onto the root of a wild rubber tree of the same genus as cassava, which gives the growing tubers more access to sunlight and nutrients (Foresta et al, 1994). That this technology has aroused so little scientific attention, and was not reported in the literature until more than 20 years after it was devised (Bruijn and Dharmaputra, 1974), may reflect the indifference among most researchers towards cassava, a low-status staple crop on which hundreds of millions of people depend for much of their sustenance. Or perhaps it reflects a lack of interest in innovations that do not come from the scientific community.

Smallholding farmers around the world at present are probably exploiting less than 50 per cent of the existing genetic potential of various crops due to less than optimal management. In many cases this is because the returns to labour are not high enough to justify intensification, but often it is a matter of not knowing how to capitalize on synergies that could raise these returns. Reducing the yield variability of traditional varieties and taking fuller advantage of their genetic potential through nutrient cycling and better soil and water management within complex farming systems could, we think, be a cost-effective strategy that complements longer-run and higher-cost biotechnological efforts being undertaken to produce new and better varieties. Increased production of other food sources, including fish culture, small animals and various indigenous plants, can augment in non-competing ways whatever nutrients are provided by staples.

Even if these alternative methods by themselves cannot achieve a doubling of world food production, they could contribute substantially to this, making up the difference that is unlikely to be produced by more modern means that are heavily dependent on inputs of energy, chemicals and water. Capitalizing on 'non-modern' opportunities will require reorientation of socioeconomic as well as biophysical thinking. It necessitates looking beyond the farm and its fields, and beyond particular crop cultivars, animal species and cultivation practices, to institutions and policies.[19]

UTILIZING THESE PRODUCTIVE OPPORTUNITIES

Doing 'more of the same' in either the so-called modern or traditional sectors of agriculture is not likely to be sufficient for meeting food needs in the decades ahead. Researchers, extensionists and policy-makers who wish to assist households around the world to become more food-secure, healthy and well-off need to consider how to make broadly-based improvements in output through evolving systems that are more intensive and more complex. These will resemble but improve upon present practices that are not fully or sustainably utilizing soil, biological and other resources.

Traditional farmers are for the most part quite resource-constrained. The technologies offered by extension services were usually developed for larger, simpler production systems that are not appropriate for the kinds of systems that the majority of farmers in the world are managing. There are wide variations in productivity within and across farming communities, with some producers tapping production potentials better than others. We look towards 'hybrid' strategies to raise production, combining the best of farmers' current practices with insights derivable from modern science to tap the power of plant and animal germplasm nurtured under optimal conditions.

There is no reason to believe that the elements of 'modern' agriculture are wrong, but neither is there a warrant to consider them (yet) complete. They offer many advantages of productivity and profit for large numbers of agricultural producers – but not for all of them, and maybe not even for a majority of farming households around the world today. Our analysis here calls into question the presumption, whether it is argued or assumed, that mainstream approaches are the best or the only way to advance agriculture in the future. For the sake of productivity and sustainability, it will be advisable to 'backcross' some of the modern varieties of agriculture, which are most suitable for advantaged producers and regions, with often more traditional methods so as to develop a more robust 'hybrid' agriculture, one that can better meet the world's needs for food, health, employment and security in this century. These considerations will be recapped in Chapter 22 with a schematic comparison of present mainstream agricultural thinking and strategy vis-à-vis that derived from an agroecological understanding of needs and opportunities.

NOTES

1   This is not a statement in opposition to research on genetic modification, a controversial subject these days. Transgenic research has some potentially valuable, legitimate and safe uses and we would not want to see it curtailed – though more oversight and regulation and a different international property rights regime would make this enterprise more defensible and beneficial. Improvements in pest- and drought-resistance, for example, if achieved through advanced technology, could be great boons, particularly for the poor. Our focus on opportunities to raise production through different, more intensive management practices aims at a diversified strategy of agricultural development, one which will include work on genetic improvements.

2   'Had the cereal yields of 1961 still prevailed in 1992, China would have needed to increase its cultivated cereal area more than three-fold and India about two-fold, to equal their 1992 harvests' (Borlaug and Dowswell, 1994).

3   One of the preeminent agricultural development projects in the 1960s and 1970s, Plan Puebla in Mexico, was set up to benefit rural smallholder households by increasing their production of maize under rainfed conditions. Maize was considered their main crop. Yet a survey in the Puebla area showed that animal production provided 28 per cent of households' income, more than the 21 per cent that came from maize and almost as much as from the sale of all crops, 30 per cent. In addition, 40 per cent of household income came from off-farm employment (Diaz Cisneros et al, 1997, p123). The project made little progress with small farmers until it sought to improve production of beans along with maize, as these crops when grown together produced more than maize grown by itself and also contributed more to family nutrition. Farmers' cooperation also increased when other lines of production were assisted by the project (Whyte and Boynton, 1983, pp36–40). A more recent survey of 206 households selected randomly in four villages in the northern Philippines found that livestock contributed almost as much to household incomes (90 per cent as much) as did their rice production (Lund and Fafchamps, 1997).

4 In a watershed development programme in the Indian state of Rajasthan, where a participatory approach to technology development was taken that aimed to capitalize on local knowledge, fodder production on rainfed common lands was increased eight- to ten-fold with corresponding improvements in soil conservation (Krishna 1997, pp261–262). While such areas usually face serious physical constraints on increased production because they have been so neglected by researchers and extension personnel, they often offer substantial opportunities, previously ignored, for raising output.

5 This is discussed by Chambers (1997, especially p47). While rice, wheat and maize have received the lion's share of research funding, at least four of the international agricultural research centres in the CGIAR system have some of these other staple crops as a central part of their mandates. There are also centres now working on animals, agroforestry and aquaculture, though the centre on horticulture has yet to become part of the system (due to political reasons). The centres responsible for working on rice (International Rice Research Institute – IRRI) and wheat and maize (International Centre for the Improvement of Wheat and Maize – CIMMYT) are increasingly undertaking research that relates these staples to the growing of other crops.

6 As with most generalizations, this has some exceptions. Some perennial crops make heavy demands on soil nutrients, and others such as pineapples can require heavy agrochemical applications. On the general value of perennials in cropping systems, see Piper (1994) and Piper and Kulakow (1994).

7 Managed fallows have been largely ignored in the existing agricultural literature. To remedy this lack, a Southeast Asian regional workshop on intensification of farming systems was held in Bogor, Indonesia, in June 1997, with over 80 papers prepared for this collaborative effort of ICRAF, CIIFAD, the International Development Research Centre of Canada, and the Ford Foundation. Documentation of these resource management systems, mostly developed by farmers, is published in Cairns (2000).

8 The German and Dutch words for agriculture, *Landbau* and *Landbouw*, are more congenial to a three-dimensional conception of agriculture as they mean land-*building*.

9 'Mechanization' as used here refers to tractorization. Other forms of mechanization such as water pumps can be very valuable for increasing production, but they are not necessarily linked to monocropping in the way that tractorization is.

10 Those who can afford tractors usually own the best-quality land, making their practice of agriculture appear better.

11 When the labour power available for agricultural production is a constraint in some countries, this often reflects the fact that the low prices paid for agricultural commodities are keeping rural wage rates correspondingly low, influenced by urban-biased national policies and/or agricultural production subsidies in industrialized countries. National policies in developing countries have generally favoured urban consumers over rural producers, leading to low prices for food. Low food prices also reflect the extent of poverty, which depresses the purchasing power of the poor who have need for more food but do not have the means (effective demand) with which to acquire it. In such situations, low wages and low labour productivity for agriculture do not reflect either a true equilibrium or an efficient use of resources in terms of meeting human needs.

12 See Steiner (1982). That monocrop yields, being single, are easier to measure has contributed to the popularity of monocropping as a subject for agricultural research and extension. More effort is required to assess polycropping precisely.

The UN Food and Agriculture Organization (FAO)'s world census of agriculture in the 1980s specifically ignored all crop mixtures, deciding to record crops only as monocultures (Chambers, 1997, p95).

13   This has been seen and documented most dramatically in Indonesia, where an IPM programme started with FAO assistance showed that rice yields would not decline, and in some instances increased, when use of chemicals was drastically cut back (more than 50 per cent), and in some cases terminated where cultural practices were changed. The key was giving farmers effective hands-on training in agroecosystem management, so that they began to diagnose problems themselves and experiment with solutions, developing alternatives to chemical dependence (Oka, 1997). Widespread use of chemicals had increased the problem of pest attacks on rice, inducing build-up of pesticide resistance in pest populations at the same time that it reduced the population of spiders and other 'beneficials' that prey on pests.

14   Recent research on rice IPM has found that maintaining the populations of 'neutral' insects in rice paddies, insects that are neither pests nor beneficials, is important. Their presence can sustain the populations of beneficials when pests have been eliminated, keeping these populations vigorous and available to deal with any new increases in pest populations. Keeping sufficient organic matter in the soil to support populations of neutrals is becoming part of an IPM strategy (personal communication, Peter Kenmore, during Bellagio conference).

15   See Fernandes et al (1997). On infertile acid soils, farmers often need to use certain chemical nutrients such as phosphorous and calcium to prime biological processes such as nutrient recycling and nitrogen fixation. Research in Costa Rica found that when cultivating beans, mulches of organic matter prevent phosphorous fertilizer from becoming bound to aluminium and other ions in the acid soil, making it more available for plant nutrition. Phosphorous applied in conjunction with organic material produced as good or better yields as when three times as much phosphorous was applied directly to the soil (Schlather, 1998).

16   There is research indicating that the application of inorganic nitrogen fertilizer suppresses potentials for biological nitrogen fixation by reducing micro-organisms' production of the enzyme nitrogenase which enables soil microbes to transform nitrogen from the atmosphere into forms usable by the roots (Van Berkum and Sloger, 1983). This suggests that naturally-occurring nitrogen can be made unavailable by the application of nitrogen fertilizers, but it does not negate the point that organic and inorganic sources of nutrients are best managed in a complementary manner. It is worth contemplating the fact that since 1950, applications of nitrogen fertilizer have increased about 20-fold (Smil, 2000; p109), while crop yields have gone up at most three-fold. While nitrogen is often a limiting factor for plant growth, if it were of overwhelming importance for plant production, we should see more proportional increases in yield, rather than such sharply diminishing returns.

17   'The importance of water-control techniques in contrast with irrigation is consistently underestimated in the literature. There is a wide range of these techniques, including those that just hold water in the sandier soils [by increasing soil organic matter] as well as a series of measures to reduce runoff where crusting is the problem. These are not just indigenous techniques. The most important ones in the next decade have large potential yield effects (when combined with inorganic fertilizers) and need to be undertaken during the crop season, generally with animal traction, and not just as emergency measures on the most degraded or most easily degraded regions (hillsides)' (Sanders, 1997, p19). On this point generally, see FAO (1994).

18 In the rice–wheat rotation systems widely used in the Indo-Gangetic Plains of South
    Asia, certain kinds of ploughing techniques, adjusted by depth and timing, can
    retain enough water from the rice season for the following wheat season, so that
    the amount of water needed for the latter crop is reduced (personal communica-
    tion, Craig Meisner, CIMMYT/CIIFAD). Seeding wheat in the standing rice crop
    towards the end of its growing season enables the wheat crop to benefit from resid-
    ual soil moisture, reducing the need for irrigation (personal communication, Peter
    Hobbs, CIMMYT).

    These low-till methods are being promoted by CIMMYT and IRRI because
    they can save water, raise yields, lower production costs, reduce weeds and herbi-
    cide use, plus reduce greenhouse gas emissions ('New Movement Among Farmers
    to Give Up the Plow Takes Root', press release from Future Harvest, The Hague,
    2 October 2001, http://futureharvest.org/new/lowtill.shtml).

19 Most of the ideas in this chapter have been prompted from the co-authors' inter-
    actions with colleagues at Cornell University and in developing countries where
    CIIFAD has been engaged in collaborative, interdisciplinary programmes since
    1990 to further the prospects for sustainable agricultural and rural development
    (Uphoff, 1996a). It is hard to know where ideas come from, and to give full or
    proportional credit where it is due. We take responsibility for presenting these
    ideas for critical consideration by researchers and practitioners, not claiming
    personal credit for all of them, and acknowledging our indebtedness to colleagues
    at Cornell and elsewhere for the stimulation and challenge they have contributed
    to this thinking. Critical review by Rainer Assé and Christopher Barrett of the
    whole manuscript was particularly helpful.

Chapter 3

Agroecological Principles for Sustainable Agriculture

Miguel A Altieri

The concept of sustainable agriculture is a relatively recent response to the decline in the quality of the natural resource base associated with modem agriculture (Audirac, 1997). Today, agricultural production does not get evaluated in purely technical terms but also with regard to a more complex set of social, cultural, political and economic dimensions. Some of these latter issues are discussed in Chapter 4. The discussion here focuses on biophysical issues and dynamics, presenting agroecology as a concept and as a strategy.

The concept of sustainability, although controversial and diffuse due to conflicting definitions and interpretations of its meaning, is useful because it calls attention to agricultural opportunities grounded in the co-evolution of socioeconomic and natural systems (Reijntjes et al, 1992). To gain a broader understanding of the agricultural context, it must be studied in relation to the global environment and social systems since agricultural development results from the complex interaction of a multitude of factors. At the same time, a deeper understanding of the ecology of agricultural systems should open new management options more in line with the objectives of a truly sustainable agriculture.

The sustainability concept has prompted much discussion and has led to proposed adjustments in conventional agriculture to make it more environmentally, socially and economically viable and compatible. Various solutions to the environmental problems created by capital- and technology- intensive farming systems have been suggested, and research evaluating alternative systems is being undertaken (Gliessman, 1998). Two main focuses are on plant protection through organic nutrient sources and integrated pest management (IPM), and on the reduction or elimination of agrochemical inputs by making changes in management that give adequate plant nutrition and better crop protection. This challenges two of the 'equations' discussed in the preceding chapter. Different soil and water management practices are also important to nurture microbiological populations in the rhizosphere.

Although hundreds of research projects have now shown the benefits from such reorientation, and many lessons have been learned from old and new practices, most development investment and research programmes still emphasize chemical or engineering solutions, seeking to suppress limiting factors or eliminate the symptoms that reflect ill-balanced agroecosystem dynamics. The prevalent view is that pests, nutrient deficiencies or other particular factors are the cause of low productivity, rather than that pest or nutrient problems reflect agroecosystem conditions that are not in a biological equilibrium (Carrol et al, 1990). The focus on specific factors that can raise productivity or on limiting factors to be overcome through new technologies remains a narrow, mechanistic one. It has diverted agriculturists from appreciating the context and complexity of agroecological processes, which in turn has led to inadequate understanding of the root causes of constraints in the agricultural sector (Altieri et al, 1993).

Agroecology is an applied science, adapting ecological concepts and principles to the design and management of sustainable agroecosystems and providing a framework for assessing the performance of agroecosystems (Altieri, 1995). When fully developed, agroecology does more than inform the selection and use of alternative practices; it helps farmers fashion and maintain agroecosystems that have minimal dependence on expensive chemical and energy inputs. Agricultural systems are supported by interactions and synergies between and among biological components that enable these systems to sponsor their own soil fertility, productivity enhancement and crop protection (Altieri and Rosset, 1995).

PRINCIPLES OF AGROECOLOGY

Just adding or subtracting certain practices or elements within present production practices will not produce a more self-sufficient and self-sustaining agriculture. This transformation requires deeper understanding of the nature of agroecosystems and of the principles by which they function. Agroecology goes beyond the perspectives of genetics, agronomy, hydrology and so on, to devise an understanding of *co-evolution* at both ecological and social levels of agricultural systems' structure and functioning. Rather than address any one particular component of the system, agroecology stresses the inter-relatedness of all agroecosystem components and the complex dynamics within ecological processes (Vandermeer, 1995).

Agroecosystems are communities of plants and animals interacting with their physical and chemical environments that have been modified by people to produce food, fibre, fuel and other products for human consumption and processing. Agroecology focuses on the forms, dynamics and functions of inter-relationships among environmental and human elements, and on the multiple, parallel processes in which these elements and their interactions are involved. An area used for agricultural production, such as a field, is regarded as a system in which ecological processes that are found also under natural conditions are occurring: eg, nutrient cycling, predator/prey interactions, competition among species, symbiosis and successional changes. Implicit in agroecological research is the idea that, by understanding these ecological relationships and processes, agroecosystems can be enhanced to improve production and to produce food, fibre, etc more sustainably, with fewer negative environmental and social impacts, and using fewer external inputs.

The design of such systems, described in Table 3.1 in terms of ecological processes to be optimized, applies the following ecological principles (Reijntjes et al, 1992):

- *Enhance the recycling of biomass*, with a view to optimizing nutrient availability and balancing nutrient flows over time.
- *Provide the most favourable soil conditions* for plant growth, particularly by managing organic matter and by enhancing soil biotic activity.
- *Minimize losses of energy and other growth factors* within plants' microenvironments above and below ground. These losses result from unfavourable flows of solar radiation, air and water. Reduction is accomplished through microclimate management, water harvesting, and better soil management and protection through increased soil cover.
- *Diversify species and genetic resources* in the agroecosystem over time and space.
- *Enhance beneficial biological interactions and synergies* among the components of agrobiodiversity, thereby promoting key ecological processes and services.

These principles can be applied through various techniques and strategies. Each will have different effects on productivity, stability and resiliency for farm systems, depending on local resource constraints and opportunities, which include the effects of market forces and dynamics. The ultimate goal of agroecological design is to help integrate components so that overall biological efficiency is improved, biodiversity is preserved, and the productivity of agroecosystems and their self-sustaining capacities are maintained. The goal is to knit together agroecosystems within a landscape unit, with each system mimicking as best it can the structure and function of natural ecosystems.

Table 3.1 *Ecological Processes to be Optimized in Agroecological Systems*

- Strengthening of the 'immune system' of agricultural operations – nurture proper functioning of natural pest control.
- Decreasing toxicity in the environment through reduction or elimination of agrochemicals.
- Optimizing metabolic functioning – organic matter decomposition and nutrient cycling.
- Balancing regulatory systems – nutrient cycles, water balance, energy flows, population regulation, etc.
- Enhancing conservation and regeneration of soil and water resources and biodiversity.
- Increasing and sustaining long-term productivity.

If synergies are correctly identified and appropriately nurtured, these should have beneficial effects on farmers' production and income, justifying (usually) more intensive management. (Sometimes agroecological analysis will point to more extensive management strategies.) Where synergies cannot be capitalized upon, agroecological initiatives will not be practical and thus not sustainable.

ADVANTAGES OF BIODIVERSIFICATION IN AGROECOSYSTEMS

From a management perspective, the agroecological objective is to achieve balanced environments with sustained yields, bolstered by biologically mediated soil fertility and natural pest regulation through the design of diversified agroecosystems and the use of low-input technologies (Gleissman, 1998). Agroecologists recognize that intercropping, agroforestry and other diversification methods correspond to natural ecological processes; the sustainability of complex agroecosystems thus derives from the ecological models they follow.

In farming systems that mimic nature, optimal use is made of sunlight, soil nutrients and rainfall (Pretty, 1995). Agroecological management aims to optimize the recycling of nutrients and organic matter turnover, closed energy flows, water and soil conservation, and balance within pest/natural-enemy populations. It exploits the complementarities and synergies that result from various combinations of crops, tree and animals in particular spatial and temporal arrangements (Altieri, 1994).

Optimal productivity within agroecosystems depends on the level and kind of interactions among various biotic and abiotic components. Functional biodiversity can initiate synergies that 'subsidize' agroecosystem processes naturally, through ecological services such as the activation of soil microbial populations, recycling nutrients, augmenting the numbers of beneficial arthropods and antagonists, and so on (Altieri and Nicholls, 1999). Today there is an increasingly diverse selection of practices and technologies available. These vary in their effectiveness as well as cost-effectiveness. Key practices are those that reinforce, through a series of mechanisms, the 'immunity' of the agroecosystem against outside assaults (Table 3.2).

Strategies to restore agricultural diversity in time and space include the following kinds of practices, which exhibit beneficial ecological dynamics:

Table 3.2 *Mechanisms for Improving Agroecosystem Immunity*

- Increased number of plant species and genetic diversity over time and space.
- Enhancement of functional biodiversity – natural enemies, antagonists, etc.
- Enhancement of soil organic matter and biological activity.
- Increase of soil cover and crops' competitive ability.
- Elimination of toxic inputs and residues.

- *Crop rotations*: by incorporating temporal diversity into cropping systems, crop nutrients are provided from one season to the next, and the life cycles of insect pests, diseases and weeds are interrupted (Sumner, 1982; Liebman and Ohno, 1998).
- *Polycultures*: cropping systems in which two or more crop species are planted within certain limits of spatial proximity can result in complementarity that enhances crop yields (Francis, 1986; Vandermeer, 1989).
- *Agroforestry systems*: agricultural systems in which trees or other perennials are grown together with annual crops and/or animals can benefit from complementary relations between components, at the same time producing multiple products from the agroecosystem (Nair, 1982).
- *Cover crops*: the use of pure or mixed stands of legumes or other annual plant species, eg, under fruit trees for the purpose of improving soil fertility, enhancing biological control of pests and modifying the microclimate (Finch and Sharp, 1976); also intercropped plant species can reduce erosion and provide nutrients to the soil (Magdoff, 1992).
- *Animal integration in agroecosystems*: high biomass output and optimal nutrient recycling can be achieved through biological processing and the return of animal manure to the soil (Pearson and Ison, 1987).

These forms of agroecosystem management, though diverse, share the following features:

- *Vegetative cover* to conserve soil and water is maintained through the use of no-till practices, mulch farming, and use of cover crops and other appropriate methods.
- *Organic matter* is supplied to the soil through the addition of compost, green manures, animal manure and/or the promotion of biotic soil activity.
- *Nutrient recycling mechanisms* are enhanced, for example, through the integration of livestock systems based on legumes.
- *Pest regulation* is promoted through the enhanced activity of biological control agents, achieved by introducing and/or conserving natural enemies and antagonists (Altieri and Nicholls, 1999).
- *Soil aeration*, critical for plant performance, is supported through both biological and mechanical processes.

Research on diversified cropping systems has underscored the great importance of maintaining diversity in agricultural settings (Francis, 1986; Vandermeer, 1989; Altieri, 1995). There are a number of reasons for stressing the value of biodiversity in agroecosystems (Gliessman, 1998):

- As diversity increases, so do opportunities for coexistence and for *beneficial interactions between species* that can enhance agroecosystems' sustainability. The contributions of micro-organisms in the soil are presently poorly understood, but their enhancement can pay large dividends.
- Greater diversity usually gives *better resource-use efficiency* in an agroecosystem. Heterogeneity of habitat leads to better system-level adaptation

with complementarity in crop species' needs, diversification of niches, overlap of species niches, and partitioning of resources.

- Ecosystems in which plant species are intermingled possess an associated *resistance to herbivores*. Also, in diverse systems there is a greater abundance and diversity of natural enemies of pest insects, which helps to keep in check the populations of particular herbivore species (Andow, 1991).
- A diverse crop assemblage often creates a *diversity of microclimates* within the cropping system that can be occupied by a variety of non-crop organisms, including beneficial predators, parasites, pollinators, soil fauna and antagonists that are important to the entire system.
- Diversity within the agricultural landscape can contribute to the *conservation of biodiversity* in surrounding natural ecosystems.
- Diverse organisms within the soil perform a variety of *ecological services* such as nutrient recycling and detoxification of noxious chemicals, as well as regulation of plant growth (Hendrix et al, 1990).
- Diversity *reduces risk* for farmers, especially in marginal areas with unpredictable environmental conditions. If one crop does not do well, the income from others can compensate.

AGROECOLOGY AND THE DESIGN OF SUSTAINABLE AGROECOSYSTEMS

Most people promoting sustainable agriculture aim at maintaining productivity over the long term through a variety of methods. This is done by:

- Optimizing the use of *locally available resources* – combining different components of the farm system, ie, plants, animals, soil, water, climate and people, so that they complement each other and have the greatest possible synergetic effects.
- Reducing reliance on *off-farm, non-renewable inputs* – in part because many have potential to damage the environment or can harm the health of farmers and consumers. Economic benefits accrue to farmers from minimizing their variable costs of production by targeting the use of external inputs more carefully.

Agroecological approaches do not assume that there will be no outside inputs, but there is a burden of proof that these will actually add to economic and environmental net benefits over multiple years, and that such benefits cannot be attained by other, less costly means. This leads to the following principles:

- Relying as much as possible and economic on *resources within the agroecosystem* – with nutrient cycling, better conservation and expanded use of local resources.
- Improving the *match-up between cropping patterns and productive potentials* – as well as matching crops with environmental constraints of climate

and landscape, to ensure the long-term sustainability of current production levels.

*   Working to enhance appreciation of and to conserve *biological diversity*, both in the wild and in domesticated landscapes, making optimal use of the biological and genetic potentials of plant and animal species.
*   Taking full advantage of *local knowledge and practices*, including innovative approaches not yet fully understood by scientists although widely adopted by farmers (Pretty, 1994; Vandermeer, 1995).

The goal of agroecological design efforts is thus to integrate components in ways that improve overall biological efficiency, preserve biodiversity, and maintain agroecosystem productivity and its self-regulating capacity. By approximating the structure and function of natural ecosystems in a given locality, an agroecosystem with high species diversity and biologically active soil promotes natural pest control, nutrient recycling, and continuous soil cover to prevent resource losses.

APPLYING AGROECOLOGICAL PRINCIPLES

Agroecological analysis provides guidelines for developing diversified agroecosystems that take advantage of the effects of the integration of plant and animal biodiversity. Such integration enhances complex interactions and synergies, optimizing ecosystem functions and processes such as biotic regulation of harmful organisms, recycling of nutrients, and biomass production and accumulation. It enables agroecosystems to sponsor and support their own functioning, with the result that farming systems are economically and ecologically more sustainable, with management systems attuned with the local resource base and operating according to existing environmental and socioeconomic conditions.

In an agroecological strategy, management components should address the conservation and enhancement of local agricultural resources – germplasm, soil, beneficial fauna, plant biodiversity, etc – by encouraging a development methodology that supports farmer participation, use of traditional knowledge and the adaptation of farm enterprises to fit local needs and match up with socioeconomic as well as biophysical conditions. The larger realm of social and institutional factors is discussed in the next chapter, and economic considerations in Chapter 5.

Chapter 4

Social and Human Capital for Sustainable Agriculture

Jules Pretty

Economic and social systems at all levels – from farms and livelihoods to communities and national economies – rely for their success on the value of the services that flow from the total stock of assets that they control. Five types of capital – natural, human, physical, financial and social – are now being addressed in the literature. Much of the recent thinking on types of capital has been prompted by the 'discovery' of social capital, which has built a bridge between economists and other social scientists.[1] While there has been intuitive understanding of social capital for many years, ambiguities in its conceptualization and measurement kept this non-material factor of production off development agendas until recently, despite its important material consequences. Now that it is clearly on the agenda, social and other scientists will find it a useful expansion upon previous analytical and policy thinking. The five types can be described briefly as follows.

Natural capital produces nature's goods and services. These are varied, including food (both farmed and harvested, or caught from the wild), wood and fibre; water supply and regulation; treatment, assimilation and decomposition of wastes; nutrient cycling and fixation; soil formation; biological control of pests; climate regulation; wildlife habitats; storm protection and flood control; carbon sequestration; pollination; and recreation and leisure.

Human capital is the total capability residing in individuals, based on their stock of knowledge and skills as well as their health and nutrition. It is enhanced by people's access to services that enhance these, such as schools, medical services and adult training. People's productivity is increased by their capacity to interact with productive technologies and with other people. Leadership and organizational skills are particularly valuable for making other resources more productive.

Physical capital is the store of human-made material resources, including buildings (housing, factories), market infrastructure, irrigation works, roads and bridges, tools and equipment, communication systems and energy and transportation facilities, which make labour more productive and better utilize natural resources.

Financial capital is accumulated claims on goods and services, built up through financial systems that gather savings and issue credit. It includes pensions, remittances, welfare payments, grants and subsidies.[2]

Social capital yields a flow of mutually beneficial collective action, contributing to the cohesiveness and cooperation among people in their respective societies. The social assets comprising social capital include norms, values and attitudes that *predispose* people to cooperate, eg, reciprocity, solidarity and trust, as well as various roles, rules, precedents and procedures that *facilitate* cooperation, which can make better use of all available resources (Uphoff, 2000).

These different kinds of assets are transformed by policies, processes and institutions to create outcomes such as food, jobs, welfare, economic growth, clean environment, reduced crime and better health and schools. Desirable outcomes, when achieved, feed back to increase the asset base in its various forms, while undesirable results or side-effects from production processes such as pollution, deforestation, school dropouts, increased crime or social breakdown reduce the asset base.

The basic dynamic for sustainable development requires that the operation of farms, firms, communities and economies add to the stocks of these five assets, thereby increasing per capita endowments of all forms of capital over time. Unsustainable systems, on the other hand, deplete or run down these various forms, thereby reducing the productive possibilities for future generations. In particular situations, one form of capital or another may be in relatively short supply, and thus increasing it will have greater payoff than adding to others. Social and human capital are particularly pivotal for making these processes accumulative rather than dissipative, and where they are lacking, productive processes are seriously undermined.

THE VALUE OF SOCIAL CAPITAL

There has been a rapid growth in interest in 'social capital' in recent years (Woolcock, 1998; Dasgupta and Serageldin, 2000). The term captures the idea that social bonds and social norms are important for attaining sustainable livelihoods. Coleman (1990) describes it as 'the structure of relations between actors and among actors' that encourages productive activities. Certain aspects of social structure and organization, supported by mental predispositions to trust other people, to value others' wellbeing along with one's own, and to expect reciprocation, serve as resources for individuals to achieve things through collective action that could not be accomplished by the individuals alone. Local institutions are effective because 'they permit us to carry on our daily lives with a minimum of repetition and costly negotiation' (Bromley, 1993).

The following kinds of social relationships are particularly important for sustainable development.

Relations of Trust

Trust lubricates cooperation, and by reducing the transaction costs between people it frees up time and resources for other purposes. Instead of having to invest in monitoring others' behaviour, individuals can be confident that others will act as expected, thereby saving both money and time. Trust also creates networks of social obligation and cooperation, in that trusting others commonly engenders reciprocal trust. There are two main types of trust: that which we have in individuals whom we know, and trust in people whom we do not know, which arises because of our confidence in known social structures and shared thinking. Both are important, but the latter is crucial for creating larger enterprises of social, economic and political cooperation. While trust takes time to build, it is easily broken (Fukuyama, 1995).

Reciprocity and Exchanges

Two types of reciprocity in exchange relationships were identified by Coleman (1990). Specific reciprocity refers to simultaneous exchanges of things of roughly equal value, while diffuse reciprocity refers to continuing relationships of exchange that at any given time may be unrequited, but over time are repaid and balanced. The latter connections in particular contribute to the formation of long-term productive relationships among people. Sustainable development depends on patterns of cooperation that support resource mobilization and investment over time that create public as well as private goods.

Common Norms, Rules and Sanctions

Mutually agreed or handed-down norms of behaviour that place group interests above those of individuals give people confidence to invest in collective or group activities, knowing that others will do so too. They encourage individuals to take initiative with some assurance that their rights will not be infringed. Accepted sanctions ensure that those who break the rules know that they will be punished. These are sometimes called the rules of the game (Taylor, 1982), the internal morality of a social system (Coleman, 1990) or the cement of society (Elster, 1989). The value and productivity of these normative orientations is made clear by the consequences of their absence: destructive conflict, lack of sharing and insecurity.

Networks and Groups

Connectedness among people is a vital aspect of social capital. There can be many different types of connection between groups: trading of goods, exchange of information, mutual help, provision of loans, common celebrations such as prayers, marriages or funerals. Relationships may be one-way or two-way, and they may be long-established (and not very responsive to current conditions) or subject to regular revision.

Connectedness can be manifested in different types of groups at local levels – from guilds and mutual aid societies, to sports clubs and credit groups, to forest, fishery or pest management groups, to literary societies and mother-and-toddler groups. It also implies connections to other groups in society, from micro- to macro-levels (Uphoff, 1993; Woolcock, 1998; Rowley, 1999).[3] Connectedness can be observed in five different contexts:

1   *Local connections*: strong connections between and among individuals and within local groups and communities.
2   *Local–local connections*: horizontal connections between and among groups within communities and between communities – these connections can sometimes become platforms for new higher-level institutional structures.
3   *Local–external connections*: vertically-oriented connections between local groups and external agencies or organizations, which can be either one-way (top-down) or two-way.
4   *External connections*: connections between and among individuals who are operating within external agencies.
5   *External–external connections*: horizontal connections among external agencies, leading to collaborative partnerships and integrated approaches to development.

Even when the value of social capital is recognized in general, it is common to find only some of these kinds of connections being attended to. For example, a government may stress integration between different sectors and/or disciplines, yet fail to encourage two-way, reciprocating vertical connections with local groups. A development agency may support the formation of local associations without any effort to build upward linkages with government agencies, though a lack of such linkage impedes their chances of success.

This analysis implies that: (a) the more linkages the better; (b) two-way relationships are better than one-way; and (c) linkages that are subject to regular revision will be more suited to current conditions and needs than historically-embedded ones. Rowley (1999) found a positive relationship between connectedness and wealth when studying social capital in sub-Saharan Africa; but the direction of causality was unclear – did well-connected people become rich, or are rich people better able to afford to be well connected? In some situations, a group might benefit from isolation, being able to avoid costly, unilateral external demands.[4]

It is advisable to keep in mind that there are multiple types of social capital that enhance capacities to solve public problems and empower communities rather than thinking and talking just about overall quantitative increases in social capital. With growing uncertainty about economies, climates, and political processes and their greater fluctuation, the capacity of people to innovate and to adapt known technologies and practices to suit new conditions becomes vital. If, as some believe, such uncertainty is growing, then the need for innovation is also growing. An important question is whether sufficient forms of social capital can be built up and sustained that will enhance capacity for collective innovation and the requisite cooperation to utilize this (Boyte, 1995; Hamilton, 1995).

MAKING IMPROVEMENTS SUSTAINABLE

Development assistance can claim a number of successes in recent decades – in education and public health, in public institution building, in technology development and extension and in sector support and reform. But for the most part, external efforts have failed to make sufficient, lasting improvements for large numbers of the people, communities and economies they were supposed to benefit. Many development initiatives appear to succeed initially, but then fade away after external support ceases. Projects that lead to short-term improvements that neither persist nor spread cannot be considered as successes.

In the agricultural and natural resource management sectors, there is much empirical evidence that failure is still very common. Reviews of more than a thousand projects funded by the World Bank, the European Commission, the Danish International Development Agency (DANIDA), the British Department for International Development (DFID, formerly ODA), and the Club du Sahel have shown that agricultural and natural resource initiatives performed worse in the 1990s than in the 1970s–1980s, and also worse than projects from other sectors.[5]

Conventional agricultural projects are unlikely to continue their achievements beyond the period when external inputs are provided. As a result, donors have been turning away from the agricultural sector.[6] Yet we know from a number of studies that agricultural development efforts can be successful and have long-term effects when people at the grassroots are well organized or are encouraged to form groups, and when their knowledge is sought and utilized during planning and implementation.[7] Thus, the human and social organizational dimensions of development have crucial implications for long-term benefits.

ELEMENTS OF SUSTAINABLE AGRICULTURE

What is understood by 'sustainable agriculture', and how can transitions in both 'pre-modern' and 'modernized' systems towards greater sustainability be encouraged? Sustainable farming seeks to make the best use of nature's goods and services without damaging the environment (Altieri, 1995; Pretty, 1995a, 1998; Thrupp, 1996; Pretty and Hine, 2001). It does this, as discussed in the preceding chapter, by integrating natural processes, such as nutrient cycling, nitrogen fixation, soil regeneration and use of natural enemies of pests into food production processes, minimizing the use of non-renewable inputs (pesticides and fertilizers) that can damage the environment or harm the health of farmers and consumers. In particular, it makes better use of farmers' knowledge and skills, thereby improving their self-reliance and capacities.

Sustainable agriculture is multifunctional within landscapes and economies, producing food and other goods for farm families and markets, while contributing also to a range of public goods, such as clean water, flood protection, carbon sequestration in soils, wildlife conservation and landscape quality. It delivers many unique non-food goods that cannot be produced by other sectors, eg on-farm biodiversity, opportunities for urban-to-rural migration, and social cohesion. A desirable end-point for both modern and pre-modern agricultural systems is to have operations that enhance both the private benefits for farm households and the public benefits accruing to society from other functions.

There are many promising technological options for more sustainable agriculture.

- Farmers can improve their agriculture by making better, more efficient use of non-renewable inputs, such as precision-farming, low-dose sprays and slow-release fertilizers.
- They can focus on better use of available natural resources, such as water harvesting (Chapter 11), better irrigation management, rotational grazing, or no-till agriculture (Chapter 15).
- They can intensify a single sub-component of farm operations, while leaving the rest alone, such as double-dug beds, digging a fish pond (Chapter 9) or adding vegetables to rice bunds (Chapter 16).
- They can diversify and strengthen the agroecosystem by adding regenerative components, such as combining agroforestry and livestock (Chapter 11), using legumes as cover crops (Chapter 14) or raising fish in rice paddies (Chapter 16).

Such innovations can be quite profitable for the farm operator while at the same time producing other streams of benefit, such as cleaner water or attractive landscapes and building up different kinds of capital – natural, human, physical, financial and social.

OLD DANGERS, NEW WORDS

A very real problem can arise, however, from such sustainable agriculture 'successes'. If the technical solutions are seen to be effective (and increasingly they are), but they are not linked to the social processes that give rise to them, then agricultural development in the name of sustainability could simply repeat the same problems of contemporary agriculture, fixated on certain technologies.

Modernist agricultural development proceeded with the conviction that certain technologies will raise production, and the challenge was to induce or persuade farmers to adopt them. Yet few farmers are able to adopt whole packages of conservation technologies without considerable adjustments in their own practices and livelihood systems, as pointed out with reference to 'the food security puzzle' that Brummett describes in Chapter 9.

Imposed models may look good at first, but they seldom have staying power. Alley cropping, an agroforestry system that plants rows of nitrogen-fixing trees or bushes between rows of cereal crops, has long been the focus of research (Kang et al, 1984; Lal, 1989). Many productive and sustainable versions of such systems, needing few if any external inputs, have been developed. They can stop erosion, produce food and wood, and can be cropped over long periods. But very few farmers have adopted alley cropping systems as designed. Despite millions of dollars of research expenditure over many years, systems have been produced that are largely suitable only for research stations (Carter, 1995).

There has been, however, some success with alley cropping where farmers were able to derive multiple benefits from it, or could take one or two components of recommended packages and adapt these to their own farms. In Eastern Indonesia, farmers have for many years planted rows of *Leucaena* along hillside contours with other crops, encouraged by the benefits of fodder production and weed control in addition to soil conservation and improved production (Piggin, 2000; also Agus, 2000). In Kenya, farmers planted rows of leguminous trees next to their field boundaries or single rows through their fields; in Rwanda, alleys planted by extension workers soon became dispersed through fields (Kerkhof, 1990). Such adaptations produced synergistic gains when interacting with particular soil, water, topographic and climatic conditions that were noticeably more beneficial relative to their cost than the benefits from using the full-cost package.

The prevailing view has been, however, that farmers should adapt their practices to the technology being offered. Evaluators for the Agroforestry Outreach Project in Haiti wrote disapprovingly:

> *Farmer management of hedgerows does not conform to the extension program. Some farmers prune the hedgerows too early, others too late. Some hedges are not yet pruned by two years of age, when they have already reached heights of 4–5 metres. Other hedges are pruned too early, mainly because animals are let in or the tops are cut and carried to animals ... Finally, it is very common for farmers to allow some of the trees in the hedgerow to grow to pole size* (Bannister and Nair, 1990).

This evaluation could be read as indicating that the project was a great success: farmers were adapting the technology to their own special needs. Yet the language of the evaluators suggests that the programme was a failure.[8] What are the implications for sustainable agriculture? The process by which farmers learn about technology alternatives is crucial. If innovations are enforced or coerced, they will not be adopted for long. Small modifications that could make the technology more beneficial will remain untapped as long as 'adoption' is the goal and criterion of success. Where the process of technology development and diffusion is participatory, on the other hand, and enhances farmers' capacity to learn about their farms and their resources, the foundations for redesign – drawing on both social and human capital – have been laid.

SOCIAL PROCESSES FOR SUSTAINABLE INNOVATION

It is critical that sustainable agriculture should not prescribe or be equated with a specific set of technologies, practices or policies. This narrows future options for farmers. As conditions change and as knowledge grows, so must farmers and communities be allowed, indeed encouraged, to change and adapt what they are doing. Sustainable agriculture is therefore not a model or a package to be introduced; it is more a process for learning (Röling, 1995; Pretty, 1995b). This process both depends on and builds up social and human forms of capital.

This process is seen in the Central American case studies reported in Chapter 13. In 1994, staff of the Honduran organization COSECHA (Associaciòn de Consejeros una Agricultura Sostenible, Ecològica y Humana) returned to communities in Guatemala and Honduras where participatory methods had been used 10 to 20 years previously to improve farming systems in poor hillside areas. They sought to evaluate changes that were made after external project support had been withdrawn (Bunch and Lòpez, 1996). The most obvious and impressive finding was that crop yields continued to increase after project termination, and that resource-conserving technologies were still being used (see also Chapter 6).

However, in both cases many of the technologies that had been considered as 'successful' during the project had been superseded by new practices. Some 80 to 90 successful innovations were documented in the 12 villages studied. In one Honduran village, Pacayas, there were 16 innovations made entirely by farmers, including four new crops, two new green manures, two new species of grass used for contour barriers supporting the growing of vegetables, chicken pens made of king grass, marigolds used for nematode control, use of lablab and velvet beans as cattle and chicken feed, nutrient recycling into fishponds, composting human wastes from latrines, planting napier grass to stabilize cliffsides and home-made sprinklers for irrigation.

Had the original technologies been poorly selected? Apparently not, because many that had been dropped by farmers in the study villages were now being used elsewhere in the country. Changing external and internal circumstances – such as market shifts, droughts, diseases, insect pests, land tenure, labour availability and political disruptions – had reduced or negated the usefulness of certain technologies. The study estimated that the half-life of a successful technology in these project areas was about six years. The technologies themselves are not sustainable, Bunch and Lòpez concluded; 'What needs to be made sustainable is the social process of innovation itself'.

A similar dynamic has been reported from the Indian state of Gujarat, where many farmers developed a variety of new technical innovations after receiving support from the Aga Khan Rural Support Programme for undertaking simple conservation measures. Farmers have started planting grafted mango trees and bamboo near embankments to make full use of residual moisture near gully traps. They have introduced cultivation of vegetables such as eggplant and okra, other leguminous crops and tobacco in the newly created silt traps. These measures increased production and income substantially, particularly in poor rainfall years. Most of these innovations and adaptations have been introduced and sustained with support from the local network of village extensionists (Pretty and Shah, 1997).

Another example comes from Thailand where, through four different phases of the Thai–German highland development project, one can see the importance of active involvement of local people (TG-HDP, 1995). The project was established to work with upland communities in Northern Thailand to support their transition towards sustainable agriculture. The resource-conserving technologies developed and adapted for local use have included hedgerows along contours, buffer strips, new crop rotations, integrated pest management, crop diversification and integration of livestock into farming systems.

The approach, however, has changed significantly since the mid-1980s (Table 4.1). In the first phase, cash incentives and free inputs were used to encourage adoption of these technologies, with high adoption rates but little or no adaptation of the technologies by farmers. In 1990, all incentives were stopped when the project adopted a participatory approach; adoption rates fell sharply, and withdrawal increased. But by 1993–1994, participatory village planning had begun to involve communities fully, and the ratio of adopters to withdrawers was now equal. Since then, the numbers of farmers using sustainable technologies has grown rapidly, but more important, farmers are now adapting these – and are innovating new technologies – to satisfy their particular needs (Steve Carson, personal communication, 1996).

LEARNING RATHER THAN TEACHING

Sustainable agriculture depends on new and more varied ways of learning about the world. Learning should not be confused with teaching, as the latter

Table 4.1 *Changing Phases in the Thai–German Highland Development Project, as Reported from 113 Villages in Nam Lang, Northern Thailand*

I 1987–1990	Cash incentives and free inputs
	High adoption of technologies, with little or no adaptation
	Adoption:withdrawal = 5:1
II 1991–1992	All incentives stopped; beginning of participatory work
	Adoption rates fell to 25% of first phase
	Withdrawal increased 3-fold
	Adoption:withdrawal = 1:2.2
III 1993–1994	Participatory village planning; communities fully involved
	Adopters and withdrawers equal in number
	Adoption:withdrawal = 1:1
IV 1995–1996	Adopters increasing as farmers adapt technologies and diversify efforts, eg, pineapple strips, lemon grass, cash crops, soil and water conservation
	Adoption:withdrawal = 3:1

Source: Steve Carson, personal communication, 1996

implies the transfer of knowledge from someone who already knows something to someone who does not know. Teaching is the normal mode of educational curricula and is central to many organizational structures (Bawden et al, 1984; Pretty and Chambers, 1993). Universities and other professional institutions have reinforced this teaching paradigm by viewing themselves as custodians of knowledge that can be dispensed or given, usually by lecture, to a recipient – a student or trainee.

Moving from a teaching to a learning style has profound implications for agricultural development institutions, as discussed further in Chapter 20. Where a problem situation is well defined, system uncertainties are low, and decision stakes are not terribly high, one may assume that standardized scientific and pedagogical methods will work reasonably well. But where problems are unavoidably ill-defined, and where uncertainties potentially affect many actors and interests, then alternative methods of learning become more promising.

We are ourselves still learning about the best conditions and approaches for engaging farmers as partners in the development and spread of more appropriate and sustainable agricultural technologies. The cases reported in Part 2 give many examples of strategies that have been successful, supporting the point that there is no single best approach. There is, however, a philosophy in common across most agroecological development efforts: one that emphasizes respect for what farmers can contribute to the process, multifaceted partnerships with a diverse set of actors, and a self-critical and continuous 'learning process' mode of operation. The desired synthesis will be not just between and among biophysical approaches or of social and learning methodologies. Rather, it will be between biophysical investigations and applications, on the one hand, and social and human processes of cooperation and learning, on the other, with a resulting wedding of science and philosophy.

NOTES

1   Contributions to this literature that illuminate 'social capital', which is our focus here, include: Bourdieu (1986), Coleman (1988, 1990), Putnam (1993, 1995), and Carney (1998). Its implications for development are addressed in Grootaert (1998), Ostrom (1998), Pretty (1998), and Uphoff (2000), with concrete applications and efforts at measurement in Krishna and Uphoff (1999) and Uphoff and Wijayaratna (2000). The following discussion is elaborated in Pretty and Ward (2000).

2   Financial capital has commonly been grouped together with physical capital, since it has material bases and can be accumulated to support expanded production. Marx's analysis included both categories of capital under this heading. However, considering financial capital separately in this framework expands the evident options available in any given context for improving the overall capital base.

3   High social capital is associated with multiple membership organizations and many links between groups. But one can imagine a situation with large numbers of organizations, each protecting and advancing its own interests with little cooperation, where outcomes are zero-sum, or even negative-sum, rather than positive-sum such as results from mutually-beneficial collective action. Organizational density may

be high, but inter-group connectedness low (Cernea, 1993). Connectedness is thus an aspect of social capital. Two categories of particular interest have been identified: *bonding* social capital that increases intra-group solidarity, and *bridging* social capital that supports inter-group endeavours (Narayan, 1999).

4   There is evidence that horizontal and vertical linkages contribute to developmental success both at the macro, national level (Uphoff and Esman, 1974) and at the micro, community or organizational level (Esman and Uphoff, 1984). While horizontal linkages contribute more than vertical ones, both are productive, and their contributions have synergistic effects. This contradicts Putnam's preference (1993) for horizontal over vertical linkages.

5   These evaluations include: Cernea (1991), Pohl and Mihaljck (1992), World Bank (1993), EC (1994), DANIDA (1994), Dyer and Bartholomew (1995) and Club du Sahel (1996).

6   See Pretty and Thompson (1996). The UN Commission on Sustainable Development (1997) reports that between 1986 and 1994, assistance to agriculture fell from US$19 billion to US$10 billion. The World Bank's financing for agricultural development fell from 30 per cent of its annual lending in the early 1980s to just 20 per cent in the early 1990s, from US$5.4 billion to US$3.9billion. The US Agency for International Development (USAID) reduced its support to agriculture in developing countries rather rapidly between 1991 and 1994, going from US$950 million to less than US$500 million, while the German development agency GTZ, and all but two other bilateral donors, similarly decreased their support to agriculture.

7   Cernea (1987), studying 25 World Bank-financed agricultural projects four to ten years after their completion, found continued success clearly associated with local institutional capacity. All 12 projects with long-term sustainability had strong local institutions. In the others, the rates of return had declined markedly, contrary to expectations at the time of project completion. Projects with no attention to institutional development and farmer participation were unsustainable. See also other studies: de los Reyes and Jopillo (1986), Cernea (1991, 1993), Uphoff (1996), Pretty et al (1995), Krishna et al (1997), Uphoff et al (1998), Pretty (1998) and Uphoff and Wijayaratna (2000).

8   For an account of this project and how it took shape, with some very impressive accomplishments, see Murray (1997).

Chapter 5

Economic Conditions for Sustainable Agricultural Intensification

Arie Kuyvenhoven and Ruerd Ruben

Large parts of the developing world have witnessed unprecedented growth in food production in recent decades. Thanks to the development of Green Revolution technologies and the extensive adoption of high-yielding staple food varieties by Asian farmers, famines in that region have been averted. Hunger and malnutrition are declining in relative terms, and many countries are basically self-sufficient now. There have been some environmental benefits too, as yield increases prevented overexploitation of marginal land and slowed the pace of deforestation.

There are reasons for concern, however. The new agricultural technologies have not been very successful in Sub-Saharan Africa, where hunger is on the increase. Important pockets of poverty remain in areas that have rainfed agriculture or fragile soils, affecting close to 1 billion people. Moreover, yield growth in high-external-input systems is slowing, and serious environmental problems have emerged. Both land and water constraints limit further expansion of irrigated agriculture. As a result, several high potential areas are showing decreasing marginal returns from further intensification, so that there are now higher potential returns from developing less-well-endowed lands elsewhere (Hazell and Fan, 2001).

A major challenge for the next decades is therefore to develop technologies and practices that enable continued agricultural growth to match growing demand for food and feed. To reduce rural poverty and hunger, the agricultural growth process needs to be equitable and to be designed in such a way that the natural resource base is maintained and pollution is controlled.

Hazell and Lutz (1998) characterize this type of agricultural development as broad based, market oriented, participatory and decentralized, and driven by new approaches to agricultural innovation that enhance factor productivity and conserve the resource base. To reduce excessive dependence on external inputs, there is growing interest in agroecological systems that create more favourable growing conditions for plants and animals as part of larger ecosystems (Altieri, 1995). Major elements in such systems include diversification of activities, interaction among cropping, livestock and forestry activities, biological control of pests and diseases, and control of soil erosion and nutrient depletion through a variety of activities that intensify agriculture.

SUSTAINABLE AGRICULTURAL INTENSIFICATION AS OBJECTIVE AND CRITERION

Evaluation of alternative approaches invariably focuses on the nature and benefits of input substitution. Green Revolution technology was characterized by embodied technical innovation via material inputs (improved seeds, fertilizers, pesticides) plus public investment in irrigation, extension and other infrastructure. Alternative approaches – for example, integrated pest and nutrient management – rely more on creating and using human and social capital, discussed in the preceding chapter. This raises important investment issues as these forms of capital take time to build up and to become effective, being usually labour- and management-intensive, and often having a large non-governmental organization (NGO) component. The use of locally available resources and enhancement of their efficiency resources is emphasized in agro-ecological approaches, with special attention paid to the resilience of the whole farming system.

Alternative approaches to more conventional agriculture have several features in common, some of which make evaluation difficult. Because systems rather than single crops are stressed, quantification and explanation of the potential of these more diversified systems is often difficult. Similarly, participatory technology development does not focus on a single technique and values the creation of capacities for flexible responses to changing circumstances through a learning process that involves local knowledge, research and extension.

There are inevitably trade-offs to be considered. To address soil fertility problems and sustain yield levels effectively, for example, the use of chemicals in combination with organic soil amendments will in many cases be appropriate (Ruben and Lee, 2000). Farmers will opt for whatever combination of inputs best serve their multiple production objectives. Since many alternative approaches require more labour, care must be taken to ensure sufficient complementary inputs, local or external, to maintain and even increase labour productivity. When this is done, attributing productivity gains to particular inputs becomes very problematic.

Benefits of alternative systems have thus far been measured mostly in biophysical terms (soil organic matter, physical yields). Less attention is usually given to their implications in terms of farm household income, consumption and labour use. We find useful the concept proposed by Pretty (1997) of *sustainable agricultural intensification* (SAI) which encompasses two key objectives: the protection and regeneration of the natural resource base with regard to soil nutrient balances, water cycles, and land productivity, and efficient combinations of production factors that improve farm household income, including returns to labour.

Because trade-offs between agroecological and welfare criteria commonly arise, we are most interested in 'win-win' technologies that give simultaneous improvement on both scores. The attractiveness of different types of natural resource management (NRM) practices in agriculture as viewed from a farm household welfare perspective is an essential concern because this affects their spread and sustainability.

The basic principles underlying SAI practices are considered in the next section. Then, economic means for assessing new SAI approaches, important for understanding their adoption by farmers, are reviewed. This points towards general conditions that should be helpful for the implementation of SAI programmes. Certain policy measures can be expected to make SAI systems more feasible, and some kinds of policy environment can accelerate the adoption of promising sustainable intensification approaches. These latter questions are not taken up in this chapter but rather are addressed in Chapter 21, after various case and country experiences have been considered.

BASIC PRINCIPLES

SAI implies that farmers attempt to increase their returns from scarce factors of production in ways that maintain the stock and quality of their natural resource base. Most agroecological approaches tend to focus on land productivity as a major indicator, with less attention given to returns to labour (Low, 1993). Farmers tend to consider yield-increasing technologies and practices based on agroecological principles from five different perspectives:

1 profitability, eg possible contributions to household income and consumption;
2 implications for input efficiency;
3 consequences for input substitution and labour use;
4 dynamic risk management; and
5 sustainability, which brings in concerns such as maintaining water supplies.

From a discussion of what guides farm household decision-making regarding sustainable technologies, we will derive a number of principles that can enhance the socioeconomic attractiveness of such technologies.

Profitability

Sustainable agricultural technologies and practices are unlikely to be adopted unless farmers attain higher and more stable income and consumption opportunities. Profitability requires both the existence of effective, accessible market outlets and favourable output–input price ratios. For example, market distortions or inefficient exchange networks may reduce incentives for investments in soil and water conservation (SWC) activities. If farmers stick to subsistence cropping and rely almost exclusively on locally available resources, agricultural intensification may become unsustainable (Lockeretz, 1989; Low, 1993).

Contrary to what might be expected, farmers are more likely to apply yield-increasing and sustainability-enhancing inputs to commercially-oriented production activities (Reardon et al, 1999; Putterman, 1995). In the cotton belts of Southern Mali and Burkina Faso, fertilizers, crop residues and animal manure tend to be mainly used for cash crops that guarantee sufficient monetary returns to warrant the costs of using them (Sissoko, 1998; Savadogo et al, 1998). Similarly, animal traction and improved tillage yield higher returns when applied on the more fertile fields where commercial crops are grown. In the Central Chiapas region of Mexico, crop residue mulching only appears to be profitable when combined with animal traction on fields devoted to intensive market-oriented cropping activities (Erenstein, 1999).

Farmers' engagement in market exchange on favourable terms is thus often a necessary condition for profitable and sustainable agriculture. Engaging in trade provides financial resources for the purchase of complementary inputs and consumption goods. Those households that have a net demand position in the food market, buying more than they sell, will benefit from low commodity prices (Budd, 1993; Goetz, 1992). Where access to formal credit services is limited, investments can be financed from income derived from off-farm employment (Ruben and van den Berg, 1999). Part of the agroecological transformation of Machakos district in Kenya, discussed in Chapter 6, is attributable to the income opportunities that residents of this rural area found in Nairobi; work there earned them cash to finance investments in terracing, livestock, agroforestry and other means for intensification. Market development commonly enhances willingness to invest, while involvement in market exchange generally improves farmers' responsiveness to price incentives. Hence, where there are market failures, policy reforms that correct a lack of access or lack of competitiveness are a first-best solution. In their absence, reliance on low-external-input technologies with low productivity tends to persist.[1]

Input Efficiency

Agroecological approaches to farming system intensification commonly substitute integrated nutrient and pest management practices for chemical inputs (Altieri, 1995). Indeed, the high costs of inorganic fertilizers and other agrochemicals often drive farmers to rely on locally available resources instead of on purchased and imported inputs. Reducing reliance on purchased inputs where these are accessible, however, implies that the right substitutes can be found, and that complementary relations between different inputs are recognized.

Prospects for sustainable agricultural intensification eventually depend on the possibilities for improving input efficiency, eg achieving positive marginal returns from additional units of organic and/or inorganic inputs. Agroecological approaches point out that nutrient efficiency (in terms of fertilizer uptake) is determined by the availability of complementary micro- and macronutrients, notably soil organic matter and phosphorous, plus active soil biology (van Keulen, 1982; also Chapter 10). Substitutes for chemical fertilizers are generally characterized by a fairly low recovery fraction due to immobilization of nutrients and slow decomposition of organic matter, however, nutrients in organic form do offer some advantages in that they enhance soil structure and biology.[2]

Nutrient recovery and the efficiency of uptake can be enhanced through soil and water conservation measures that enhance soil nutrient retention capacity, and nutrient applications timed to match the crop growth process, eg, shortly after sowing and with sufficient rainfall. Both activities are highly labour-demanding and not very amenable to mechanization. Moreover, mechanical or animal tillage speeds up nutrient release from the soil.

Agricultural yields are held down by whatever is the most limiting growth factor in the particular situation, and can only be increased when input combinations are made available with adequate *complementarities* between different growth-enhancing inputs, ie, nutrients and water, phosphorous–nitrogen and carbon–nitrogen ratios. Studies regarding input efficiency refer to the functional relations between soil carbon content and nitrogen supply to prevent the immobilization of nutrients, and the proportional relationship between nitrogen and phosphorous to guarantee a beneficial rhythm of organic matter decomposition (Penning de Vries and van Laar, 1982). This implies that input efficiency will be low when complementary inputs are not available at the right time or in sufficient amounts.[3]

Farmers have commonly learned how to time and combine different productive activities to generate positive synergy effects. Organic and chemical inputs are not full substitutes, and combinations of locally available resources with selectively applied external inputs often yield the best results (examples are given in Chapters 7, 8, 9 and 10). In practice, farmers hesitate to refrain completely from the use of purchased inputs because this permits better timing of activities, reduces the demand for labour in critical periods, and often contributes to a better appearance of the produce in the marketplace. Where soil nutrient content is low and the nutrients available from organically produced fertilizers (green manure, mulch, dung, compost) are insufficient or too slowly released, use of chemical fertilizers will continue to be necessary.

Since organic matter decomposition takes time, as does building up biotic activity in the soil, optimal results are more likely from gradual reduction in levels of fertilizer application rather than abandonment. The attractiveness of inorganic nutrient sources will be affected by how great an increase in production they can in fact contribute to when used in association with other practices. When yields can be doubled or more with agroecological practices, as reported for rice, maize, beans and potatoes in Chapters 12, 13 and 14, farmers' willingness to use more labour-demanding inputs can be substantially changed.

Nitrogen derived from cover crops through biological fixation can be made more effective if sufficient phosphorous is available. Since tropical soils typically have shortages of this nutrient, applying phosphate fertilizer or rock phosphate can be very helpful in increasing overall input efficiency (Kuyvenhoven et al, 1998a). Similarly, nitrogen requires a minimum amount of water and organic matter to become effective. Where exclusive reliance on local inputs impedes nutrient efficiency, selective application of complementary external inputs should be encouraged (Triomphe, 1996; Buckles et al, 1997).

Similar complementarities are found in integrated pest management (IPM) programmes where improved nutrient application can be a means for controlling pests and diseases. Farmers who use small amounts of chemical fertilizer may suffer less crop loss from competition or infestation. When no fertilizers are applied, some diseases can more easily penetrate into fields, although the incidence of diseases or weeds often increases with high doses of fertilizer.[4]

Factor Substitution

Most analyses of sustainable agriculture practices devote much attention to short- or long-run yield effects, but generally do not assess labour requirements and returns to labour in any detail. Implicitly, family labour is thus considered an abundant resource. While technical efficiency is usually evaluated against the background of the most limiting factor for yield increase, whether water, nutrients, energy, pests or diseases, economic efficiency should be understood according to the critical factors that determine farm household income: land, labour, capital and knowledge, as well as natural resources. In particular, limitations on the scope for substituting labour for external inputs should be recognized.

Most sustainable agroecological practices tend to be more intensive in their use of labour. Physical soil conservation measures promoted in the Central American hillsides and West African lowlands have resulted in yield increases, but with large amounts of labour for construction and maintenance and substantial costs for the purchase and transport of materials (Stocking and Abel, 1989). Given their high labour intensity and greater gestation period, the returns to labour with such measures are critical considerations for adoption (Lutz et al, 1994; de Graaff, 1996). Similarly, green manure practices and crop residue mulching require additional labour for harvesting, transport and ploughing-under (Ruben et al, 1997; Erenstein, 1999). This is why synergistic effects – if they can be achieved – are so important in the adoption of an agroecological system, because they repay several benefits from a single cost or achieve proportionally higher outputs.

Most mixed cropping and agroforestry systems demonstrate lower returns to labour due to high establishment, maintenance and harvesting costs (Current et al, 1995). Production of fodder crops for livestock feeding improves the availability of manure for arable cropping and enables farmers to recycle their crop residues, but both activities demand additional labour (Breman and Sissoko, 1998). Labour requirements for integrated pest and disease management are similarly high due to the substitution of manual for chemical operations. For most of these NRM practices, mechanization is not a feasible option due to strong terrain slopes and the small scale of operations.

For systematic evaluation of the attractiveness of any practice from the farm household perspective, returns to land and labour need to be compared simultaneously (Reardon, 1995). Attention has to be given to their *marginal*

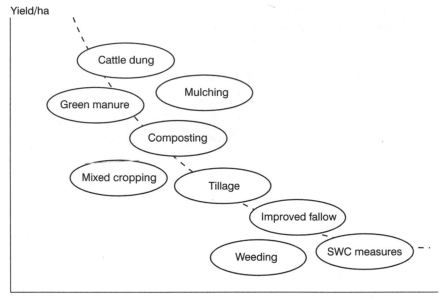

Figure 5.1 *Factor Intensity and Yield Effects of Major NRM Practices*

returns compared with other activities, ie, off-farm employment or the hiring out of land. When sustainable agroecological practices improve nutrient stocks and soil organic matter content, the improvement in yield should be superior compared with the additional inputs requirements, as seen in the case studies. This can be explained by the fact that labour is accomplishing the timely availability of nutrients to the cropping system. Generally, returns to labour will be higher for technologies that utilize external inputs in ways that capitalize on benefits from input complementarity.

Figure 5.1 provides an overview of major NRM practices, taking into account expected yield effects and labour requirements. The final selection of NRM practices made by the farmer is likely to depend on the labour opportunity cost–output price relationship. Soil fertility-enhancing measures give best results on both scores, followed by mixed cropping and minimum tillage. Soil and water conservation measures and intensive weeding are only attractive for cropping activities with a high value added, or where labour costs are relatively low.

The higher labour intensity of most NRM practices needs to be considered as a major limiting factor for their adoption. Labour tends to be scarce in semi-arid areas, particularly during the periods of soil preparation, weeding and harvesting (Fafchamps, 1993), and competition for labour occurs when mulching, manuring or crop residue recycling are introduced. Resource-poor farmers are likely to derive part of their income from off-farm activities that have to be reduced when labour-led intensification of their farming system takes place (Reardon et al, 1988). Certain NRM practices, notably physical soil conservation measures, can be executed during off-season periods when economic opportunity costs are low, but they will take up leisure time that could be reserved for social or communal purposes.

Risk Management

Resource-poor farmers are inclined to rely on fairly diversified patterns of activities to maintain acceptable levels of risk. Diversification of cropping and livestock production and their integration with agroforestry, aquaculture and improved fallow practices can reinforce the resilience of farming systems through processes of nutrient recycling, biodiversity management, and integrated pest and disease control (Muller-Samann and Kotschi, 1994). Consequently, yield levels tend to be more stable, and dependency on purchased inputs can be reduced.

It is increasingly recognized, however, that risk management can also take place through farmers' engagement in non-farm and off-farm activities (Reardon et al, 1994). The revenue streams derived from these activities are far less dependent on weather conditions, which vary, and thus provide insurance against co-variate shocks (Udry, 1990). Besides diversifying cropping systems, diversification into non-agricultural activities can be considered a promising risk-management strategy. This becomes more feasible when labour demand for agricultural activities can be reduced, and household members possess sufficient skills and knowledge for entering into wage labour or self-employment (Reardon, 1997).

Another issue in short-term risk management is farmers' capacity to adjust their input use under changing weather or environmental conditions. Adaptive behaviour strongly depends on the capacity for learning that enables prompt reactions to unexpected events (Fujisaka, 1994). Although most agroecological practices have been developed through participatory and horizontal extension methods, eg farmer-to-farmer approach or farmer field-schools, there is often little understanding of the dynamics of production systems. An example is the disadoption of maize-cover crop systems in Honduras, documented by Neill and Lee (2000). The abandonment of a previously attractive leguminous cover-crop technology by thousands of farmers can be explained in terms of inadequate response to weed invasion and the subsequent abandonment of 'companion technologies' like live barriers, contour cultivation, crop-residue recycling and reseeding, with influences also coming from external factors like changing land tenure rules and competing employment opportunities. As economic conditions and opportunities changed along with biophysical processes, the combination of practices that were perceived as best serving household needs and interests did too.

Sustainability

SAI implies that the production capacity of the resource base can be maintained in the long run. This does not necessarily mean that agroecological balances must be strictly maintained at each moment in time. In principle, farmers may allow some resource depletion in the short run while investing in its recovery in subsequent periods. This concept of 'weak' sustainability (Pearce and Turner, 1990) can be applied in the economic analyses of landuse systems.

Typical examples of 'optimal depletion' can be found in traditional fallow systems that are allowed to recover after some period of permanent exploitation. Similar natural regeneration can occur for wildlife species, fisheries and forestry systems (Bulte, 1997). For given prices and discount rates, there should be some optimum composition of the stock of renewable resources that satisfies intertemporal welfare optimization criteria. Consequently, it can be economically rational to reduce stocks in the short run and to earmark investment funds for their recovery in subsequent periods (though this may, in fact, not occur).

Farmers' preference for weak sustainability can be explained from a trade-off perspective. When comparing current and future costs and benefits, discounting procedures are used that reflect farmers' relative time-preferences. People facing more risk tend to maintain a higher discount rate, reflecting a preference for immediate revenues. Investment activities with long gestation lags are especially sensitive to high discount rates, as Current et al (1995) have demonstrated is the case with agroforestry investments.

A second type of trade-off occurs when farmers assess the welfare and sustainability implications of alternative technologies (Kruseman et al, 1996). Farmers' adoption of sustainable practices can only be expected when positive welfare effects are expected. In practice, however, methods intended for agro-ecological sustainability can involve a short-term sacrifice in terms of income or consumption objectives as soil systems adjust to the new system of management. Moreover, production systems that may be sustainable at lower system levels (field, farm) can encounter negative externalities when operating at higher system levels (village, region). In such cases, certain policy instruments can be helpful to overcome adverse trade-offs as discussed in Chapter 20. Suitable incentives need to be identified that permit simultaneous improvements to both welfare and sustainability – 'win–win' scenarios (Kuyvenhoven et al, 1998b).

APPRAISAL METHODS

Empirical studies evaluating sustainable practices and technologies tend to focus on yields and resource balances, as we have noted. Positive returns to land are usually considered as an indication of financial feasibility. However, economic evaluation of their attractiveness from a farm household perspective requires taking a variety of criteria into account, as presented in Table 5.1. Based on the criteria used for the socioeconomic appraisal of agricultural technologies and production systems, different combinations of analytical methods can be recommended (Ruben et al, 2001).

The profitability aspects of agricultural intensification can be measured in a rather straightforward manner, making use of conventional cost–benefit analysis (CBA). However, attaining profitability is only a necessary condition for adoption; it does not take into account various non-income farm house-

Table 5.1 *Available Analytical Procedures for Appraising NRM Practices*

Criteria	Analytical procedures	Examples of empirical studies
Profitability	Partial farm budgets and cost–benefit analysis	De Graaff (1996), Current et al (1995), Lutz et al (1994), Erenstein (1999), van Pelt and Kuyvenhoven (1994)
Input efficiency	Production functions	Mausolf and Farber (1995), Heerink and Ruben (1996)
Factor substitution	Farm household modelling	Singh et al (1986), De Janvry et al (1991), Fafchamps (1993)
Risk management	Portfolio analysis	Reardon et al (1994), Reardon (1997), Scoones (1996)
Sustainability	Bio-economic modelling	Kruseman and Bade (1998), Barbier and Bergeron (1998), Kruseman et al (1996)

hold objectives. CBA provides an appraisal of average costs and revenues at prevailing prices, usually in a partial equilibrium framework. CBA is most often applied in the appraisal of specific NRM practices, like soil and water conservation (de Graaff, 1996; Lutz et al, 1994), crop residue mulching (Erenstein, 1999), or agroforestry systems (Current et al, 1995). Objectives other than income can be taken into account by extending CBA to provide a multi-criteria analysis (MCA) (van Pelt and Kuyvenhoven, 1994). Its partial character is, however, normally retained.

To make a thorough appraisal of input efficiency, information regarding marginal returns to factors of production is required. For this, production function analysis (PFA) provides an appropriate analytical framework (Heerink and Ruben, 1996; Mausolff and Farber, 1995). This can be used to estimate marginal returns to land and labour for agroecological and conventional production technologies, enabling one to identify the range of input–output price ratios within which conversion is likely to take place. Moreover, typical farm household characteristics associated with the adoption of certain sustainable technologies can be revealed.

A full analysis of the economic attractiveness of sustainable technologies considering prospects for factor substitution requires reliance on farm household modelling (FHM) as explicated by Singh et al (1986). Farm household models explicitly consider complementarities between inputs and provide an analytical framework for the simultaneous evaluation of production and substitution effects. Differences in supply response between tradable and non-tradable commodities are recognized and can be assessed (de Janvry et al, 1991). Further extensions towards village-wide modelling can include market linkages and general equilibrium effects (Taylor and Adelman, 1996). FHM offers procedures for policy simulation by assessing farmers' likely supply response to different types of economic incentives.

Aspects of risk management can be included in programming models and econometric procedures. However, explicit appraisal of farmers' risk behaviour and their coping strategies requires a separate treatment. Portfolio analysis can be used to assess the variability among different household income categories (farm, non-farm, off-farm) and to identify major strategies for consumption-smoothing (Deaton, 1992). In such analysis, attention is given to linkages with non-agricultural sectors, with differences in the supply response between food-deficit and food-surplus households being accounted for.

Finally, to make a comprehensive analysis of the sustainability implications of different production technologies or strategies, bio-economic modeling is recommended. Such models permit appraisal of both current and alternative (more sustainable) technologies and their contribution to farm households' welfare and agroecological sustainability (Kruseman and Bade, 1998; Barbier and Bergeron, 1998; Deybe, 1994; Kruseman et al, 1996). Trade-offs between both objectives can be established, and policy instruments to enhance the adoption of sustainable practices can be identified and assessed. Further consideration of policy measures to further these objectives is deferred until the concluding chapter.

CONDITIONS FOR IMPLEMENTING SAI

A major constraint for the adoption of agroecological practices is their economic feasibility. Returns must be sufficiently attractive compared with income derived from off-farm employment, and sustainably-produced products must be competitive in the market to be economically sustainable. Even when cost-benefit appraisals yield positive results, farmers carefully consider other factors and risks. Given the frequently high labour requirements of most agroecological practices and the existing limitations on factor substitution, returns to land and labour have to increase simultaneously. *Additional* reliance on purchased inputs may be a preferred mechanism to maintain farmers' incomes and improve food security prospects, at least in the short run, knowing that everyone needs to live in and through the short run.

Despite widespread efforts by non-governmental organizations and local development programmes to promote shifts towards agroecological practices, adoption often remains limited to farmers who receive direct technical or financial support. Without such assistance, these practices are soon abandoned, indicating that their underlying economic feasibility is not always apparent to farmers.[5] Consequently, at least three conditions should be satisfied to make sure that both farm productivity and household incomes can be improved through SAI.

First, the economic viability of agroecological practices can be strongly enhanced when *public investment and services* are made available to farmers in remote regions. Without such alterations in opportunity sets, low-input technologies tend to be restricted to medium-sized farmers who are only marginally engaged in market exchange. Market development and reduction of transport costs are usually the most important requirements for agricultural intensification, since exchange relations favour access to complementary inputs and provide incentives for investment. Improving poor farmers' access to physical infrastructure thus represents a major condition for equitable and sustainable rural development.

Second, sustainable intensification requires improved *access to and information about factor and commodity markets* in order to reduce uncertainties and permit flexible responses to changing production and exchange conditions. Substantial increases in agricultural productivity can only be reached when internal farm household resources are combined with selectively applied external inputs.[6] Considering the requirements of input efficiency and factor substitution, greater agricultural yields strongly depend on the possibilities of overcoming critical input constraints, whatever they are. Certainly, the availability of complementary inputs and an adequate supply of labour to guarantee their timely application are required.

Third, the adoption and maintenance of sustainable production systems critically depend on policy measures that enable farmers to invest their resources in *better integrated farming systems*. Even when land and water conservation practices, improved tillage systems and better nutrient management offer wide prospects for enhancing productivity, to reduce poverty the availability of financial services, marketing outlets and off-farm employment opportunities are equally important. While structural adjustment policies have generally improved market prices for farmers, input costs have remained high and delivery systems inefficient (Kuyvenhoven et al, 1999; Reardon et al, 1999). Access to inputs has proved to be strongly dependent on individual characteristics such as education and on community networks. Therefore, investments in both human and social capital can be particularly important to enhance the adoption of sustainable practices and technologies.

Concerted action in the field of participatory technology development and dissemination combined with market and institutional reforms will remain important for the adoption of agroecologically sustainable and economically efficient landuse practices and technologies. Economy-wide market liberalization policies will not be successful in supporting SAI in the absence of appropriate public investment in marginal and remote areas, and without local initiatives that assure farmers access to markets and information and provide them with sufficient input purchasing power – and a degree of market power generally. Usually, such efforts require rather solid social networks that involve all or most of the relevant rural stakeholders and that nurture linkages with other regions and non-agricultural sectors. Economic analysis, focusing particularly on household incentives and capabilities, thus needs to intersect with social analysis and action as well as biophysical potentials.

NOTES

1   One caveat needs to be borne in mind here. Market penetration, such as occurs with improved roads, can open up an area to commercialization and the extractive exploitation of soil, timber and other natural resources that makes long-term agricultural productivity unsustainable unless there are legal or social controls. We are assuming here some degree of local or external regulation of resource use so that profitability does not undermine the natural resource base.

2   Some questions are being raised about whether present estimates of plant nutrient requirements, derived from measurements made with inorganic nutrient application, may be too high, given examples of the good and even superior results that are possible with smaller amounts of nutrients, provided slowly but continuously (Bunch, 2001). This is discussed in the next chapter and Chapter 12.

3   A major continuing concern in agricultural research is the very low rates of nitrogen-use-efficiency and the diminishing returns when nitrogen is applied in an inorganic form. The worldwide average efficiency for uptake of inorganic nitrogen by cereal crops is about 33 per cent, and often as low as 20 per cent (Kronzucker et al, 1999; Ladha et al, 1998). Concerns about the timing and efficiency of nutrient application are thus not limited to agroecological practices.

4   In the case of rice intensification reported in Chapter 12, where chemical fertilizer is not necessary for high yields and is seldom used, the Madagascar farmers report fewer problem with pests and diseases because of the plants' vigorous growth. There is growing evidence that fertilizer application without sufficient soil organic matter contributes to greater crop vulnerability through pest and disease losses.

5   An example is the indigenous technology of 'raised-beds' revived in Bolivia and Peru in recent years. This method of ensuring and enhancing production of potatoes and other crops at very high altitudes, by heaping up soil on 'platforms' (known as '*suko kollo*' in Quechua) for growing crops with channels filled with water running around them, was being promoted by NGOs in the early 1990s with considerable subsidies. By the mid-1990s, however, the rate of abandonment of raised beds was matching the rate of their new construction, as the increases in yield were not sufficient to cover the cost of building and maintaining the raised fields (CIIFAD, 1997).

6   We refer to 'selectively applied inputs' rather than prescribe any particular amount because there remains debate as to how extensive external inputs must be for significant yield increases, given differing conditions in sites and crops. Proponents of agroecological approaches, having seen major increases that do not depend on large amounts of external inputs as reported in Part 2, challenge the view held by many agricultural scientists that such external inputs need to be substantial to raise production. This is an empirical question about which there is too little systematic evidence to reach firm general conclusions.

Chapter 6

Can a More Agroecological Agriculture Feed a Growing World Population?

Mary Tiffen and Roland Bunch

Even though more people in more countries are making decisions that favour smaller families, the existing demographic structure ensures that the world's population will continue to increase for at least three to five more decades before it stabilizes. These additional people will require more food and other goods and services, and more and more of these people will be living in towns. We need to consider, therefore:

- What fraction of the additional food production required can be delivered by improved, low-external-input ecologically-oriented agricultural systems?
- Can these systems achieve the extra production required without eliminating the remaining natural forests of the globe?
- Can they provide food to the people needing it and where it is required, including to the expanding urban population of the world?

To know where and how technological change can play a successful role, we need to understand the processes of change at work in society. This chapter begins by mapping out three stages that, broadly speaking, agriculture and human societies have passed through as they have developed. This analysis places agroecological innovations in a historical context, tracing the influence of demographic and biophysical relationships, though the same patterns may not always apply in the future.

We then consider the opportunities for increasing production by methods that can be characterized as agroecological, with examples from Africa, Asia and Latin America, focusing on the second historical stage. We discuss where current second-stage societies are located, how many people live in them and their potential contribution to future food needs. Finally, we take up the questions posed above. Throughout we emphasize the need to understand and respond to the interests of food producers, as these interests are understood by rural people themselves.

The process of sustainable agricultural intensification as discussed in the preceding chapter has two basic requirements:[1]

1   *Maintaining soil and water resources* in productive condition over the long term by: (a) replacing depleted nutrients over a period of years, if not necessarily annually; (b) maintaining the structure and biological qualities of the soil in productive condition; and (c) maintaining a supply of water that is adequate in quantity and quality for humans, livestock and plants.
2   *Providing an adequate return to the labour and physical capital inputs invested.* Farm families expect a rising standard of living in line with the norms for their country, or they or their children will quit farming when other options are seen as accessible.

Everywhere, households want to be able to sell their products remuneratively in order to be able to provide for their many non-food needs. This makes marketing and other infrastructure very important. The importance of access to markets is seen in the two African cases reviewed in the next chapter. In many parts of the world, the present rewards to agricultural labour, whether working as a small farmer or as an agricultural labourer, are scanty, even to the point in some places of starvation. The central aim of development policies must be to provide more abundant returns to labour. The two requirements cited above are closely linked since the maintenance and, even more importantly, the improvement of soil and water conditions require the investment of labour and/or capital.

DEMOGRAPHIC FACTORS SHAPING ECONOMIES AND SOCIETIES

The nature of agriculture – and of the culture and economy of a population dependent on it – has always been deeply influenced by population density, which in economic terms translates into the relative abundance of land and labour. Boserup (1965) has provided insights into the pressures and incentives that have increased labour intensification and innovation in response to the needs of a growing population.

Population density has been intimately related to markets and the possibility of specialization. A small, scattered population has to be self-sufficient because the markets where it can purchase the things that it needs but cannot produce itself are distant. With few opportunities for exchange, there are few specialists who can meet particular needs with skill and innovation. Such a situation is not enviable and indeed perpetuates poverty and deficiencies.[2]

However, as populations have grown, a variety of specialists could be supported, and the number of markets to which farmers can sell and from which they can buy expands with the growth of towns. This process affects not only land–labour relationships, but permeates society, changing the way that children are brought up, for example in gendered labour roles. It has influenced as well various features of farming systems, such as the preferred mode for ensuring the land's continued fertility, and the role of livestock within farming.

Table 6.1 identifies these many inter-related changes from a generalized perspective on historical processes of change. In previous centuries, when population growth was slow with many set-backs, these changes took hundreds of years. In the past century, population growth has been rapid, so these changes have been telescoped into decades.

SUSTAINABLE AGRICULTURE MUST BE A CHANGING AGRICULTURE

Sustainable agriculture, as noted in the introduction, must be an agriculture that changes over time – in its products, in its technical methods, and in its combination of the factors of production: land, labour and capital. No system remains reliable for generating income and opportunity unless it can adapt to changing external circumstances and incorporates innovations. Changing circumstances are of two kinds.

Relatively Slow, Long-Term Trends

These include, particularly, increasing rural population density and growth in the proportion of urbanized people engaged in industry and services who need to buy all their food. Such trends affect the scarcity value of land, labour and capital. Three broad situations have been identified in Table 6.1, and differentiating situations include the following.

Land is Plentiful: Labour and Capital are Scarce

This occurs where population density is low. Shifting cultivation with long fallows and free-ranging domestic livestock is often appropriate for a land-abundant situation. Even farmers who once used intensive methods will adopt this type of farming if they migrate to land-rich areas, as reported in the Kofyar region of Nigeria (Netting, 1965; Netting and Stone, 1996), as well as in the newly-settled United States.[3]

Land Begins to be in Short Supply Relative to Demand

In this second stage, farms become smaller and are more intensively worked. Labour is still relatively plentiful, and capital remains scarce and difficult to accumulate. Over time, old production methods fail either to meet people's welfare needs or to maintain the land's productivity because of ever-shorter fallow periods. Boserup (1965) has explained how the pressures in such a situation eventually lead to the introduction of more labour-intensive methods

Table 6.1 Changes in Agriculture Associated with Changing Population Density

	Extensive agriculture	Conditions associated with: Intensifying agriculture	Intensive agriculture
Climatic zones	Arid, semi-arid or humid	Semi-arid or humid	Semi-arid or humid
Population per km^2	Low densities: <30/km^2	Medium densities: 30–100/km^2	High densities: 100–600/km^2
Agricultural modes of production	In arid areas, pastoralism; in semi-arid areas, agropastoralism and shifting cultivation; in humid areas, shifting cultivation, with no animals if diseases are present	Fallows shorten; annual cultivation begins when population density reaches 50–70km^2; crops and animals become more integrated	Double or triple cropping common; varying extents of livestock husbandry, which can include aquaculture; more penned livestock
Soil fertility management	No manuring since animals graze freely; long fallows practised or opening of new land	Manure used; animals are herded and penned; mixed cropping or rotations including legumes practised	Manure scarce in relation to demand; chemical supplements are purchased
Land availability	Less than 10% of potentially cultivable land is farmed	Percentage of cultivated land in older-settled areas steadily increases; new land of lesser quality is cleared	No unoccupied or unclaimed land; land has high market value
Typical problems	Few serious problems: opportunistic use of pastures can lead to 'over-grazing' until herds move or die, then grasses recover; forest clearance may leave soil temporarily unprotected, but regrowth soon occurs	Diminishing fertility as fallows shorten; loss of vegetative cover contributes to erosion; conflicts between herders and farmers as the latter expand into new semi-arid and valley areas; erosion results if farmers move up slope	Soil erosion common; pollution from agro-chemicals occurs; urban pollution new phenomenon; growing competition for increasingly scarce water
Investments	No land-related investment	Investments made in new land clearance; soil conservation begins after a delay; start tree protection and planting	Land improvement and water conservation increase; new investment in irrigation and in tree crops

Tenure	Clearer has temporary use rights; community may exclude others from its territory; other rights are unimportant since land is plentiful	Families establish firm rights to cleared and cultivated land, then to stubble, then to private grazing land; rights first gained by inheritance, later by other means	Land can only be acquired by inheritance, purchase or lease; active markets for land exist
Market access	Marketing is difficult; transport costs high; self-sufficiency is necessary	Market access slowly improves; imported food is still costly, so farmers produce own food crops and sell some	Market access good; farmers buy more of their own food needs and also specialize
Division of labour	Few specialists; distance makes access to them difficult	More specialists become available as large villages and small towns develop tool-makers, teachers, builders, artisans, traders, etc	Large non-farm sector providing goods and services, and an expanding market; increased urbanization
Technology and information	Low technology level	Technology improved by experimentation, learning, market access	Access to information is good; technology rapidly improving
Education	Education by parents	Primary schools cheaper and nearer	Secondary education and training common
Family size preference	Large families preferred; high child death rate	Large labour force needed; increasing rate of child survival	Costs of raising children increase; small family preferred

Source: Adapted from Tiffen (1998)

for raising output and to the creation of land-based capital.

If prices and markets are attractive, capital may be imported from other areas, such as through the remittances of men who have migrated (as in the Machakos case discussed in the next chapter), or by investments of entrepreneurs who have made their money in trade. (This was frequently the case in the 18th-century Britain described by Adam Smith and Samuel Johnson.) Typical innovations are labour-intensive, depending mostly on local and biological resources, eg, digging drains or irrigation channels, terracing, planting hedges, and confining animals at night to make use of their manure. This stage is often associated with increased animal inputs both for traction and for manuring. Farmers and scientists also experiment with breeding more productive crops and animals.

Beyond this, the future of agriculture is intimately linked with the path of the non-agricultural sector. Where there is significant growth in economic specialization in industry and services, accompanied by urbanization, rural labour becomes scarce, capital more plentiful, and land relatively more cheap because it is now less in demand as a source of livelihood. Farms become larger as holdings are consolidated. To remain profitable, agriculture adapts to the new situation by mechanization and by labour-saving methods for restoring soil nutrients and dealing with pests and disease, such as the application of chemicals. There may be no labour available for the daily care of livestock. This is the current situation in the United States, most of Europe and some parts of Asia.

There can be a different path, when economic opportunities outside agriculture do not develop significantly and rural population remains dense. Then the value of land remains high because of high population–land ratios. This is unfortunately the situation in many parts of the developing world, where a dualistic agriculture has emerged. In this situation, a small number of large-scale, highly capitalized operations coexist with a multitude of small, fragmented holdings. The situation need not become as extreme as that described by Geertz (1963) for Indonesia, a condition of 'agricultural involution' where returns to labour become pitifully low. Such a situation usually arose because the free movement of labour and land transactions were distorted by colonial or other governments. The cure, as in Indonesia, involves changes in policy: in that case, lifting the requirements that farmers should grow certain kinds of crops. Alternatively, governments can refrain from policies that remove the incentives or means to invest in farm improvements and new technologies. Response to such policy changes will be gradual as they represent another, possibly countervailing, influence among various long-term trends.

Short-term Changes

These must occur for farming to remain financially sustainable. To meet the continuing needs and rising expectations of rural people, agriculture has to adapt to changes in its environment. Many kinds of change can occur, some favourable and others adverse, eg:

- New transport infrastructure may make new markets accessible, and/or bring in competing goods and services.
- A change in a distant country can raise local prices for a globally traded product, eg, frosts in Brazil can hike the price received for coffee produced in Kenya.
- The growth of a town or a mining industry can create new demand in its vicinity for perishable products such as vegetables, fruit and milk.
- A change in fashion can create new demands, eg, exotic fruits and vegetables have become accessible in Europe due to new means of transport and are made fashionable by cookery writers; or organic foods, made popular by environmental writers, may attract a premium price.
- New agricultural practices can make old practices non-competitive, and therefore obsolete.
- Insect pests and diseases coming from other areas and, increasingly, from other continents can invade farmers' fields and require alterations in present farming systems to cope with them.
- New laws having to do with land tenure, forest management, the prohibition of certain chemical inputs, international tariffs, non-tariff trade barriers and many other actions that affect agriculture directly or indirectly can force farmers to change their systems of production.
- Changes in average family landholding size and labour availability will make certain technological changes more attractive and certain current practices less so.

Like their counterparts in industries and services, agriculturists need to respond to changes in the relative scarcity of factors of production, market demands and other factors that impinge upon them from the outside. Technological change is what enables us to countervail the economic law of diminishing returns to additional inputs of labour or capital.

THE ROLE OF AGROECOLOGICAL METHODS IN RAISING FOOD PRODUCTION

As in all agriculture, the possibilities for production and the productivity of different technologies will depend particularly upon the human and natural resources available.

Prospects for Low Population-density Areas

Quite large areas of the globe still have population densities under 30/km2. If one excludes very arid areas where agriculture is impossible without irrigation and areas with long cold winters, most of the remaining low-density areas are partly under humid or dry forest. While these areas are large, their populations are small, and usually they have problems with marketing. Accordingly:

- No substantial increase in the world's food production is likely to come from these areas, represented in Chapters 11 and 14. The best that can be hoped is that they can more or less feed themselves. The drier an area is, the more difficult this will be.
- The natural tendency of people in these areas, which are labour-short, is to use long fallows for fertility regeneration, and to cater for an increasing population by expanding the cleared area. In the first decades of settlement in forest areas, people make use of stored fertility and are unlikely to be interested in intensive methods. Where there are no forest resources, production is accordingly more difficult.
- These areas will not contribute much to meeting urban demand, except in the case of very high value-per-weight items, such as the spices widely produced by scattered farmers on some of the islands of Indonesia.

Possibilities for Medium Population-density Areas

What can and should be done in these areas, where much larger numbers of people live with concentrations of 30–100 people per km2? These areas represent a growing fraction of the total world population, and thus deserve more attention. It is common to see density quickly increase when a road is built through or adjacent to a low-density area. This facilitates the marketing of products, and settlement begins to thicken up adjacent to the road. The addition of social facilities such as schools and clinics makes it even more attractive for people to concentrate, and in such situations farmers will be under some pressure to intensify their agriculture, moving into the second stage.

In many of these areas, agroecological methods of intensification will be appropriate and cost-effective and could make a substantial difference to yields and output. This will be true particularly where people can market easily and in areas where the main factor limiting small farmer systems is soil fertility, rather than water, so we discuss it first and at greater length. Water is, of course, essential for plant and animal growth, so we need to consider potentials that it creates or constrains.[4]

Dealing with Soil Fertility Problems

Low fertility was the main constraint for small farmers in the African and Central American cases reported in Chapters 7 and 13. There is nothing unusual or unique about these areas. As programmes around the world learn better how to deal with deteriorated soils in tropical areas, it should become increasingly possible to get elsewhere the same kinds of results that are documented in those chapters.

The first response made to falling fertility – when farmers become unable to maintain their fallowing systems because land availability has become a constraint – is usually to increase the inputs of animal manure by integrating crop and animal production more closely. Manure may be applied by systematic grazing of animals on fields, or by penning the animals, at least at night, and carting their manure to fields. If more animals are kept by collecting and preserving crop residues and by improving feed resources in various other ways, this adds to the resources for soil improvement. Livestock-crop integration has the advantage that in addition to improving soil fertility via manure, it may also provide draft power for improving tilling, weeding and various soil and water conservation techniques, with further income from sales of stock and manure.

In some areas, however, increasing the animal–cropland ratio may be problematic, for climatic, disease or other reasons. In any case, farmers are well advised to seek considerable quantities of plant material that can be mulched or turned into the soil. Even farmers with livestock may reach a point where their manure has to be extended with additional vegetative material to meet soil fertility needs.

Two promising kinds of innovation that are gaining acceptance around the world will make fertility maintenance easier than was possible with the technologies available in the 1970s and 1980s. These should not be seen as new technologies in that they have been used by farmers for centuries, even millennia. However, they are now being used more intensively and systematically.

Agroforestry systems, often in conjunction with the use of *green manures and cover crops* (GMCCs), have proved themselves capable of improving eroded and deteriorated soils dramatically in a variety of countries with differing soils and climates, and at relatively low cost. Improved fallows and dispersed trees are related kinds of agroforestry that can build up soil capacity (Cairns, 2000). Here we focus particularly on GMCCs, as agroforestry is considered in Chapter 8. These have shown themselves capable of producing prodigious amounts of organic matter, even more than 25 tons of green matter per hectare per year in areas with reasonable rainfall.

Contrary to the conception of green manures as plants that are monocropped on a piece of agricultural land and then incorporated into the soil, GMCCs can be grown in many tropical contexts with little or no opportunity cost. They can, for instance, be intercropped with regular crops, grown during the dry season or grown on land too depleted to grow food crops. They are often left on top of the soil to decompose, rather than being incorporated, which reduces labour requirements. Mulch as a particular set of agroecological practices is discussed below and in the case study in Chapter 15.

In the more than 70 developing countries where GMCCs have been introduced or are already used by small farmers, this technology has shown great power to raise yields by acquiring nutrients and providing organic matter for the soil. Because GMCCs have been utilized traditionally, they are often looked down upon as 'not modern', but in fact they represent a technology capitalizing upon the genetic potentials of plants, and their blending in cropping systems can be quite scientific.

There are quite a variety of plant species and methods that can help farmers get sufficient inexpensive organic material to improve soil performance. Leguminous species such as peanuts, tephrosia, bahunia and canavalia can grow, and often even grow well, in semi-arid areas. Sahelian farmers have a long tradition of mixed cropping, cultivating leguminous crops such as peanuts and cowpeas alongside their cereal crops. This has two advantages: the legumes enrich the soil by fixing nitrogen from the atmosphere, and their leaves provide fodder for animals that in turn provide manure. These farmers have been able to maintain soil fertility at a stable, albeit low level, even when population densities have grown to the point that prohibits fallowing (Harris, 1996; Mortimore, 1998).

In some places in Africa, improved short-season varieties of dryland crops such as millet, sorghum and maize have enabled farmers to maintain or even improve yields despite unfavourable rainfall, using a mix of their own selections and varieties coming out of research stations (for examples, see Tiffen et al, 1994 on Machakos; and Faye et al, 2000 on the Diourbel region of Senegal; see also Chapter 11). Unfortunately, less research effort has been put into these crops than into rice, wheat and varieties of maize. Many farmers in these drier areas already know the value of certain trees such as *Faidherbia albida* for returning nutrients to the soil, and the importance of others as browse for their livestock. (This example shows how fine can be the line between agroforestry and GMCCs). Generating and validating new practices in this area will benefit from participatory approaches that mobilize indigenous knowledge.

In areas with low population density, farmers have little motivation to adopt such technologies since land is cheap and abundant, and fertility can be more easily restored by fallowing. Conversely but with similar consequences, land use in high-density areas can become so intense that there is no space for GMCCs and the opportunity costs of labour are higher. Then, improved fallow systems evolve into systems similar either to GMCC systems that are biologically improved, or to traditional orchards or tree plantations that are economically improved (Cairns and Garrity, 1999).

One consideration affecting the spread of GMCCs is whether local or national consumption of legumes will increase enough to maintain attractive market prices for the food products (grains) of these legumes. When small farmer agriculture is intensified, farmers are likely to maintain fewer animals, and dietary protein will have to come increasingly from legumes, a process already observed in many developing nations. If there is sufficient increase in the production of food legumes to lower the market price for beans and pulses, farmers will begin to use more of these as a quality animal feed.

The long-term impacts of these factors on the use of GMCCs around the world are difficult to predict. Nevertheless, scores of GMCC, improved fallow and dispersed tree systems are now known, with hundreds of others yet to be documented. In humid and sub-humid areas with medium population pressure, there is experience showing that enough organic matter could be grown or collected to enable small farmers to double or triple agricultural production, provided there are no introduced disincentives.

If we add to agroforestry and GMCC systems some additional simple technologies that make better use of sources of organic matter now being wasted – crop residues, tree litter, coffee pulp, sugarcane bagasse, animal manure, commercial chicken manure, urban waste, etc – there should be reasonably abundant organic matter available for soil enrichment in most systems. Some of these materials will not be used because transport costs make them financially unattractive, but these costs can often be reduced once the materials are used on a larger scale. Incentives to use compost or other organic matter as a source of nutrients can also be increased by making their use more productive by promoting better conjunctive practices or raising factor productivity (Chapter 12). One needs to think always about how to use *sets of resources* rather than single resources.

Capitalizing on Organic Matter Interactions with Chemical Fertilizers
Under a variety of conditions, large amounts of organic matter can provide higher yields than are achieved with the normally recommended applications of chemical fertilizers, especially on the kinds of highly degraded soils where so many of the poorer farmers operate (Subedi, 1998; also Chapters 11 and 14). Conventional agronomists might respond that these organically achieved yields are not sustainable, because of depletion of phosphorous when locally grown GMCCs are used. However, we are not talking about agriculture where there are no external inputs; rather we are contrasting low-input agriculture with high-input production systems. Small amounts of chemical fertilizer can be beneficially used to maintain phosphorous levels over time. As noted already in Chapter 2, research increasingly shows positive effects from using nutrients derived from both organic and inorganic sources in a complementary manner. This need not be a zero-sum relationship. However, given the current neglect of organic sources, we call attention here to their potentials.

Acid and Nutrient-poor Soils
Much has been written about the very difficult problems of growing crops on the extremely acid, phosphorous-deficient and often aluminium-toxic soils of the humid tropics. Nevertheless, several technical approaches beginning to be tested and applied suggest that production can be raised fairly cheaply even on these soils. Such practices could make a huge difference in the feasibility of sustainable agriculture in 'marginal' areas.

One of these practices is to use, either as mulches or in compost, plants that are known to be *nutrient accumulators*. Certain plant species, most notably the wild sunflower (*Tithonia spp*) and cassava, are able to access significant amounts of phosphorous in soils where other plants suffer from phosphorous insufficiencies. These plants can take up soil phosphorous that is normally unavailable to plants and then deposit it on the surface in forms that other plants can access.[5] Such techniques need always to be assessed from farmers' perspectives, since unless their benefits exceed their real costs in terms of additional labour and finance, uptake will not be sustained. However, considerable investment of labour can pay off if substantial increases in productivity can be obtained this way, like the examples for *Tithonia* use given in Chapter 8.

Mulch-based agriculture offers a second set of practices that draw on agroecological principles such as maximizing in situ biomass production, keeping soil covered, zero tillage and feeding the plants with the resulting mulch. This can be practised together with agroforestry and/or GMCC practices, so it is not something entirely separate. Such combinations of practices show evidence of being able to increase farmers' yields as much as reported from Honduras in Chapter 13 with soils that are quite deteriorated and acidic. These practices can be used on a large scale, as seen with no-till cultivation in Brazil (Chapter 15).

A possible explanation for the productivity of low-input agroecological systems comes from research reported by Primavesi (1980, p49). She describes experiments in which maize plants grown hydroponically – in solution with just 2 per cent of the concentration of nutrients usually considered necessary for plant growth, ie, a 'normal' solution – grew as well as plants grown in a 100 per cent solution, provided that the nutrients in the weak solution were replaced every other day, to maintain their constant availability in small amounts. (The roots of the plants grown in 2 per cent solution, it should be noted, grew to be eight times greater in mass than those in the 'normal' solution.)

This suggests that acceptable crop productivity can be achieved even in poor tropical soils with very low cation-exchange capacity, provided that the nutrient supply is continuous and constantly replenished. There appears to be greater benefit to plants from any given amount of nutrients if these are released slowly, as provided by compost or mulch compared with chemical fertilizer. This suggests that if soil structure is good and plant roots have easy access to nutrients in or on the soil, crop production can be quite satisfactory even with lower nutrient levels than now thought necessary (Bunch, 2001).

The age-old, worldwide process of achieving soil recuperation through long-term fallowing in systems of shifting cultivation did not restore soil productivity by pumping nutrients from sub-surface horizons. Rather, there was a significant increase in organic matter in and above the root zone that brought about continual replenishment and improvement in the structure of soils that were still very low in total nutrient content. Large amounts of biomass were created and sustained with fairly low quantities of soil nutrients.

The dynamics of such a process can be imitated by farmers, with good results, from manuring and composting or through the GMCC and improved fallow systems mentioned above. More research into these dynamics will be required before firm conclusions can be drawn, but data suggest important interactions between the qualitative and quantitative aspects of soil and plant nutrition.

Dealing with Water Shortage

In semi-arid areas and sometimes in sub-humid areas without conventional irrigation, water – and especially the distribution of rainfall over the growing season – is the limiting factor for small-farmer production systems. Outside of the arid regions of the world, rainfall distribution is a more serious constraint on agriculture than the overall lack of water. Sadly, the problem of irregular rains appears to be increasing. Water stress not only lowers productivity directly, but also affects it indirectly by making crops and animals more susceptible to attack by insects and disease. Water stress thus increases the demand for chemical pesticides.

Irrigation is understandably the first line of action against the shortage or irregularity of water supplies, as noted in Chapter 2. Nevertheless, as irrigation has been expanded dramatically over the last half-century, the areas where large-scale irrigation projects can still be built at costs that are economically and ecologically acceptable have dwindled. Large-scale irrigation will play a lesser role in increasing world food supplies in the next 50 years than in the past five decades.

On most sloping fields, anywhere from 30 to 60 per cent of the rain that falls runs off. Often, less than half actually filters into the soil and becomes available locally for plant and animal use. There is need for a major push in development circles to find ways of *harvesting and storing rainfall* for later use, especially for supplemental irrigation when periods of drought occur during the normal growing season. Some village-wide systems and farmer-field systems for such harvesting already exist and have been documented in areas where land per household is plentiful (see cases in Reij, 1996).

There is a widespread need to develop and diffuse knowledge of simple, inexpensive systems that farmers can use even in areas with very small landholdings. Other continents can learn something from Africa in this regard since farmers there have had to develop a variety of appropriate, sustainable water-harvesting and -conservation techniques over large semi-arid areas. Terraces, tied ridges and zai holes, for example, have been shown to be very good for capturing and holding water in farmers' fields (Tiffen et al, 1994; Wedem et al, 1996; Ouedraogo and Kabore, 1996; Schorlemer, 1999; Chatterji et al, 1999; see also Chapter 11).

Several NGOs in Honduras are now experimenting with *microcatchments* for small-scale water harvesting and *home-made filters* for recycling greywater. Enough experimentation has been done with small farmers that we can see how these technologies could substantially increase incomes in drought-prone areas and are quite acceptable to poorer farmers. Water can be held for six months in micro-catchments whose total cost is only US$15 per cubic metre of capacity. Such technologies should be improved, adapted and spread widely rather than through a slow process of diffusion.

With low-cost water harvesting techniques, average yields in the semi-arid and sub-humid areas of the tropical world could probably be doubled on those areas on which the water is concentrated. Another productive use of concentrated water is the farm pond, which supports aquaculture and complements a home garden as discussed in Chapter 9. Once farmers have a more or less guaranteed water source, soil conservation and soil improvement practices become more economically attractive, especially for high-value crops.

People who solve their problems of water availability will likely move on to soil improvement for the less valuable crops, thereby increasing their yields further (see Chapters 10 and 11). The technologies are fairly simple, though often cooperation among farmers is required to permit larger structures that are more water-efficient, and such cooperation is not always forthcoming. This brings agroecological practice into the domain of social capital discussed in Chapter 4.

Until now, the emphasis of most agriculturalists has been on soil conservation measures as a means to prevent soil loss. However, some recommended anti-erosion techniques also conserve soil moisture, and farmers have found that a major benefit from certain types of terracing, ridging, etc has been moisture conservation, which dramatically affects yields and incomes. For example, in Machakos district of Kenya, farmers in the 1950s adopted a type of terrace where soil was thrown uphill, even though it was more labour-intensive than the officially recommended contour ditches, for which soil was thrown downhill. Why was the more labour-intensive technology adopted? Because it conserved water better, as discussed in the next chapter. Another technique much used there was channelling road runoff via a cut-off drain to an area where it could be useful on the farm (Tiffen et al, 1994).

Dealing with Pests

While integrated pest management (IPM) by itself may not increase yields very much, it has proven potential to halve pesticide use while at least maintaining yields (Chapters 16 and 17). Where farmers are cultivating improved soils with higher organic matter content and with more consistent water supplies, plants are healthier and better able to resist pest and disease attacks. As we learn more about the use of additional biological control measures and as these become commercially available, reduced pesticide use and enhanced incomes through savings brought about by this reduction should become more widespread.

Tillage Practices

There is increasing acceptance both in developing-nation agriculture and in richer countries of various forms of *reduced*, *minimum* or even *zero tillage*, discussed in Chapter 15. In Brazil and Central America, GMCC systems are rapidly following in the footsteps of conservation and zero-tillage systems, if they have not preceded them. This allows farmers to control any weed problems created by decreased tillage through the increased use of cover crops. At the same time, the spread of IPM practices and conservation tillage among 'conventional' farmers is already moving many of these farmers towards a much more ecologically-based agriculture.

Areas with High Population Densities

Larger and larger proportions of the world's populations will live, or already live, in rural areas with densities over $100/km^2$. In many cases they are near rapidly growing towns with expanding manufacturing and service sectors that compete with agriculture for labour. In such areas, very labour-intensive methods of restoring fertility or of water harvesting are unlikely to be accepted, unless they are for a crop or a product that yields very high financial returns. Farmers will need to find methods of restoring fertility that combine ease of use (probably inorganic fertilizers) with good management of the soil to assure good structure and biological health, for which up to now farmers have generally used manure and compost. This will be a challenge. In some cases, as in Kenya, the answer will be intensive methods of livestock-keeping, as seen in the next chapter.

Farmers are always going to do what seems best to them in their particular circumstances, with their combination of land, labour and capital, their climate and their market opportunities. Practices will be altered as circumstances change. Agroecological techniques, which may either be worked out locally or introduced by outside agencies, will be adopted only if, and to the extent that, they suit local circumstances, and for as long as they suit these circumstances. Farmers will change their practices, sometimes frequently, in order to survive and prosper.

CAN THESE TECHNOLOGIES BE INTRODUCED WIDELY AND AT LOW COST?

Evidence from a number of development programmes run by different nongovernmental organization (NGOs) and government agencies around the world in greatly differing conditions indicates that the extension of low-input technologies can be efficient and sustainable. Some of these programmes have achieved significant and sustainable increases in productivity with a total investment of less than US$1000 per household, including transportation, salaries, administration and other programme costs. While this may sound like a lot of money, many times this amount have been spent in numerous agricultural development projects around the world over the last three decades, with far fewer results to show for the expenditure.

This level of efficiency cannot be achieved through traditional systems of extension focused on technology transfer and adoption. The kinds of programmes that can reach reasonable levels of efficiency and effectiveness with the complex and diverse systems of the world's poorer farmers will be ones that:

- involve farmers as extensionists;
- focus on a limited number of innovations during the first years of the programme;
- teach farmers to experiment and innovate;
- avoid artificial and unsustainable incentives, such as subsidies that reduce prices (different from investments that are expected to return long-term benefits); and
- solve marketing constraints where farmers have monetary as well as survival needs, so that they can get real financial benefit from their work and innovation.

The main factors impeding widespread adoption of low-input technologies are:

- the unwillingness or inability of organizations, public or private, to use the above principles of participatory technology development and diffusion; and
- the small amount of funding that is available to programmes that follow this approach.

Therefore it is important that donors become convinced of the value of these alternative approaches and of the value of the new methods for technology development and diffusion now validated in many countries. Major advances could be achieved in not only feeding the world's still increasing population, but doing it in ways that have at least some chance of being sustainable economically, agronomically and ecologically. Scaling up represents the largest current challenge for all agricultural technologies, whether requiring many or few external inputs. The latter technologies should not be more difficult to extend, especially if developed and popularized with farmer involvement.

CAN A MORE ECOLOGICAL AGRICULTURE FEED A BURGEONING POPULATION?

In most areas of low population density in developing countries, it appears that people will continue, as they have in the past, to feed themselves but barely, not producing much of a surplus for income generation. Population in these areas is usually low because of soil, climatic or other constraints on production. In areas of medium population density, on the other hand, which are often semi-humid, it should be possible to double current low yields – or to increase them by even more except where there are severe climatic constraints – using presently known low-input technologies if these are widely disseminated. (Some presently low-density areas may move into this category as key production constraints are alleviated.) Such yield increases should meet the food needs of the growing population in these areas and also provide some surplus for food-deficit areas, including towns and cities. These potentials are repeatedly shown in the cases reported in Part 2.

Semi-arid areas will continue to present more of a challenge, but there are ways to increase yields of cereals and legumes even here, even with decreased fallowing, and to improve incomes by techniques of water harvesting for valuable crops. Scientists could help farmers in these areas who have already proved their capacity to make good economizing use of their resources and to integrate crop and livestock activities, by giving more attention to breeding short-season cereals and legumes, using GMCCs, harvesting water and combating animal diseases.

In areas of high population density, especially around towns and cities, some of the food will probably have to come from outside the developing countries. These populations are, fortunately, the ones most likely to have the incomes necessary to take advantage of the globalization of trade, buying cheaper grains from abroad while consuming vegetables and some fruits from their own highly productive rural areas.

Happily, we are seeing in some places the rise of a new (or previously ignored) phenomenon, urban agriculture, though its contribution to total world food supply will remain marginal because there will be better competing uses for people's labour and for land close to urban centres. While densely-populated areas cannot become proverbial 'breadbaskets', they can contribute to food requirements more than they do now. This was seen in parts of Machakos district in Kenya, discussed in the next chapter, where over a 60-year period, while population density was increasing five-fold, the value of agricultural production, including in some semi-arid areas, increased 11 times.[6]

Thus, while the major towns and cities will depend more on imported foods in the future (in fact, imports of food will probably push at least some locally produced foods out of these markets), the burgeoning populations of medium-density areas should be quite able to feed themselves with food produced by local farmers. Even high-density areas using intensified production systems, which require and can remunerate more labour, should be able to rely on more local food than present trends indicate.

How Well Can High-input Technologies Meet the Future World Food Needs?

While this book focuses on productive potential of low-external-input technologies, it is appropriate to consider at least briefly the possibility that high-input technologies are themselves going to run into serious difficulties in maintaining their present advantages. Ecological concerns that toxic chemicals and chemical fertilizers are damaging our water supplies and our soils and are creating a series of other environmental problems have been well documented (Pretty, 1995).

No certain predictions on the future price movements of petroleum and its products such as fertilizer can be made, but they are unlikely to become relatively more cheap in the decades ahead. Within the next 15 to 30 years, even allowing for major efforts at conservation (which remain to be instituted), petroleum production is unlikely to keep pace with demand spurred by both population and economic growth, so prices are indeed likely to rise. Already many farmers are finding their past levels of fertilizer and energy use, which had previously been heavily subsidized by governments, uneconomic. If these prices increase further, the advantages of petrochemical inputs will diminish, making it cost-effective to invest more in labour and to depend more on organic sources of nutrients.

The world will not suddenly run out of petroleum supplies, and other sources of energy will be increasingly tapped in the future. But once we no longer have the luxury of abundant and relatively cheap petroleum, which has subsidized food production for the last 40 years, the prices of energy and energy-intensive farm inputs will increase, and so will the cost of growing food with high-input technologies. As these prices rise, 'modern' technologies will become still less accessible to a majority of farmers. The world does not just need more food; it needs food produced in ways and at a price that the poor can afford. If farmers cannot themselves afford the means of production, this aggravates the extent of poverty and hunger that we are concerned about.

While there may be some doubts about the ability of low-input, ecological agriculture to feed the world's growing population, there are equally good reasons to wonder whether conventional modern agriculture can feed the world's poor any better in the future than now, especially if we remember that reducing hunger is not simply a matter of supply. Food produced in Europe and North America is not going to prevent hunger among the marginalized populations of the developing countries if they cannot afford to buy it. Moreover, transport costs are an important factor in moving bulky commodities such as grains to the points where they are needed.

This means that poorer countries will need to maintain their own supply capabilities, although this does not mean that their policy should be the autarkic one of self-sufficiency. Because food security cannot be attained entirely through market transactions, farmers in developing countries need production technologies with inputs they can afford, and which are low in risk. Impressive productivity of high-input agriculture in the rich countries, where these technologies may still perform satisfactorily, will only make the plight of hungry people elsewhere all the more ironic.

LOOKING AHEAD

This discussion and the case studies that follow point to courses of action that expand upon the options of the presently predominant modes of agriculture. The basic elements of a strategy for moving in these new directions are as follows:

- Increase greatly the research being done on low-external-input technologies, especially on organic means of soil improvement, microscale water harvesting and non-toxic alternatives to pesticides, cures for animal diseases, and crops suited to semi-arid areas.
- Allocate development funds differently. Not only should donors become convinced of the value of researching and supporting low-external-input technologies, but they need also to realize the importance of farmer-to-farmer extension and of teaching farmers to experiment in a process of participatory technology development.
- Evaluate past agricultural development efforts more rigorously. Studies of the impact of development programmes five to ten years after termination are especially important for understanding what can be expected to be 'sustainable' in agricultural development. Such studies are few. There is reason to question the long-term sustainability of the high-input agriculture that is promoted on a large scale in many parts of the tropics.
- Terminate the use of incentives that in effect 'bribe' farmers to use agricultural innovations. These should be productive enough that their adoption can be justified in terms of market prices (provided, of course, that other

prices also reflect the market value of these inputs, goods or services). There are many reasons to move towards a more market-oriented, less subsidized agriculture.

The cases that follow provide evidence that low-external-input agriculture using agroecological practices can make an important contribution to feeding the world's burgeoning population over the next 30 to 50 years, after which time we have reason to expect that the total numbers requiring food will begin to diminish. If food insecurity and malnutrition persist and even increase, it will not be because technologies are inadequate or insufficient.

Rather, the problem will be more one of inadequate political support for innovations now available. Those who have the money that could fund these new directions will either use it for other ends or fail to use it effectively, if we are to be guided by an understanding of past experience. Many institutions, and some political leaders in developing countries, afraid of empowering rural people, may refuse to accept people-centred approaches. Powerful interests, such as fertilizer and chemical companies, may oppose low-input agriculture and perhaps may complicate farmers' already difficult economic situations with initiatives such as the current attempts by multinational companies to corner many of the world's most important seed research and production resources.

At the same time we hope for continued efforts to breed more productive varieties, varieties that can resist important pests and diseases, and ones that can tolerate stressful conditions like drought and salinity, as these could greatly benefit farmers in presently marginal areas. Commercial interests are likely to concentrate on the crops grown in the great grain baskets of the world, so there is need for local, national and international research that deals with the special needs of the poorer areas we have been describing. Our comments should not be read as opposing a private sector role in agricultural development since we recognize that a wide-ranging and more genuinely competitive private sector can have the effect of empowering farmers by giving them a wider range of options than they have in more state-controlled agricultural sectors.

The key factor in any changes will be human resources, which are more than just labour. If policies create a push from rural areas matched by urban pull, and few people are able and willing to engage in agricultural production, there will be food shortages of quite a different origin than those that worry scientists and policy-makers now. The agricultural strategy proposed here is concurrently a human resource development approach that seeks to upgrade people's capacities as decision-makers and managers. Strategies are needed that do not de-skill rural people by treating them like plantation labourers. Rather, the approach should engage the knowledge and the critical and inventive faculties of farmers of both genders to improve their own productivity and that of others.

For the growing millions of people living in large towns and cities, the main staple production areas are likely to continue to be irrigated areas and temperate zones. In the agriculture of both regions, a higher proportion of purchased inputs is likely to be appropriate, although agroecological methods can make contributions also here to maintaining soil health and water quality. How far the spread of agroecological methods will go cannot be predicted. Twenty years ago, few would have predicted that 'no-till' agriculture would become a preferred practice in North American, European and Latin American (Southern cone) agriculture, or that IPM would have become so widespread as it is now.

Natural resource endowments and the relative cost and productivity of labour have been driving forces affecting agricultural change over decades and centuries. At the same time, regulatory measures and policies seeking to achieve public benefits that would not otherwise be produced reliably by individual private calculations have retained or reshaped forms of agriculture.

Entrenched ways of thinking as well as vested interests create inertia, which slows changes that could apply available resources better to serving human needs. Rather than proposing actions and incentives that promote agroecological approaches, it is more appropriate to level the present playing field for agricultural development. Making it open to the opportunities that agroecological and participatory approaches offer will give farmers every-where a broader range of choices that can benefit their families, communities, countries and the natural environment.

NOTES

1   Our discussion focuses primarily on sustainable crop-based agriculture, though we consider the contributions made by animals. In Table 6.1 and in the rest of this chapter, we are not considering those very arid areas where only livestock keeping is possible.

2   In pondering the causes of poverty in Northwest Scotland in the 18th century, Dr Samuel Johnson said: 'Men, thinly scattered, make a shift, but a bad shift, [having to do] without many things. A smith is ten miles off: [so] they do without a nail or a staple. A taylor [sic] is far from them: [so] they'll botch their own clothes. It is being concentrated which produces high convenience' (quoted in Boswell, 1993, p169).

3   A Swedish traveller (Kalm) wrote of English settlers practising agriculture in North America in 1749: 'They make scarce any manure for their corn fields ... when one piece of ground has been exhausted by continual cropping, they clear and cultivate another piece of fresh land; and when that is exhausted, [they] proceed to a third. Their cattle are allowed to wander through the woods and other uncultivated grounds, where they are half-starved, having long ago extirpated almost all the annual grasses by cropping them too early in the spring' (reported in Smith, 1991, p204). These settlers came from a country that was well known at the time for manuring and other productivity-enhancing techniques.

4   The *amount* of rainfall is not the only factor that is important for crop production; its distribution and intensity also influence the contribution that rainfall can make to agriculture. One should distinguish: (a) semi-arid farming areas with seasonal rainfall of 250–600mm, usually with only one rainy season and great rainfall variability, as in the Sahelian regions of West Africa and parts of Southern Africa and India; (b) semi-humid areas that have rainfall in the 600–1200mm range; and

(c) humid areas with rainfall above 1200mm, with perhaps two cropping seasons, even without irrigation.

5   Another example is wild ginger (*Aframomum angustifolium*), which is starting to be used in composts in Madagascar. This plant's leaves and stems contain about 0.2 per cent phosphorous, and initial evaluations on farmers' plots suggest that while typical compost can double yields of poor farmers, equal amounts of compost made with wild ginger can triple the yield of maize, potatoes, beans and other crops (CIIFAD, 1997, p76).

6   Food output grew more or less in line with population while higher-value crops proliferated. Various techniques for concentrating water on certain areas were crucial to this transformation.

Part 2

Experiences from Africa, Latin America and Asia

Chapter 7

The Evolution of Agroecological Methods and the Influence of Markets: Case Studies from Kenya and Nigeria

Mary Tiffen

Many parts of Africa have been and continue to be in the second of the three stages identified in Chapter 6, when land is becoming scarcer and agroecological methods of improvement become more worthwhile. In these circumstances, such methods can contribute to a substantial rise in output and welfare, partly because land formerly under scrub woodland and rough pasture is put into cultivation, and partly through the adoption of methods that lead to higher output per hectare. However, for such intensification there need to be incentives and rewards for hard work and enterprise, plus access to markets where farm goods can be sold.

In areas with very erratic rainfall, farmers can safeguard household incomes when the rains fail by encouraging at least one family member to have an off-farm job, either in the dry season or permanently. The non-agricultural sector of the economy will grow as urbanization and specialization proceed, but to get good employment in it requires qualifications. Hence, farmers have to juggle their cash income carefully to improve their farms, provide for the education of their children, and support family-related non-farm activities.

This process of change can be seen from two case studies, one in East Africa and one in West Africa, which are followed by some comparisons with Senegal that elaborate the analysis. These reflect basic patterns of agricultural change seen across Africa and, with some modifications for historical and other reasons, elsewhere.

THE MACHAKOS EXPERIENCE IN KENYA

Changes over a 60-year period in the Machakos district of this country, from the 1930s into the 1990s, have been documented and analysed in a previous study that illuminated the process of agricultural development in a poorly-endowed region in East Africa (Tiffen et al, 1994). Back in 1937 this district was described by knowledgeable observers from the Kenyan Department of Agriculture as suffering from over-grazing and poor husbandry, writing that 'the inhabitants of [Machakos] are rapidly drifting to a state of hopeless and miserable poverty, and their land to a parching desert of rocks, stones and sand' (Maher, 1937). Yet, over a six-decade period, during which time the population grew from 250,000 to 1.5 million, the value of agricultural output per hectare was increased 11-fold, with a three-fold increase per capita.[1]

Environmentally, the land had recovered from severe erosion in its cultivated parts, and much of the unproductive tsetse-infested bush in its uncultivated parts had been transformed by households into terraced and cultivated land, on which little erosion took place, or into privately-owned, managed grazing areas and woodland on which erosion was diminishing. Planted trees, many of them useful exotics, had greatly increased in numbers, providing fruit, timber, windbreaks and amenity. Farmers attributed the improvement on cultivated land largely to their activities of manuring and terracing. The latter conserved valuable rainfall in an area that had inadequate and erratic rains, and it also ensured that manure and other nutrients stayed where they were put.

The substantial increase in the value of output per hectare and per head was achieved by several innovations and investments, the most notable of which are outlined below.

The Conversion of Unterraced Cultivated Area and Rough Grazing Bush to Terraced Land

This is done to boost production per hectare on smaller farms. The extent of this practice, farming in three dimensions as suggested in Chapter 2, can be seen from Figure 7.1, which was built up from 1948, 1961 and 1978 air photographs. Terraces in the long-settled areas were originally built mainly by compulsory communal labour. But once some farmers were seen to be making good incomes by growing high-value crops such as tomatoes and onions for the thriving town of Nairobi in the adjacent district, terracing was also used for other crops. Tomatoes and many other fruits and vegetables could not be grown under Machakos conditions without terraces that conserved water. These were a simpler and more appropriate technology than irrigation would have been for dealing with the water constraint.

The type of terrace first recommended, introduced from the United States by Maher in the 1940s, was not very good for conserving water, as the soil dug to make a contour ditch was thrown downslope. Farmers came to prefer

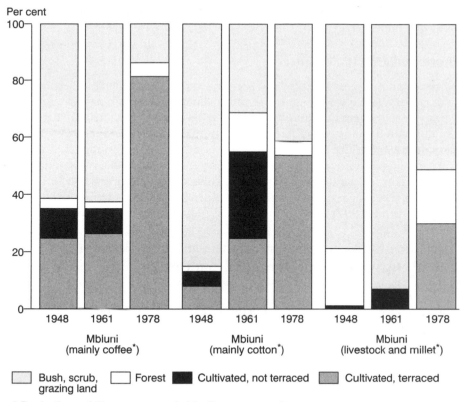

Figure 7.1 *Land Use in Three Areas, Machakos District, 1948–1978*

the more labour-intensive method of throwing soil from the contour ditch uphill, forming a bench terrace locally called *fanya ju* ('throwing upwards'). A former soldier recruited from the local population who had seen this technology in India during his service in World War II started growing onions on this type of terrace in 1949. By 1952, others were using it very profitably for tomatoes. The Agricultural Department then began promoting it for coffee, a crop that a limited number of farmers in the wetter parts of Machakos were permitted to cultivate (Tiffen et al, 1994).

Because this terracing was productive under local conditions, farmers began to adopt it also for their maize and beans, the staple food crops. At the same time they began extending their cultivated areas into formerly grazed areas, and migrating downslope into the thorny, tsetse-infested scrub that covered most of the drier areas of the district. The first stages of this process did not look sustainable as it involved burning and slashing the bush. In new areas, new land was cultivated each year, as this was the accepted way in which the settler could establish land rights. Terraces were put in as people could afford them, some years later, and this tended to stabilize separate cultivated and grazing areas, since cattle could not be allowed to trample a painfully built terrace.

Increased Use of Manure

Since common-access pasture land was constantly diminishing, investments in hedging or fencing were required in areas claimed as private pasture, to create *bomas* (pens) where the cattle could be kept overnight, with manure carried to the cultivated fields. As grazing land diminished still further, crop residues were more and more frequently collected and stored for fodder, and grasses were cultivated on terrace edges for cutting. In the most densely settled parts of Machakos, in the slightly wetter hills, cows now spend the whole day in the *boma*, except when led to water, and fodder is carried to them (Tiffen et al, 1994; Murton, 1997).

The Use of Ox-ploughs for Cultivation and a First Weeding, Better Organization of Time, and the Breeding of Shorter-season Maize Varieties

These changes enable farmers to plant and harvest two crops per year more frequently. Even so, on average, farmers can expect to get a harvest in only 60 per cent of seasons, ie, four crops in three years, given the vagaries of weather. If long-season sorghum and maize were still in the field when the long rains began, a second crop planted hastily, broadcast without further land preparation, yielded little. The ox-plough was a crucial innovation that enabled earlier planting and weeding, giving short-season maize crops a better start and a chance of getting more rain. The ox-plough was initially introduced on European farms and then spread by traders to African farmers.

The Concentration of Water

This often involves constructing a pond, ditch or road cut-off on a small area of high-value crops such as fruits and vegetables, planted according to market demands. At the time of our 1990–1991 study, one recent innovation was planting very fine French beans, specially grown for the high-class restaurant market in Kenya and Europe, and fetching a high price. By 1996, Kenyan beans (of slightly lower quality) had become a common supermarket product in Europe. While their price had become lower, more farmers were supplying the enlarged market, which enhanced household incomes.

Prompt Response to Market Signals

The way in which farmers have changed their composition of output over time is shown in Figure 7.2, which summarizes information received from older village leaders on the main products marketed at different times. Because water is a limiting factor, the greater part of land has had to remain under crops demanding less water, such as grains and pulses. The effect of better varieties,

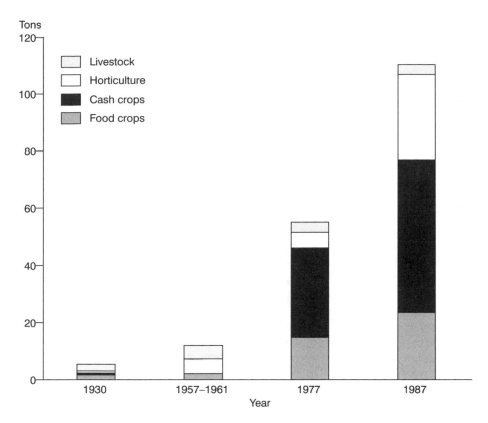

Figure 7.2 *Output per km², in Maize Equivalents, at 1957 Values*

more effective water conservation by terracing and early planting, manuring, improved weeding, etc can be estimated from the rough yield data we have compiled. Because rains and yields are so variable, it is better to look at what can be produced over a three-year period.

Average annual grain production seems to have increased from 1500kg over three years (three crops of 500kg each) to 5100kg for such a period (an average of 850kg/ha per season during 1974–1988) if all land is planted twice a year. This estimate is possibly an overstatement since in some very dry seasons, the extent cultivated will drop. (In really bad seasons, with late and scanty rains, farmers may not plant at all.) However, it seems likely that output per hectare over three years had at least trebled, to about, 4500kg.[2] This increase in output was all the more impressive since new farmers were being forced by land shortage to farm in progressively more arid areas of the district. Yields in the 1930s were from land of higher average natural potential than was being utilized in the 1980s.

The cumulative effect of these changes on production is shown in Figure 7.2, which gives output per head stated in terms of maize values in constant 1957 prices. Of course, the exchange ratio between maize and other products has varied over time. Coffee prices were high in maize terms in the 1950s and again in 1977–1980 (the effect of a harvest failure in Brazil), but low in 1987. This resulted in reduced maize purchasing power between 1977 and 1987, though this remained considerably higher than the 1957–1961 level. As coffee prices fell, there was a switch into fruit and vegetables. In welfare terms, we estimated that incomes per head had increased more than these diagrams indicate because of the growth of the rural non-farm sector. By 1983, about half of rural incomes were coming from non-farm sources. With more of their product being marketed, farmers were able to purchase an increased amount and variety of goods and services, and there was a corresponding growth of non-farm jobs.

At the time of our investigations, farmers purchased relatively little fertilizer. Its price was high compared with the benefit, particularly in the drier areas with high risk of a poor harvest due to lack of rain. What fertilizer was purchased was applied mainly in the wetter areas to coffee and vegetables (Murton, 1997). Coffee farmers also purchased other chemical inputs such as sprays. These areas had the highest population density, the smallest farms, and therefore, a minuscule proportion of grazing land when compared with drier areas. The number of cattle per hectare was higher than in the 1930s, however, because most of the small farms now had at least one cow. Manure was bought from ranches in the lowlands, and town rubbish was collected. A combination of organic and non-organic inputs enabled households to raise coffee yields considerably in years when they thought that price ratios justified the investment of effort. The yield variation according to price is shown in Figure 7.3.

An investigation of soil conservation in other parts of Kenya in 1996 found that in most of the well-watered, densely populated parts of Kenya, farmers were conscious of falling yields because they could no longer afford to buy fertilizer. Most of these farmers kept a cow or two in *bomas* for milk production and placed as much manure or compost as possible on their fields. They knew this was inadequate in quantity to replace nutrient losses in their soils, but they could no longer buy fertilizer because the milk marketing board was in chaos and owed them several months of back payments for milk. The private sector and cooperatives have since stepped in with new distribution and processing networks, and milk prices at the farm gate are now higher than they were under monopoly conditions; still, in 1996 these new institutions were only getting established.

In more remote areas, farm families still relied on male out-migration and remittances to supplement the inadequate returns from small, underfertilized farms. One woman told me during a visit to her compound that she used the manure she had wasting in an unprotected heap 'when she had time'. If she had been living in Central Kenya she would have made the necessary time, because with easy market access for her products, her labour would have been well rewarded. Farmers in Kenya are not oriented to organic farming on principle; they use a variety of methods to keep their land in good health, varying their investment of cash and labour according to market conditions for inputs and outputs.

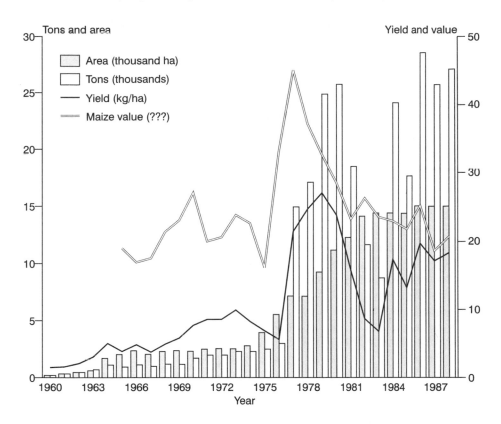

Figure 7.3 *Area, Output, Yield and Maize Value of Coffee, 1960–1988*

We have now begun studying the recent agricultural history of Makueni district, formerly part of Machakos district but with different endowments and lying at lower elevation.[3] We want to understand how agriculture has developed in the driest and most disadvantaged parts of the region since 1990. In much of this area, average yearly rainfall is only 600–800mm, divided between two seasons, and with great variation from season to season.

Sequences of bad seasons are not uncommon. Many farm families there, particularly in the very driest villages, are now without even a single cow, through a combination of drought and disease effects. Nevertheless, in the slightly better areas that are closer to markets, some farmers have made a remarkable investment in cross-bred cattle, because milk becomes profitable. This also induces investments in fencing, improved pastures, the creation of ponds and improved storage facilities for crop residues.

Yet while some farmers have several cattle, many families are without any cattle and have a shortage of manure, even if they have goats. They are well aware of the benefits of organic material, not only to improve fertility, but also for improving the soil's water retention. Composting in the dry areas is not easy due to a lack of moisture. A less hilly landscape means proportionately fewer farmers are able to concentrate water on fruit and vegetables, though in certain areas there is some irrigation. Fertilizer is unattractive because of its price and the risk of crop loss. It is apparent that these farmers face very difficult problems due to the erratic nature of the rainfall. They invest heavily in education in order to qualify some members of their family for non-farm jobs. When there are good rains and good harvests they invest in their farms; when the rains fail they rely on non-farm income and help from relatives. These areas of low and erratic rainfall cannot always feed their population, let alone feed the towns, and in really bad years, the government has to assist them with food-for-work programmes (Mbogoh, 2000).

What has become clear from interviewing farmers is that the type of farming practised in the 1990s required much more judgment, knowledge and information than the older subsistence-oriented system of the 1930s. Education has long been a main focus of family investment, in both the arid and the less arid areas, mainly so that children will be better able to find non-farm employment, but also so that literacy and numeracy can help improve their farming and marketing. As one farmer told us: even a herd boy has to be able to read these days – instructions on veterinary supplies and supplementary feeds are important.

Teaching is one of the service industries showing spectacular growth rates in the rural areas. The availability of, and attraction of, off-farm employment reduces the time available for labour-intensive practices and daily livestock care. Despite the fact that most rains in the 1990s have been poor, farmers have made huge efforts to ensure their children's education. Almost all children go to nursery schools (where fees are required) to give them what is regarded as the essential foundation for eight years of primary school – which is nominally free except, of course, for text books and materials, uniforms, contributions to encourage teachers, etc.

Parents regard primary schooling as the essential minimum. About 30 per cent go on to the considerably more costly secondary education, where fees are required. At the moment, slightly more girls than boys are enrolled in secondary schools (Nzioka, 2000), reflecting farmers' judgment of the current demands of Kenya's intensely competitive labour market. At a workshop in Makueni in November 1998, farmers were asked how they chose between farm, non-farm and educational investments, given their general shortage of cash. Education, we were told, was not a choice now but a necessity.

It is difficult to say what proportion of income comes from crops, from livestock, from local non-farm jobs or from remittances from husbands, sons and daughters working away from home, because amounts vary from year to year. In a bad year, animals are sold, crops bring nothing and relatives give help. In a good year, on the other hand, when crops are bountiful, farmers are able to sell a surplus, buy new animals, invest in a small dam or in a building to rent out in the nearest market centre, and possibly put another child into secondary school, knowing that this latter course of action brings few rewards unless it can be maintained for four years.

EXPERIENCE IN GOMBE, NIGERIA

A study of long-term change, covering the period 1900–1966, was previously carried out in this region of northeastern Nigeria (Tiffen, 1976). Gombe is surrounded on three sides by the Gongola River, which loops round it. While this allowed for some limited river transport to the south even many years ago, the area had limited road access to the west and the north. With an average rainfall of 900–950mm concentrated in one season, the area is much less limited by arid conditions than is Machakos, although Gombe has a long annual dry season.

In 1900, there was only a small and scattered population in the district, predominantly of settled Fulani who combined cattle raising with the cultivation of sorghum for subsistence. Population density in 1916 was only around 10/km². Only the fields most conveniently situated near the village were manured, by penning the cattle on them. Outer fields were fallowed when they lost fertility, as new fields could be opened freely in the plentiful bush.

For a combination of economic, political and religious reasons, Gombe attracted large numbers of immigrant farmers after 1920, and especially between 1945 and 1960. It contained a famous centre of pilgrimage, and many pilgrims decided to settle when they saw that there was good land, economic opportunity and relaxed taxation by chiefs who welcomed newcomers. By 1963, the southern district in Gombe that was the first area to attract settlers had a density of 74/km². An adjacent district that was also attracting new colonizers had reached 54 people/km². Thus, this part of Gombe was well into what was described in Chapter 6 as the second stage of agricultural development. The northern part, with less favoured soils and poorer rainfall, still had less than 30 inhabitants/km².

Initially, only cattle and manpower could be marketed because roads were almost non-existent. Cattle could be trekked to distant markets, and men could do spells of work in the tin mines 200 miles to the west around Jos. As river navigation developed after World War I, peanuts (groundnuts) could be sold to the south. In 1946, an agricultural officer described Gombe as 'this groundnut division'. By 1948, lorry traffic, no longer impeded by wartime shortages of vehicles and tyres, was increasing along a road that had finally been completed in 1939, after several unsuccessful attempts to find a good bridging point. Foodstuffs as well as manpower and cattle could then be sold in Jos, the regional market centre.

After the war, a cotton ginnery was built. The national seed-cotton price was raised first to four pence in 1948–1949 and then to six pence per pound in 1951–1952. This proved attractive to Southern Gombe farmers who had soils well suited to the crop. They competed to get the scarce ox-drawn ploughs that the agricultural department first introduced in 1946. The department endeavoured at first to allocate ploughs only to farmers who would build pens for their oxen and who would cart manure to their fields (to avoid 'soil-mining'). But this required heavy additional inputs of labour, which farmers regarded as unnecessary. Local blacksmiths began to make ploughs out of any scrap metal at hand, and painted them blue to match the imports. Also, traders busily brought in second-hand ploughs from areas where this equipment was less popular. As a consequence, farmers were able to ignore any necessary linkage between ploughs and manure collection and use. By 1960, there were an estimated 4000 ploughs in use, almost all of them in the southern part of Gombe.

As in Machakos, a lot of the additional production initially came from clearing new land from the bush, which was plentiful. The necessary burning and slashing was done both by new immigrants and by indigenous farmers expanding their holdings. For this, the plough was crucial since it not only enabled farmers to plant more land but also permitted early planting and reduced weeding labour. (Additional labour was required for de-stumping fields.) The agricultural department, which used Gombe as a cotton seed multiplication area, sent an energetic extension officer to the area to promote good methods of cotton cultivation, including early planting, early weeding, rotation with sorghum, and the annual cutting down and burning of cotton residues to reduce pest carry-over.

Fertility was not a problem at first on the newly cleared land as farmers were 'mining' the stored nutrients by burning the bush. But as land became short, farmers had to find methods of maintaining productivity. Initially, the preferred option was *kraaling* their cattle on the fields after harvest, and also inviting in visiting pastoral Fulani herds. However, as time went on, farmers in Southern Gombe found it more and more difficult to maintain a large number of cattle, because the bush was being increasingly converted to cultivation. They maintained their plough oxen, sheep and goats, but reduced their cattle numbers. There was less manure to be purchased from pastoral Fulani because more of these decided to settle and kept their manure to use on their own fields.

Northern Gombe, with soils less suited to cotton, received few immigrants and, still having plentiful bush, farmers there expanded their cattle holdings and maintained *kraaling* of animals. In 1967–1968 interviews (the sample was unavoidably biased toward older and richer farmers), choices among fertilizer, manure or the use of neither differed significantly between the north and south, as shown in Table 7.1. The first official fertilizer campaign was in 1963, with sales of 540 sacks; this number rose quickly, reaching 4200 sacks by the first part of 1966.

Gombe farmers were at the same time increasing their sale of foodstuffs to Nigeria's expanding urban markets. Most farmers practised the recommended sorghum-cotton rotation, and grain output apparently increased in line with the expansion of cotton, but we were not able to get data on grain yields. There is some evidence on what was happening to cotton production and yields, because the amount of cotton seed distributed and the amount of cotton bought were recorded. The database is less reliable on the number of people who were paying taxes to the local authority, which gives some idea of the growth in population through natural increase and immigration, which caused a growing shortage of the best types of farm land. With some idea on when and how cotton production increased, we can calculate whether this was due to the use of more land, or to the use of better and mainly biological techniques up to the 1963 fertilizer campaign, and the role of fertilizer after that date.

Table 7.1 *Fertilizer and Manure Use, Selective Sample of Gombe Farmers, 1967–1968*

	North (N=60)	South (N=132)
Counted tax payers per mile^2 (land scarcity measure)	4.8	15.4
Counted cattle per tax payer (manure scarcity measure)	3.2	1.1
Percentage in sample using: manure	36.7	25.8
fertilizer	6.7	37.6
neither	60.0	47.9

Source: Tiffen (1976, p117)

From the first introduction of cotton, there were some limited chemical inputs as the seed delivered to the farmers was already dressed, and farmers soon began purchasing seed-dressing for their food crops. Farmers waxed almost poetical as they described the benefits of seed-dressing and the vigour of healthy green shoots pushing up.

Cotton seed issues rose from 450 tons per annum in 1951–1953 to 1100 tons between 1959–1961.[4] This gives an idea of how much additional land was put under cultivation. Assuming that farmers followed a constant rate of seed application, roughly 2.5 times more land was put under cotton. However, sales to the marketing board rose from 3584 tons to 15,291 tons, an increment of more than four times. Thus almost half the increase in sales would have come from more tons of cotton produced per ton of seed distributed. This means that about half of the increase was on the extensive margin (using more land) and the other half came from intensification, raising yields.

In turn, about 40 per cent of the intensification could be attributed to research supported by the government and to breeders' success in introducing better varieties; the other 60 per cent gain over eight years would have been due to better farming techniques, including early planting, early weeding, more use of manure, crop rotation, etc (Tiffen, 1976, p88). There is reason to believe that similar techniques, plus seed-dressing, enabled farmers to raise their food crop yields, but we do not have enough data on this for even a rough calculation.

By the 1990s, cotton had become an unattractive crop. Population density had increased and farms had become smaller. A large dam on the Gongola River had enabled some land to be irrigated to allow the growth of two crops a year (though there had been considerable loss of rainfed land to build the reservoir). There is no indication that farms in Gombe had become unproductive. On the contrary, the area had become known as a major exporter of maize to other parts of Nigeria, together with other foods crops such as rice, beans and cowpeas (information from traders in a wholesale food market near Kano, in 1995).

Between the 1960s and 1990s, there had been considerable improvement in the roads so that access to Nigeria's rapidly growing cities was cheaper and swifter. In 1967, maize was a minor crop, grown mostly for eating as green cobs. An early agricultural development programme in Gombe spread the use of new maize varieties and subsidized fertilizer to increase yields in the 1970s; by the time of a survey in 1989, fertilizer was used by almost all farmers in Gombe, as it was in other parts of northern Nigeria (Goldman and Smith, 1995). From this we can deduce that by the 1980s and 1990s, farmers in Gombe found it necessary to increase their inputs of chemical fertilizers in order to raise the output of the new range of crops that they were producing for a changing market situation. Land had undoubtedly become more scarce, as farms were subdivided as inheritances. Agroecological methods which led at first to big advances in yields were later sustained by additional inorganic inputs. As population density continued to rise, and as roads improved and made Nigeria's rapidly expanding towns more accessible, so Gombe began moving into stage three.

Gombe illustrates the possibilities for an area with a reasonable rainfall regime and good access to markets to accomplish respectable increases in agricultural productivity. It also illustrates a change in response to changing markets, having moved in the course of the century from cattle to groundnuts, then to cotton and sorghum and now to maize and a range of other food crops.

THE IMPORTANCE OF MARKETS AND EDUCATION

A common factor in both the Machakos and the Gombe cases is the importance of markets and education for bringing about change in the agriculture practised. Infrastructure for physical access is tremendously important. In the Gombe case, bridging the Gongola River and building the road to Jos were crucial to the agricultural development that followed. During Nigeria's oil boom years, the road to the north was tarred, and a second bridge over the Gongola brought Kano, the major city of Northern Nigeria, within easy reach. However, market growth is also shaped by factors like the taxation regime and the existence (or absence) of monopolies – usually, in Africa, state monopolies.

In Senegal, where parallel research is being done, local food markets have not developed to nearly the same extent as in Nigeria, despite the existence in Senegal of big cities like Dakar and Touba, and despite a good transport infrastructure. This is partly because state policy for many years favoured the import of cheap rice for consumers, which discouraged greater domestic production, and also because state monopolies controlled not only the export of peanuts but also trade in rice and even the local millet. This discouraged the growth of local markets, which officials felt might distract farmers from producing the peanuts that the state depended on for revenue. Only in recent years have farmers been able to obtain freer market opportunities and to respond to market signals. In the area we are currently studying in Senegal, farmers have done this by greatly intensifying their management of sheep and goats for the urban market – and thereby increasing also their access to manure for agriculture.

Approaches to education have also been different between countries. Rural people in Senegal have considered the type of education provided by the state (with instruction in French language from the first year) poorly suited to their local realities or to their religious beliefs. Accordingly, educational investment and attainment have lagged. Like their counterparts living in the drier areas of Kenya, rural Senegalese have sought off-season urban work or have migrated permanently to the towns or overseas. But they have not qualified for the better paying urban jobs, and this has limited the flow of funds back to their rural relatives.

A recent example from Kenya of the importance of market outlets is a non-governmental organization (NGO) project, which began by encouraging groups of farmers to increase the productivity of their soils by double-digging their vegetable gardens and by making and using compost. Farmers found that indeed this did improve their soils, but they did not adopt these practices because they saw no purpose in producing more vegetables than their families could consume, unless they could market these for a good price. The project has now been re-oriented to build up brands that can be sold at premium prices to satisfy the growing urban demand for organic products. This is encouraging more farmers' groups to adopt ecological methods, managing their soil more effectively rather than depending on purchased inputs for its fertility.[5]

Kenyan farmers are now better able to assess the market and to manage group activities because of their good levels of numeracy and literacy. They also manage the flow of funds between family members in farming and those in non-farm jobs. The high level of education not only contributes to the growth and diversification of urban activities, but also enables rural children to find skilled, and therefore relatively well-paid, work in the expanding formal private sector. In Senegal, by contrast, there has been less widespread investment in education, and most of the urban jobs that are open to rural youth are unskilled, low-paid activities in the informal sector. Unfortunately, the links between education and farming have been little researched. Experience in many countries shows, however, that the productivity and hence the acceptability of agroecological methods is critically linked to factors beyond the farm, and that economic and human resource considerations need to mesh with biophysical ones for new practices to be adopted and sustained.

NOTES

1 This study, done under the auspices of the Overseas Development Institute and the University of Nairobi, acquired data from oral histories and documentary and photographic evidence. To control for price changes, the value of all produce was translated into its purchasing power in terms of maize, the main staple commodity that all farmers either sell or buy. Adam Smith has suggested that the price of the main grain staple is the best indicator of value, since it is closely related to the value of labour.

2 Agricultural officers add together the areas planted and the amounts harvested each season and then calculate the average yield for the year. The average of 850kg therefore takes account of bad seasons, as in any one year there could be a yield of 2000kg/ha in one season while the next season could yield just 300kg/ha. While

agricultural estimates and statistics are always subject to error, they do give inform-
ative orders of magnitude.

3   This work is being conducted by Drylands Research, a partnership set up by Mary
Tiffen and Michael Mortimore to do research on development in semi-arid areas.
Current research in Kenya (Makueni district) and Senegal (Dourbel region) as well
as in Niger and Nigeria is being undertaken with national researchers to examine
the effects of national policies on development in four zones that have similar
agroecological features, and especially to assess the impacts of policy on dryland
farmers.

4   Three-year averages were taken for these calculations, to even out the effects of
variable rainfall.

5   The Association for Better Land Husbandry was established some years ago with
the aim of helping farmers in Western Kenya to make their soils more productive
using a variety of agroecological methods. However, participating farmers soon
made the project realize that: 'To get out of poverty they must be involved beyond
the soil alone and conservation farming. To relieve poverty on a large scale, it is
necessary for them to market products – some processed to add value – with a
higher value than maize. The effectiveness of cash as a spur is demonstrated by the
fact that a canning factory within one year of start-up had attracted 5000 farmers
producing green beans for canning for export.' The second phase of the project is
concentrating on providing marketing and processing facilities to farmers' associations and other NGOs (Cheatle, 1999).

Chapter 8

Benefits from Agroforestry in Africa, with Examples from Kenya and Zambia

Pedro A Sanchez

Agroforestry – integrating trees and other perennials into farming systems for the benefit of farm families and the environment – is an ancient practice that began moving from the realm of indigenous knowledge into agricultural research only about 25 years ago (Bene et al, 1977). During the 1980s, agroforestry was promoted widely as a sustainability-enhancing practice with great potential to increase crop yields and conserve soil and recycle nutrients, while producing fuelwood, fodder, fruit and timber (eg, Steppler and Nair, 1987; Nair, 1989). At that time, agroforestry was considered almost a panacea for solving landuse problems in the tropics. Many development projects pushed agroforestry technologies that were without foundations in solid research. During the past decade, however, agroforestry studies have become more empirical, based on process-oriented research (Sanchez, 1995; Young, 1997; Buck et al, 1999).

Agroforestry is now recognized as an applied science based on principles of natural resource management (NRM) (TAC, 1998; Izac and Sanchez, 2001). The application of such principles includes the following practices:

- participatory, multidisciplinary and analytical approaches;
- technical and policy research;
- working at and across different spatial and temporal scales;
- beneficiaries identified at the community, national and global levels;
- working along the whole research–development continuum;
- working in partnership with governmental and non-governmental organizations (NGOs);
- moving rapidly into on-farm research with a decreasing degree of researcher control;

- assessing impacts in economic, social and environmental terms; and
- being a credible partner in development.

Agroforestry is in fact a very widespread practice, found from the Arctic to the southern temperate regions, but most extensive in the tropics. Approximately a fifth of the world's population (1.2 billion people) depend directly on agroforestry products and services in rural and urban areas of developing countries (Leakey and Sanchez, 1997).

Agroforestry products include fuelwood, livestock fodder, food, fruits, poles, timber and medicines. Agroforestry services include erosion control, soil fertility replenishment, improved nutrient and hydrological cycles, boundary delineation, poverty reduction and enhanced food security, household nutrition, watershed stability, biodiversity, microclimate enhancement and carbon sequestration. Many agroforestry systems are superior to other landuse systems at global, regional, watershed and farm scales because they optimize trade-offs among increased food production, poverty alleviation, and environmental conservation (Izac and Sanchez, 2001). Being complementary to rather than competitive with arable or pastoral practices makes agroforestry an important part of strategies to produce sufficient food in the decades ahead in ways that meet both human and environmental needs.

While the original impetus for agroforestry was very practical and empirical, it is supported increasingly by scientific foundations that permit its extension and extrapolation across the tropics. The complex agroforests of Indonesia are one example (Michon, 1997). Research based on the principles of competition for light, water and nutrients (Ong and Huxley, 1996) or the complexity of interacting socioeconomic and biophysical factors (Sanchez, 1995) has led to new agroforestry components that increase the sustainability and profitability of existing farming systems.

Much of the information for determining the biophysical performance, profitability and acceptability of agroforestry comes from on-farm trials (Franzel et al, 1998). At the same time, farmers have become increasingly interested in on-station work and have often suggested technologies that address constraints and identify opportunities showing promise upon further investigation. Farmers' informal and formal visits to research station work has led to keen interest for collaboration between researchers and farmers in testing and assessing agroforestry technologies on-farm.

Three broadly defined partnerships for on-farm trials have been adopted by the International Centre for Research in Agroforestry (ICRAF): those that are researcher-designed and -managed (type 1), researcher-designed and farmer-managed (type 2) and farmer-designed and -managed (type 3) (ICRAF, 1995; Franzel et al, 1998). The suitability of these different kinds of trials depends on the objectives; no type of trial is intrinsically 'better' than another type. Which type should be preferred depends on the objectives of the participants (facilitators and farmers) and on the particular circumstances.

The collaborating farmers for these on-farm trials are selected through locally-based institutions such as extension services or farmer groups. Farmers are asked to volunteer and are selected to represent the range of different farmers in the study area, eg, large to small sizes of holdings, male and female, different wealth categories. Farmers choose among several alternative technologies to test, and except for receiving planting material and information, they are not provided with any further incentives for participating in trials with researchers.

Researchers are involved mainly in technical backstopping for farmers in trials of types 2 and 3, and they help lay out type 2 trials. There is great variation in the number of trials and the number of farmers involved, but in most cases small numbers of farmers start type 2 trials, on average about ten; after modification during the first year, the number is expanded up to about 50. The number of farmers involved in type 3 trials can be much larger. In Eastern Zambia in 1997, for example, extension services and NGOs were helping 2800 farmers test improved tree fallows in type 3 trials (Franzel et al, 1998).

Two agroforestry components that have been extensively researched and adopted are the domestication of indigenous trees (Leakey et al, 1996; Leakey and Tomich, 1999) and soil fertility replenishment (Buresh et al, 1997). In the process, some previously promoted practices such as alley cropping that have not met science-based tests are no longer advocated on a large scale. There are, however, some indigenous alley cropping systems in Indonesia that are many decades old and popular with farmers where they fit certain ecosystem niches (Piggin, 2000; Agus, 2000).

Some agroforestry innovations that can be applied to meet the particular agricultural challenges in Africa are discussed here – how to assure food security, reduce poverty and enhance ecosystem resilience at the scale of thousands of smallholder farmers. There are many examples of successful agroforestry innovations in other parts of the world that could be cited (Buck et al, 1999).

REDRESSING SOIL FERTILITY PROBLEMS

When smallholding farmers throughout the sub-humid and semi-arid tropics of sub-Saharan Africa, hereafter referred to as Africa, are involved in diagnosis and design exercises, they invariably identify soil fertility depletion as the fundamental reason for declining food security in this region. Scientists concur. No matter how effectively other constraints are remedied, per capita food production in Africa will continue to decrease unless soil fertility depletion is effectively addressed (Sanchez and Leakey, 1997; Sanchez, Shepherd et al, 1997; Sanchez, Buresh and Leakey, 1997; Pieri, 1998).

During the 1960s, the fundamental cause of declining per capita food production in Asia was the lack of rice and wheat varieties that could respond efficiently to increases in nutrient availability. Food security was only effectively addressed with the advent of improved germplasm in this region for higher-yielding varieties. Then other key aspects of agricultural development that had been previously less important – enabling government policies, irrigation, seed production, fertilizer use, pest management, research and extension services – came into play in support of the spread of new varieties.

The need for soil fertility replenishment in Africa is now analogous to the need for Green Revolution germplasm in Asia three decades ago.[1] A full description of the magnitude of nutrient depletion, its underlying socioeconomic causes, the consequences of such depletion and various strategies for tackling this constraint are described elsewhere (Buresh et al, 1997; Sanchez and Leakey, 1997; Sanchez, Shepherd et al, 1997; Sanchez et al, 2001). Fortunately, strategies for soil fertility enhancement and agroforestry can be combined based on much research (Sanchez and Leakey, 1997).

Nitrogen and phosphorous are the most severely depleted nutrients in smallholder African farms. Although such constraints can be alleviated with imported mineral fertilizers, economic, infrastructural and policy constraints make the use of mineral fertilizers extremely limited in such farms. However, Africa has ample nitrogen and phosphorous resources – nitrogen in the air and phosphorous in many rock phosphate deposits. The challenge is to get these natural resources to where they are needed and in plant-available forms. For nitrogen, this can be achieved through biological nitrogen fixation by leguminous woody species utilized in fallows. For phosphorous, there can be beneficial direct application of reactive, indigenous rock phosphate combined with biomass transfers of non-leguminous shrubs.

Two-year leguminous fallows, leaving land uncultivated but with selected leguminous species growing on it, can accumulate 200kg of nitrogen per hectare in plant leaves and roots. Incorporating these into the soil, with subsequent mineralization, provides sufficient nitrogen for two or three crops. Resulting maize yields can be two to four times higher (Kwesiga and Coe, 1994; Kwesiga et al, 1997; Kwesiga et al, 1999).

The greatest impact of this work so far has been in Southern Africa, where about 10,000 farmers are now using *Sesbania sesban*, *Tephrosia vogelii*, *Gliricidia sepium* and *Cajanus cajan* in two-year fallows followed by maize rotations for two to three years (Rao et al, 1998). The species used in such fallows produce nutrients that would cost US$240/ha for an equivalent amount of mineral fertilizer, well beyond the reach of farmers in this region who make less than US$1 per day.

The provision of nutrients through such plant and soil management methods, which require hardly any cash, repays the labour invested very well. The results of such practices were summarized by one farmer, Sinoya Chumbe, who lives in Kampheta village near Chipata, Zambia, when he stated: 'Agroforestry has restored my dignity. My family is no longer hungry; I can even help my neighbours now' (interview, 12 April 1999).

In many high-potential areas of East Africa, smallholder farms are depleted of both nitrogen and phosphorous, necessitating the combined use of organic and mineral sources of nutrients (Palm et al, 1997). Short-term improved fallows of up to 16 months' duration using *Tephrosia vogelii*, *Crotalaria grahamiana* and *Sesbania sesban* are an effective and profitable way of adding about 100kg of nitrogen per hectare and of recycling other nutrients in the nitrogen-depleted soils of Western Kenya. With fast-growing trees, fallows as short as six months have tripled maize yields in villages where farmers are now practising a fallow-crop rotation every year in a bimodal rainfall environment (Niang et al, 1998; Rao et al, 1998).

In phosphorous-deficient soils, Minjingu rock phosphate from Northern Tanzania has proven to be as effective as imported triple super-phosphate, as well as more profitable for small farmers (Sanchez and Leakey, 1997; Niang et al, 1998; Sanchez et al, 2001). Basal applications of 125–250kg of phosphorous per hectare as a capital investment are beginning to be used by farmers with an expected residual effect of five years. In addition, biomass transfers from hedges of wild sunflower, tithonia (*Tithonia diversifolia*), have shown tremendous effects on yields of maize and high-value crops such as vegetables in Western Kenya (Gachengo et al, 1998; Jama et al, 2000).

Tithonia biomass has high concentrations of nitrogen, phosphorous and potassium and decomposes very rapidly in the soil (Palm et al, 1997). Given the large additions of soluble carbon and nutrients to the soil when tithonia leaves decompose, it appears that these processes enhance phosphorous cycling and therefore the conversion of mineral forms of phosphorous into organic ones (Nziguheba et al, 1998).

Combining the application of tithonia biomass with phosphorous fertilizer has been shown to be particularly effective (Rao et al, 1998). Tithonia grows abundantly along roadsides and in farm hedges at intermediate elevations throughout sub-humid Africa, making it an easily available natural resource that can be utilized to replenish soil fertility.

About 4000 farmers are currently trying these techniques in Western Kenya. Most of the dissemination work has been done at the village level as a pilot development project (Niang et al, 1998). An assistant chief of Barsauri sublocation in the Siaya District of Nyanza Province, Hosea Omollo, summarized the results of this agroforestry technology as follows: 'For the first time there have been no hunger periods in this village. Only two ears of maize have been reported stolen this year' (interview, 7 July 1998).

Many farmers who have adopted tithonia biomass transfers to their fields have shifted now from maize to high-value vegetables, which can be readily sold in nearby towns, effectively bringing them into the cash economy. One farmer, Charles Ngolo of Ebuyango in Vihiga District, Western Province, reported to ICRAF that his annual cash income had increased from US$100 to US$1000 through the sale of kale, locally known as *sukuma wiki* (*Brassica oleracea cv. Acephala*). He commented: 'My wife and I are living the tithonia life. I built a new house with a tin roof, and we are going to be able to send our children to school' (interview, 4 June 1997).

These people now farming on replenished soils have achieved food security, and Mr Ngolo is an example of a smallholder who is beginning to work his way out of poverty. Economic analysis has shown high net present values for these technologies (Sanchez and Leakey, 1997). The potentials of agroforestry practices can no longer be discounted as hypothetical. The question now is how to scale-up their delivery, from thousands to millions of farmers, a major challenge facing national governments and international agencies (ICRAF, 1998).

Enabling policies at the national, district and community levels are beginning to emerge in support of technological advances. These include increasing the availability of phosphorous fertilizer and high quality seeds, providing microcredit and levying fines on farmers who let their cattle eat their neighbours' sesbania fallows (ICRAF, 1998; Sanchez and Leakey, 1997; Sanchez, Shepherd et al, 1997). The Kenyan government has established and funded a pilot project on soil fertility recapitalization and replenishment for Western Kenya. Further, the government as a member of the Consultative Group on International Agricultural Research (CGIAR) is providing financial support for ICRAF to conduct strategic research to underpin replenishment efforts. Technological and policy research are both needed in agroforestry. Their joint impact can enable soil fertility replenishment through biological means to make a major contribution to food security in Africa.

OPPORTUNITIES

It would be unwarranted to generalize that all agroforestry interventions will have similar degrees of success. But there are surely many more innovations still to be identified and evaluated. Agroforestry is not the best landuse option for all tropical areas, and some practices have met with widespread failure when they were not based on solid technical and policy research. Science-based agroforestry pursued in cooperation with farmers can, on the other hand, assuredly produce economically, socially and environmentally sound results. These examples and many others that are emerging and spreading throughout the world, where trees and other perennials are integrated with other farming components and practices, can raise productivity and security for several billion people who will benefit from combining ancient practices and modern science.

NOTE

1   Two of the 'fathers' of the Green Revolution agree with this analogy: Norman Borlaug (Borlaug and Dowswell, 1994), and M S Swaminathan (personal communication, July 1998).

Chapter 9

Realizing the Potential of Integrated Aquaculture: Evidence from Malawi

Randall E Brummett

In an integrated farming operation, the wastes from different farming activities are recycled into other enterprises, thus raising overall economic and ecological efficiency. Aquaculture has often played an important role in the development and functioning of integrated farming systems for smallholders. This is due to farm ponds' particularly effective role in processing waste materials without creating some of the problems that can be associated with mulches and green manures, such as weeds or insect pests.

The classical image of an integrated farm comes from China where a wide variety of integrated systems, eg duck–fish, rice–fish, mulberry–fish, chicken–pig–fish, etc have evolved over nearly 2000 years (Kangmin and Peizhen, 1995). These systems when studied for their economic efficiency have been found to have many desirable features. They have consequently sometimes been packaged by development agencies for widespread transfer to smallholding farmers, including those in sub-Saharan Africa.

Unfortunately, such integrated systems have not produced the expected results. In fact, they have seldom been adopted for the simple reason that smallholders do not normally make their adoption decisions solely on the basis of a technology's economic performance (Brummett and Haight, 1997). Smallholdings must first and foremost produce food all year round, being at least minimally productive in an often-harsh environment and within a complex village social system. Many factors lead over time to the establishment of what can be termed the 'food security puzzle', as suggested in Figure 9.1.[1]

According to this way of understanding decision-making processes, the range of social, ecological and technical conditions surrounding a smallholding combine to produce a resilient and not easily modified structure. Of course, every farm will have its own mix of puzzle pieces: some of these will be common to other farms within a particular area, and some are unique to that farm. As long as the solution that a household has worked out for this puzzle

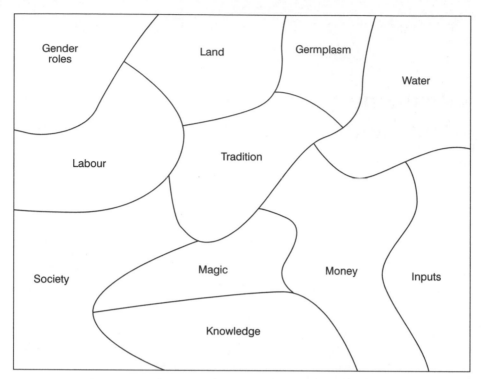

Figure 9.1 *The Food Security Puzzle for Smallholders*

enables it to meet its basic needs, smallholders will be hesitant to change the system. This is partly a matter of what is called risk-aversion, but there are also elements of habit, inertia and comfort level. Farmers are unlikely to dig up any substantial part of their farms and install a pond solely on the basis of an extension agent's advice.

We start our analysis by noting that these puzzle-farms encompass many different crops and activities. As mentioned above, the farm must provide food year-round. A mix of crops ensures that food will be available each month. In case of adverse weather, certain crops may still succeed when others fail. Mixed cropping systems are commonly viewed by farmers as more or less independent enterprises conducted within seasons rather than as inter-connected resource systems with potentially synergistic links to each other (Lightfoot and Minnick, 1991). Given the complexity of these systems and the difficulty of detailed comparisons, a decision-making criterion of 'satisficing' (Simon, 1957) often applies. It is hard to know how 'sub' a sub-optimal strategy really is.

Having a mix of crops can be a fertile bed for innovation, however, if an acceptable approach to transforming farmers' perspectives can be found to move towards more integrated and productive strategies (Lightfoot and Noble, 1993). The International Centre for Living Aquatic Resource Management (ICLARM) believes that such an approach has been established in Malawi, through what is called the Farmer–Scientist Research Partnership (FSRP).

PARTICIPATORY TECHNOLOGY DEVELOPMENT AND DIFFUSION

The concept behind the FSRP process is common to the development of agriculture in many countries: a working relationship between farmers and researchers which ensures that the problems being addressed by researchers are really the ones that farmers face. The technologies that address these problems should evolve over time on the farm itself, so that farmers thoroughly understand and are comfortable with them.

In industrialized countries, farmers have a relatively high level of education and there is a cash basis to their agriculture. This makes communication between farmers and scientists easier, and they are normally agreed that the main objective of the farm is to produce money income. In poorer tropical countries, the situation is different. Most farmers have little or no education and are not comfortable with quantitative analyses or explanations presented in terms of soil chemistry. They regard income generation – remember the puzzle – as only one among many factors when deciding whether or not to adopt a new technology or practice.

The FSRP starts from a participatory resource assessment, of which an example is given in Figure 9.2. Diagrams that identify all production activities and any connections between them and the market put the researchers and farmers on common ground.[2] This process of consultation sidesteps many of the socioeconomic factors that impede adoption decisions by letting the farmers themselves decide which technologies would be suitable for their farms. A range of methods should be presented to farmers, and these must be sufficiently easy to understand so that farmers can make informed choices (not guesses) about which would suit them best. Making available a range of technological options not only increases the chances for adoption of at least one improved method: it also takes into consideration the variation among farms that can usually be found within even a small area.

Generating a variety of technologies from which farmers might select requires some preparation. Some techniques, such as green manuring and composting, are more or less universally relevant and can be obtained through literature review. Others that are more suited to a specific locale, such as breeding techniques for an endemic fish species, may need to be generated at the experiment station.

To enhance the chances that new technologies can make real contributions to farm livelihoods, they must be 'appropriate.' Such understanding can be enhanced by a systematic characterization of farming systems in the area conducted prior to starting the programme of research. Researchers must avoid the temptations of proposing capital- or knowledge-intensive technologies that cannot easily become available to farmers.

Once a starting point has been selected, farmers are encouraged to improve their farming system incrementally over several seasons, thus giving them a chance to learn all the details of the new technology. The nine phases of the process are outlined below:

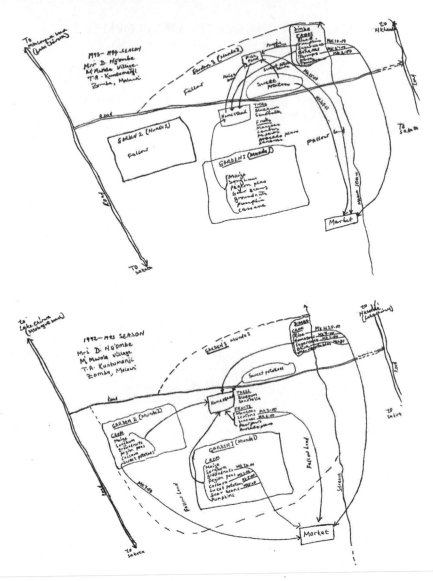

Figure 9.2 *Resource-flow Diagrams Depicting Qualitative Changes in a
Malawian Smallholder Farming System*

1   Careful study and analysis of farmers' goals and farm-level constraints
    and potentials.
2   Design or adaptation of a range of simple integrated technologies, consid-
    ering existing practices and also ease of integration and adoption, rather
    than simply aiming for the optimization of production right away.
3   Introduction to farmers of the integration concept through discussions and
    resource flow diagramming.
4   Selection by farmers of an introductory technology for on-farm trial.

5   Monitoring of progress and performance by research staff.
6   Parallel on-station experimentation, matching real farm conditions and resources, with a view to improving production incrementally.
7   Discussions among farmers and researchers of their respective results.
8   Choice of an improved technology developed through step 7.
9   Repetition of steps 5–8 until the innovations have been adjusted and fully accepted.

This process can lead to high rates of integration and intensification. It is up to farmers to decide which innovations they want to make and evaluate. There are many combinations of practices assembled by farmers once this process begins, as outlined below:

- Rice–fish systems are easy for farmers to adopt and result in increased rice yields as well as fish for sale and home consumption.
- Azolla produced on farm ponds is used as a fertilizer and soil amendment.
- Bananas can be planted on pond banks at higher than normal densities with no loss of productivity per plant.
- Off-season vegetables can be grown using pond water as necessary to counter seasonal rainfall shortages, producing high profits.
- Ponds located near households provide a wide variety of services.
- More complex systems emerge over time, as illustrated by an azolla–vegetable–fish enterprise.

Farm ponds and their associated practices have been quite attractive and successful by capitalizing on well-known processes of nutrient cycling and complementarity among practices. Of Malawian farmers who have been exposed to integrated aquaculture technology through the FSRP methodology, 86 per cent have adopted at least one of the technologies presented for consideration; 76 per cent adopted at least two; and 24 per cent adopted four (Brummett and Noble, 1995). Moreover, the innovations presented and adopted have been sustained over time.

All of the farmers with whom ICLARM staff have worked who have access to permanent water supplies are continuing to grow fish and improve their production. Among those farmers who have rainfed fishponds, 36 per cent dropped out for one reason or another (40 per cent of these did so because of family deaths or illness rather than for any agricultural reason). Those remaining with the programme have continuously improved their ponds and production. For example, their average pond size has increased 37 per cent, from 64m^2 to 88m^2, with new gardens being planted around the ponds (Brummett and Chikafumbwa, 1995).

Once introduced in a rural community, the technologies have spread and evolved without further extension support, indicative of net benefits to the households adopting them. A survey found that within six months of a field-day in May 1990 explaining the new opportunities, 46 per cent of farmers in the target area adopting an aquaculture practice had learned about it from other farmers; a third of these farmers had adopted two or more technologies from their neighbours. By the end of 1992, almost 80 per cent of the farmers who were practising integrated rice–fish farming in Zomba District had never witnessed first-hand an extension demonstration (Chikafumbwa, 1994). In Zomba East, where ICLARM worked initially with 34 farmers from 1991 to 1995 (ICLARM-GTZ, 1991), there were within six years 225 practising fish farmers (Scholz et al, 1997), more than a six-fold increase. Thus farmers are learning from other farmers, facilitated by whatever social and human capital exists at community level (Chapter 4).

IMPROVEMENTS IN PRODUCTIVITY

Average fish productivity of integrated Malawian smallholdings is 1350kg/ha/yr in rainfed areas and 1650kg/ha/yr in springfed areas. This is 50 to 83 per cent more than the average production level achieved by the 48 most productive commercial fish farms in Southern Malawi, about 900kg/ha/yr (Chimatiro and Scholz, 1995). The difference in productivity stems from the variety of inexpensive inputs that are available as pond inputs and the location of the ponds close to other farm enterprises, as well as from more intensive management.

To supply a typical farm pond properly, a farmer needs about 522kg of dry matter (Brummett, 1997). On integrated farms, ponds are generally located within or next to vegetable gardens; or, as often happens, vegetable gardens develop around the fishpond to take advantage of its water for emergency irrigation when needed. Wastes from the garden are used to feed fish, and typically these wastes amount to some 3700kg of dry matter per year, much more than is required. That the material is generated in close proximity to the pond minimizes the work involved in transportation.

Farms that have not integrated their operations, on the other hand, usually use maize bran as fish food, which is recommended by the extension service. Household production of maize bran averages only around 192kg of dry matter, however, only 37 per cent of the amount needed. Moreover, the bran is produced in the house, often far from the pond, which adds to the labour required to operate the pond. Also, we note that maize bran is sometimes needed as an emergency food for humans, whereas vegetable garden wastes are typically just burned if they are not used in a pond. So there are no opportunity costs for the wastes from gardens used to feed the fish.

On a continent where an estimated 80 per cent of the population is presently rural, the potential impact on food security of integrating aquaculture components into household farming systems could be enormous. Using very conservative figures, the United Nations Food and Agriculture Organization (FAO) has estimated that 31 per cent of sub-Saharan Africa (parts of 40 countries, covering 9.2 million km2) is suitable for small-scale integrated fish farming (Kapetsky, 1994).

If production levels from relatively recent projects (1300–2300kg/ha/yr) are used for projections, 35 per cent of Africa's increased fish demand up to the year 2010 (580,000 tons) could be met by small-scale fish farmers using only 0.5 per cent of the total area potentially available (Kapetsky, 1995). Also, enabling local producers and production systems to meet local needs obviates the problems inherent in long-distance marketing (Brummett, 2000).

ECONOMIC GROWTH

Economically, such integrated farms produce almost six times as much cash as is typically generated by Malawian smallholders (Chimatiro and Scholz, 1995). The integrated pond–vegetable garden is an economic engine for these farms, generating almost three times more annual net income than the staple maize crop and the homestead combined. The vegetable–fish component contributes, on average, 72 per cent of annual cash income for participating households (Brummett and Noble, 1995).

On a per unit area basis, the vegetable-garden–pond resource system generates annually almost US$14 per 100m^2 compared with US$1 and US$2 for the maize crop and the homestead, respectively, from an equivalent area. If this level of economic return is sufficient to overcome recurrent cash flow problems of smallholding farmers and to give them enough cash to reinvest in their farms (something not yet proven), then integrated farming could contribute significantly to real economic growth in rural communities.

Such a farm-level economic impact could produce wider economic growth. A review of results from Burkina Faso, Niger, Senegal and Zambia found that 'even small increments to rural incomes that are widely distributed can make large net additions to growth and improve food security' (Delgado et al, 1998). Winkleman (1998) has identified interventions that lead to improved incomes at the level of the rural farmer and resource manager as 'having a larger impact on countrywide income than increases in any other sector'.

Circumstantial evidence indicates that ponds also have the potential to improve the stability of small farms. All of the farms involved in integrated pilot research were badly affected by a drought that was very serious from 1991 through 1995. Yet in all cases, even though maize crops failed and farmers suffered economic losses, the pond–vegetable systems kept operating and sustained the farms during these years of stress.

By retaining water on the land, ponds have enabled farms to maintain their food production and to compensate for losses on seasonal croplands. For example, in the 1993–1994 drought season, when rainfall was only 60 per cent of normal, the average net cash income accruing to a study group of rainfed integrated farms, thanks to their fish, vegetable and other intensively managed outputs, was 18 per cent higher than comparable non-integrated farmers in an area with some of Malawi's severest poverty (Brummett and Chikafumbwa, 1995).

Due to the incremental nature of FSRP these benefits accrue over time, as shown in Figure 9.3.[3] In contrast, as most development professionals are aware, attempts to transfer complete technology packages or modules directly result in unsustained adoption and in subsequent declines in productivity as projects phase out. This is partly due to not having considered farmers' 'food

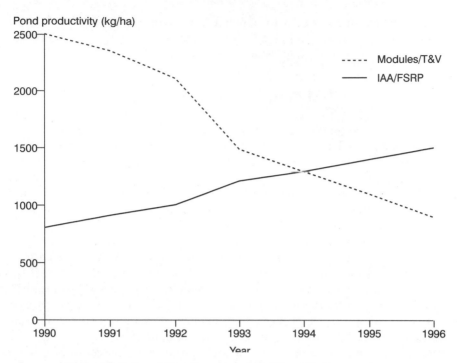

Figure 9.3 *Pond Productivity over Time: A Comparison*

security puzzle' (Harrison et al, 1994). FSRP starts with direct discussion and simple methods so that farmers can understand and become comfortable with the technology. Then, in partnership with researchers, the farmers themselves work to improve their output over time. This takes time and patience, but unlike many other approaches, it works and lasts.

ENVIRONMENTAL AND SOCIAL SUSTAINABILITY

Small-scale integrated farming systems are more efficient at converting feeds into fish and produce fewer negative environmental impacts than purely commercial fish farms (Figure 9.4).[4] They also have the advantage of not using one human foodstuff to produce another. Some analysts have predicted that the widespread adoption of integrated aquaculture might actually improve local environments by reducing soil erosion and increasing tree cover (Lightfoot et al, 1993; Lightfoot and Pullin, 1995), although this remains to be demonstrated on the ground.

At some point in the evolution of their integrated farms, farmers will need to begin importing nutrients to replace those that are exported to the market. This, in the face of increasing population pressure and traditional landholding practices that allocate land among all the families in a village regardless of a family's ability to farm, is inevitable. Land productivity per unit area must

A
To support a 1m^2 **cage system** for raising tilapia, one needs:

- 21,000m^2 of ocean to grow fishmeal for inclusion in fish feeds.
- 420m^2 of cropland to grow grains for inclusion in fish feeds.
- 60m^2 of green plants to produce oxygen for consumption by fish.
- 115m^2 of benthic community to assimilate waste phosphorous.

Ecological footprint = **21,700m^2**
= 6g fish produced per m^2 of footprint.

B
To support a 1m^2 of **waste-fed integrated fish pond** for raising tilapia, one needs:

- 0.9m^2 of additional benthic community to assimilate waste phospohorous.
- 0.9m^2 of green plants to produce oxygen for consumption by fish.

Ecological footprint = **1.8m^2**
= 264g fish produced per m^2 of footprint.

Figure 9.4 *Comparative Efficiency of Two Alternative Aquaculture Systems*

increase substantially if the next generation of Africans is to be fed (Brummett, 1995). This imperative often leads planners to the false assumption that larger-scale commercial farms would be better investments, believing these will produce larger quantities of fish per unit area, and use capital more efficiently due to economies of scale. The empirical basis for these assertions is insubstantial. As seen above, larger operations in Malawi have lower output per unit area, in addition to requiring more expensive inputs.

Moreover, large-scale commercial farms seldom return much economic benefit to the local community, so they usually have little positive impact on rural food security or poverty. Reliance on jobs and food from large agricultural estates has a long and often tragic history in Africa. For rural economic development to take place, the inevitable growth of agribusiness mentioned above would best be based on the evolution of integrated smallholdings into larger and more prosperous integrated farming operations. Intensively-managed operations, using natural, human and physical capital synergistically, are more likely to result in development that is both environmentally sustainable and that returns long-lasting benefits to rural communities.

NOTES

1   A range of social, cultural, economic, ecological and technical factors create a complex conjunction of objectives and activities that renders smallholder farms resistant to rapid change. Proposed modifications that can be contained within a single piece of the 'puzzle' are easier to get adopted. For example, introducing new germplasm often requires increased use of fertilizer and/or pesticides. This transition involves at least the germplasm and the money pieces in Figure 9.1. When germplasm introduction does not require new inputs, and hence is contained within a single piece of the 'puzzle', it is easier to transfer.

2   The diagrams in Figure 9.2 show resource and product flows before and after
    integration of a farm pond into a particular farming system. Prior to integration
    (a), all resource flows take the form of food consumed by the household or sold
    for cash in the market. With integration engendered by the addition of a fishpond
    (b), there is a 60 per cent increase in resource flows that result from linkages estab-
    lished between and among farm enterprises.

3   This graph compares the yields from ponds integrated through the FSRP approach
    with ponds on farms to which integration modules were introduced through a
    Training and Visit (T&V) strategy in Southern Malawi. The T&V production
    target of 2500kg/ha was arrived at by extrapolating Malawi's national fish produc-
    tion need to the land area available for aquaculture. The technology promoted was
    too complex for most farmers to fully understand and/or adopt, which resulted in
    declining production as T&V extension support waned. The FSRP entry-level
    technology is much simpler, with less productivity initially. But it evolves and
    improves on-farm as farmers who understand the technology are able to manipu-
    late it more efficiently to suit their individual situations.

4   This analysis evaluates the efficiency of a system according to the concept of
    'ecological footprint', which is the quantity of environmental goods and services
    consumed by a food production system in the generation of external inputs and
    the processing of wastes (Berg et al, 1996). The integrated pond uses agricultural
    by-products as inputs to fuel natural processes that generate the bulk of the food
    for the fish. This process treats what are waste products in a cage system as inputs
    for an integrated system, with a consequent reduction in pollution.

Chapter 10

Management of Organic Inputs to Increase Food Production in Senegal

Amadou Makhtar Diop

In the Sahelian countries of Africa, the major constraints on food production are related to soils as well as water. Most soils are sandy and deficient in organic matter, as well as in clay content, cation exchange capacity (CEC), moisture-holding capacity and natural fertility. Those that are heavier and better in quality are invariably subject to intensive use and are thus vulnerable to various types of erosion. At present, soil erosion and degradation threaten large areas devoted to crop production not only in Senegal but also in most developing countries of Africa. A recent study by Scherr (1999) identifies soil degradation as probably the major threat to developing-country food security in the near future.

The constraints that farmers in Africa face are multiple. Soil structures are commonly weak, and many soils are naturally acidic, which impedes the uptake of certain nutrients that are in the soil but become unavailable. Under natural conditions, scarce nutrients such as phosphorous are recycled for plant use. However, in recent years much of the vegetation has been cleared for cultivation. Continuing loss of vegetation and the resultant loss of soil organic matter are primary contributing causes of soil degradation (Charreau, 1974; Rodale Institute, 1989).

Recognizing the need to find effective means to reverse this process, a Regenerative Agricultural Research Centre (RARC) was established in Senegal in 1990 by the Rodale Institute to operate a multidisciplinary, participatory programme that integrates applied research and community action.[1] This centre works with national government organizations and non-governmental organizations (NGOs) to develop and spread innovative production systems that utilize renewable resources with substantial farmer participation. The programme assists families to secure healthy food and increase their incomes by using and improving local resources while protecting the environment.

In Senegal, regenerative agriculture initiatives in the so-called Peanut Basin have already resulted in positive biophysical-environmental and socioeconomic impacts. The primary cropping system in the basin is a rotation of millet, the staple grain, and peanuts, both a cash crop and a food crop. Before cultivation, fields are cleared by burning any residues, and they are then ploughed, primarily with shallow tillage using animal traction. Fallow periods have shortened dramatically in recent years, and the majority of fallow is now involuntary, often a consequence of seed shortages. Inorganic fertilizers and pesticides are rarely used these days following the removal of government subsidies over the past two decades.

Actually, inorganic fertilizer applications in the Peanut Basin generally do not produce higher yields unless the soil's organic matter content is concurrently improved (Freeman, 1982). When soil organic matter and clay particles are not present in sufficient quantities to hold nutrients in the soil-rooting zone, the first heavy rains of the wet season remove through runoff and leaching any minerals that have been added to the field. Further, if fertilizers are added at the time of planting, microbial activity and weed growth are major sinks for available nutrients and compete with crops for any mineral amendments.

Given the short growing season, turnover of immobilized nutrients in these sinks may not occur until well after crop harvest. Research at the Senegalese Institute for Agricultural Research (ISRA) has shown that organic amendments to the soil system should accompany inorganic fertilizer applications to make optimal use of any chemical treatments (Dancette and Sarr, 1985). Incorporation of straw along with additions of nitrogen–phosphorous–potassium fertilizer helps maintain millet production levels, given adequate rainfall levels of at least 400mm/yr. Certain tillage practices, crop rotations and micronutrient applications may also be needed, however, to make best use of nitrogen–phosphorous–potassium amendments. Centre staff and others have developed the following conceptual framework for understanding and promoting the process of regenerating degraded soils in the Peanut Basin (Rodale Institute, 1989).

MODELLING PROCESSES OF SOIL DEGRADATION AND REGENERATION

When the soil is in a state of severe deterioration, as depicted in Figure 10.1, land is in need of regeneration. After years of intensive cultivation, one finds that soil organic matter is reduced to negligible levels by the effects of mechanized tillage and consequent wind and water erosion of fine material from the topsoil. The soil loses its ability to retain moisture and essential nutrients such as nitrate (NO_3) and phosphate (P_2O_5) due to low clay content and loss of soil organic matter binding capacity. With the first heavy rains, high erosion and surface runoff transport out of the root-soil system much of the nitrogen, phosphorous and potassium applied in inorganic fertilizers. This severely limits the utility of chemical inputs for improving plant nutrient uptake and crop yields.

Microbial populations, which have been inactive during the dry season, undergo an explosion of biological activity once the rains come and there is

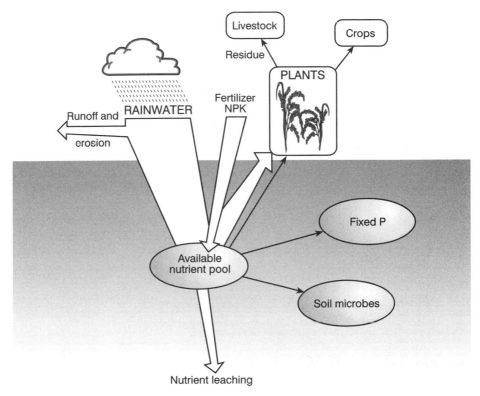

Figure 10.1 *Model of Soil Degeneration*

moisture in the soil. These highly opportunistic micro-organisms compete efficiently with the young crop plants for the nutrients that are suddenly liberated from the small amount of residual organic matter that is rapidly decomposing in the soil. Thus, nitrogen and phosphorous are mostly immobilized during periods that are critical to plant growth and development. Available soil phosphorous is kept from uptake by the plant roots due to fixation with aluminium and iron complexes that are abundant in the acidified soil profile.

Crop yields decline in a deteriorating cycle of nutrient extraction from the soil system. Very little organic material is returned to the soil since crop residues are either burned or are removed and fed to livestock. The overall moisture-conserving capacity of the soil is lost along with the soil organic matter, and crop yields become highly dependent on annual rainfall amounts. Drought years are likely to result in crop failures.

The cycle of degradation can be reversed, with regeneration of soils as is being done in the Peanut Basin following the model shown in Figure 10.2. The rehabilitation process focuses on building up the levels of organic matter in the soil in order to increase moisture infiltration into the soil and its water-holding capacity. This also reduces surface runoff and erosion. Construction of soil and water conservation structures such as windbreaks, living hedges

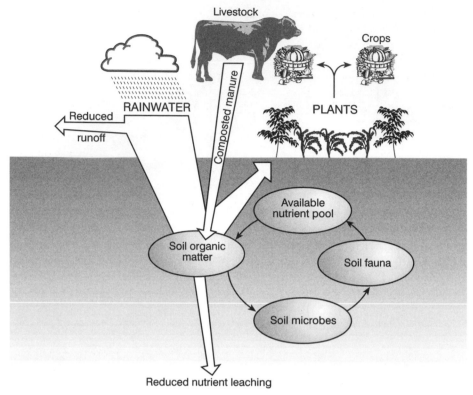

Figure 10.2 *Model of Soil Regeneration*

and rock walls in and around fields should accompany initial amendments of organic matter to the soil to complement biological dynamics with physical support.

As trees and leguminous crops add their organic debris to the soil, and as livestock are integrated into the recycling process to accelerate nutrient mineralization, greater moisture in the soil favours the development and maintenance there of an active biological community. This completes the cycling and turnover of essential plant minerals such as nitrogen and phosphorous. Such a process of soil and productivity improvement is documented in the following chapter on integrated soil, water, plant, animal and nutrient management in Mali. The biology, physics and chemistry of this process are well explained for persons not acquainted with this body of scientific knowledge in Smillie and Gershuny (1999).

Microbial populations of bacteria and actinomycetes rapidly decompose active soil organic matter components, and some make nutrients such as phosphorous available. In contrast to the immobilization of scarce nutrients in a microbial biomass sink as shown in Figure 10.1, under these alternative conditions, the development of populations of protozoa and soil fauna that feed on micro-organisms and excrete mineral waste products make nitrogen and phosphorous available for plant root uptake.

After several seasons of rebuilding the soil ecosystem and nutrient cycles, crop yields increase and still more organic material is available to be invested in the soil to intensify the regeneration process. Rainwater runoff, soil erosion and nutrient leaching are all greatly reduced as water infiltration rates improve.

Crop yields can in this way be decoupled, at least to some extent, from the annual amounts of rainfall. Droughts, while having a negative impact on yields, do not result in crop failure as often or as severely when soil systems are vigorous.

These processes have long been known and are well documented in the literature, however, they have been overshadowed by overconfidence on the part of scientists, extensionists and policy-makers that inorganic fertilizers can raise yields by themselves. In soils already rich in organic matter, the payoff from fertilizers can be high, though they can also suppress some of the soil biological community and its capacity to support plant growth. Where soils are poor because they lack biotic activity, adding inorganic nutrients will not make them rich. Only biological activity can do this.

FROM MODEL TO FIELD: FARMER PARTICIPATION

In farming systems research, on-farm trials play a key role in the development and verification of appropriate technologies. Involving farmers who must ultimately make decisions concerning adoption or rejection of innovations provides opportunities for them to become the primary developers and evaluators of these technologies.

There is an 'information gap' between what has been learned by agricultural researchers working primarily with ISRA and smallholder farmers who need appropriate solutions to problems associated with declining soil fertility and erratic rainfall. In the Peanut Basin in particular, the traditional farm fallow practices that allow for natural regeneration of soil fertility have almost disappeared. This has resulted from increased population pressure on the land and from the widespread availability of animal traction equipment.

Throughout the 1970s and into the 1980s, the Senegalese government invested heavily in certain regional parastatal institutions to extend not only agricultural technologies developed by researchers, but also credit and inputs for crop production as well as buying produce, particularly peanuts, from farmers for export. When the government accepted structural adjustment recommendations from the International Monetary Fund (IMF) in the mid-1980s, it began to disengage itself from the agricultural sector. This meant ending subsidies to farmers and reducing support to parastatals. As a result, a number of NGOs began to assume some responsibility for agricultural extension services to farmers.

In 1990, the RARC and ISRA began jointly conducting soil regeneration research on farmers' fields. The aim was to help farmers integrate better natural resource management into food and fibre production systems and to enhance farmers' capacity to participate in and contribute to technology development and dissemination. Various NGOs were brought into the process, playing roles similar to that of the RARC. Subsequently, other groups such as consumers, young people and small traders have been drawn into partnerships designed around Rodale Institute's philosophy of 'healthy soil, healthy food, healthy people'.

Seeking to strengthen the relationships depicted in Figure 10.2, the centre started work in 11 villages. Cooperating with various NGOs and government organizations, RARC staff talked with farmers in these villages to identify what farmers perceived to be their primary soil-based constraints. In some villages, soil erosion was identified; in others, declining soil fertility drew the most attention. Initially, work concentrated on these two components of the model: holding the existing soil in place, and improving the quality of the conserved soil.

SOIL REGENERATION IN THE PEANUT BASIN OF SENEGAL

From the initial set of villages engaged in soil regeneration experimentation, the effort expanded to include more than 50 villages in nine of the ten regions of Senegal by 1998. Much of the western half of the country is now covered by activities of the RARC; the eastern half is sparsely populated.

In sub-Saharan Africa as a whole, there are more than 60 million small-holder farmers (World Bank, 1998). Poor and declining soil fertility is a major impediment to increased food production in this region, and in the Sahelian countries in particular. Aggravating this problem is an apparent decline in rainfall in recent years. Although the causes of this are still in dispute, and possibly this is more a short- to medium-term trend than a long-term one, farmers' current problems of reduced precipitation are aggravated wherever soil fertility and organic matter are low.

Rainfall figures for the Peanut Basin over a ten-year period show a statistically significant trend of decreasing rainfall levels. The severe land degradation in arid and semi-arid regions of Africa has rendered some areas no longer able to sustain crops and other vegetation. In Senegal, in the village of Tatene in the Thies region, almost half of the arable land has not been cultivated for several decades now. In many cases, all productive soil has been washed away either by wind or water erosion, exposing rocks on the surface. These areas desperately need regenerative agriculture to halt the losses and where possible, to reverse present trends.

The regenerative system adopted in the Thies region consists of:

• conserving the remaining soil by preventing both water and wind erosion;
• introducing leguminous trees and shrubs as windbreaks and as sources of organic materials for the soil;
• intercropping or rotating leguminous and cereal crops; and
• adding manure and/or compost to increase fertility.

Successful soil regeneration was possible in Senegal, with accompanying increases in yield, when these practices were integrated.

Soil Conservation and Improvement

RARC staff work with local farmers to get rock walls built along field contours as barriers to runoff erosion. Leguminous trees are planted as windbreaks to reduce wind erosion. Additional steps are taken to improve the soil's organic matter content and fertility. With these technologies, farmers see increased crop production and can reclaim land that had once been unusable. The use of stone barriers and trenches along contours improves the soil's water-holding capacity, retains significant amounts of top soil, enhances seed germination of indigenous plant species that were believed to be lost and puts back into cultivation seriously degraded lands that have not been used for several years (Table 10.1).

After three years, the establishment of stone barriers resulted in the retention of 10–12cm of soil particles on the most developed sites, with a soil water-holding capacity of almost 20 per cent after the second year compared to 2.2 per cent on unimproved sites. Today, crops are being planted on the site, and farmers are able to graze their animals in those fields after harvest. After four years, approximately 50 hectares of land had been put back into production. Most of these lands were reclaimed by their original owners after they saw the improvement made on-site. A significant amount of food is being produced from this area without using external inputs.

Table 10.1 *Efficiency of Stone Barriers at Tatene, 1990–1992*

Treatments	Water-holding capacity (%)		Soil particle retention (cm)	Biomass production (kg/m^2)	
	1990	1991	1992	1991	1992
I	2.2	2.2	0.3	0.180	0.120
II	3.5	10.5	1.3	0.152	0.896
III	3.6	14.2	1.8	0.101	1.588
IV	3.8	19.3	2.7	0.139	2.500

Note: Treatments: I = plot without stone barriers; II = plot with 1 stone barrier; III = plot with 2 stone barriers; IV = plot with 3 stone barriers

Integration of Livestock and Crop Production

Cereal-based farming systems are the most common type in Senegal. The integration of livestock into such systems can be very challenging because grazing animals can interfere with the establishment of forage species introduced as windbreaks, and can destroy trees planted as barriers to erosion. Mixed cropping has been the strategy used by many farmers to minimize their risk of crop failure. Cowpea intercropped with millet provides additional fodder for animals, contributes to reduced erosion, increases the organic matter content of the soil and, more importantly, provides communities with at least one crop to be harvested when millet does not reach maturity because of insufficient rainfall. When both cowpea and millet are harvested under normal circumstances, there is evidence of total yield benefit from intercropping. Even though the respective yields of these two component crops may be slightly reduced compared to their yields when grown separately, the total yield from these two crops in intercropped plots is usually higher per unit land area (Koenig, 1990).

Integrated crop–livestock systems reduce risk, contribute to the sustainability of smallholder farmers, improve household diets through the addition of protein, increase income opportunities and contribute to the restoration of soil organic matter, as discussed in Chapter 2 and seen again in the Chapter 11. To improve the nutritional security of rural Africans, low-input sustainable production systems like those described here are needed. Such systems can both maintain the fertility of soils and promote social equity and community wellbeing.

EXPERIENCE FROM NDIAMSIL

In this village in the Diourbel region, about 120km from the capital Dakar, farmers fatten cattle, goats and sheep to increase their household incomes and the availability of manure. The RARC created a small fund to allow participating farmers to make an initial livestock purchase. In the first year, 1995, five out of six farmers in this community were able to sell their livestock profitably, and four were able to reimburse the community fund completely. (One family experienced a financial crisis that made it impossible to repay at the end of the year.) We could see that most such investments could pay off within one year, so there was the possibility of designing a programme that could be self-sustaining.

Stall-feeding cattle is a logical component of the farming system for better utilization of off-farm resources. However, feed shortage during the long dry season can present a challenge. The RARC has been helping farmers to establish mixed-crop plots of millet and cowpea. Not only do farmers see an increase in yields and an improvement in soil fertility, but they can produce more fodder for livestock and increase their incomes.

Farmers have tested different techniques to accelerate the animal fattening process using locally available resources. One technique used in Ndiamsil that works well is to mix 500g of ground millet grain with water and feed this mixture to the stabled cattle twice a day, morning and night. This has resulted in an average daily weight gain of 0.94kg, which is excellent. In another community (Baback), a feed mix consisting primarily of leaves of wild plants, peanut hulls and dried cowpea residues averaged 0.84kg weight gain daily, which was also very good, especially from feeding materials that have no competing human uses.

As part of the cattle-fattening programme, feed gardens have been established by many farmers in family compounds as well as in waterlogged areas of their farms that do not produce anything of value. Such gardens alternate rows of grasses (*Andropogon gayanus*) with fast-growing leguminous trees (*Gliricidia sepium* and *Leucaena leucocephala*). Three years of results from feed-garden harvesting in one community in the Thies region (Samba Dia with

450–600mm of rain) suggest that planting seedlings rather than direct seeding of legume tree species will result in better biomass production of individual plants, at least three times greater. *Gliricidia sp.* outperformed *Leucaena sp.* by at least 300 per cent, regardless of the planting method used. When labour is a constraint, the direct seeding of Gliricidia to establish forage feed gardens can be recommended.

Farmers report that with such increases in available fodder, they are better able to stable their cattle, which makes manure collection easier. Those who compost the manure before applying it to their fields achieve more improvement in soil fertility, as shown below. One farmer from Ndiamsil has found that stable-feeding just one cow for four months can provide sufficient manure for improving one hectare of cropland.

Use of Manure with Natural Rock Phosphate

From 1990 to 1995, the RARC conducted research trials in Ndiamsil to test the benefits of using natural rock phosphate in combination with animal manure. The trials showed that manure is very effective when used in combination with natural rock phosphate on millet crops and on intercropped millet and cowpea. As a result, the company that processes rock phosphate in the Thies region has now made this soil amendment accessible to farmers at a reasonable price through its fertilizer sale outlets. In 1998, the government of Senegal launched a nationwide programme to support application of rock phosphate as a capital investment to help farmers create more favourable conditions for soil regeneration. Studies conducted in the Thies region have indicated that the availability of manure of good quality and in sufficient amounts is the key factor affecting the spread of these practices (Sagna Cabral, 1988). If there is adequate supply of manure, there will be considerable demand for and use of rock phosphate.

Composting manure and crop residues increases nutrient concentration, improving the quality of the composted end-product. It also reduces the huge amount of raw manure that is needed for soil regeneration and getting substantial yield increases (10t/ha every two years). Raw manure can either be beneficially amended with rock phosphate or it can be composted with crop residues.

While manure amendments surely enhance concentrations of phosphorous and potassium in the soil in the Peanut Basin, it is not certain whether the timing of early nutrient release from organic fertilizers coincides with periods of peak plant demand (Freeman, 1982). If nutrients are being flushed from the entire soil system with the first heavy rains, this could explain why millet yields are not increased by putting manure on the surface and then ploughing it into the soil. The timing and manner of application are obviously important for the greatest benefit.

Rock phosphate used with composting has been shown to increase the available phosphorous for plant uptake to levels comparable to single superphosphate responses (Batiano et al, 1986). Substantial sources of natural rock phosphate fortunately exist in a number of West African countries, eg, Mali,

Niger and Senegal. More research is needed on the best use and combination of manure and chemical inputs for the soils and climatic conditions of the Peanut Basin. The objective of such studies should be to increase nutrient-use efficiency within cropping systems, as depicted in Figure 10.2.

Two methods of manure management have been tested in the Senegal Peanut Basin by the RARC. A first study conducted at Ndiamsil in collaboration with ISRA evaluated the effects on crop yield and soil conditions of adding natural phosphate rock (37 per cent P_2O_5) to animal manure. Each of the seven farmers participating in this study applied the same set of treatments to millet and peanuts in a completely random block design with three replicates. The four treatments were:

1   check plot without manure;
2   two tons of manure per hectare every two years;
3   two tons of manure per hectare plus 30kg P_2O_5 (as rock phosphate) every two years;
4   farmers' practice – roughly 2t/ha every two years, the exact amount depending on manure availability.

Plot sizes were 100m² (10m x 10m) for millet and 90m² (10m x 9m) for peanuts.

Results obtained after the first cropping season indicated a substantial increase in yield for both crops when natural phosphate rock was added with the animal manure, compared with manure alone or farmers' practice (Table 10.2). Probably the decomposition process for the manure had not yet reached a level that could release nutrients and make them available for uptake by crop plants. However, manure alone applied on the soil surface is expected to provide a better environment for individual seeds. Better crop stands were observed in all plots that received it. Managing nutrient concentrations and manure handling are very important for getting optimum benefits from its application.

Table 10.2 *Peanut Yields as Affected by Applications of Manure and Rock Phosphate, Senegal, 1991*

Treatments	Peanut hay (kg/ha)	Millet yield (kg/ha)	Combined yield (kg/ha)
Control plot	500	340	840
Manure: 2t/ha/2yrs	505	485	990
Manure: 2t/ha + 30kg P_2O_5	580	680	1260
Farmers' practice	590	440	1030

COMPOSTING

Composting is an old practice still being used in a traditional way by African farmers. The use of compost has great potential in developing countries, particularly in Senegal where farmers cannot afford chemical fertilizers now that these are no longer subsidized. Several compost-making techniques have been experimented with in Senegal. Women, particularly, are interested in learning to make compost. So far, they have been pleased with the results. A woman in Gade Khaye, a village near Thies, reported to RARC staff that fertilizers were 'burning' the vegetable seedlings in her garden and killing them; when she used compost instead in 1993, she did not lose her crop again.

Composting has become the most popular topic for training sessions held by the RARC in Senegal, with more than a third of the trainees being women. They are interested in growing vegetables to eat and/or sell, to benefit their families from improved nutrition and extra income. In addition, producing vegetables at home saves women the long trip to distant markets several times a week, and money previously spent on fertilizers can now buy extra seeds or can be used in times of emergency, like when a child falls ill.

Money earned from vegetable production is often held by women's groups as a pool of credit for individual women. They have opened up small stores, started craft businesses or bought millet-grinding mills. In the village Keur Banda, women found that making compost using Rodale's methods was too difficult for them. Most of the men in the village worked elsewhere and could not help the women dig the four-foot-deep pit required to hold compost materials, usually millet stalks, ash and manure. The women worked with RARC staff to figure out a method requiring less heavy labour. Modified composting methods now allow women working together to grow more vegetables, improve their health and augment finances for themselves, their children and their communities.

Farmer-managed trials were conducted at seven different locations in the region of Thies, to evaluate the response of millet and peanut crops to the application of manure and of compost that was made from manure mixed with plant residues (millet stalks, grasses). The average yields obtained for millet and peanuts, over a five-year period, showed increases of 54 to 59 per cent with the addition of two to four tons of manure per hectare. But yields were much higher with the addition of just two tons per hectare of composted manure; there was a tripling of yield (Table 10.3). It was very interesting to see that a doubling the application of composted manure to four tons per hectare gave little or no further increase. This suggests that the *volume* of nutrients added was not so important as the compost's effect in nurturing active microbiological processes in the soil.

Yield increase under the conditions of the experiment could also be due to the creation of a protective physical barrier on the surface by applying composted material. This reduced wind erosion during the dry period prior to the rainy season, and improved soil moisture conservation needed for good crop establishment. This was observed in the previous experiment when animal manure was applied. However, application of composted manure and crop residues should be made at a time when plant nutrient uptake is at a maximum level to avoid leaching and volatilization.

The compost-making process can be time consuming and labour intensive, in part because it requires a large amount of water to support the microbial

Table 10.3 *Peanut and Millet Gain Yields in Ndiamsil, 1991–1995*

	1991	1992	1993	1994	1995	Average increase	Average % increase
Peanut yields (kg/ha)							
Control	469	236	383	170	455	342	
2t/ha manure	736	360	652	502	870	62	483
4t/ha manure	676	361	671	527	933	63	385
2t/ha compost	1014	668	1327	848	1384	1048	207
4t/ha compost	992	577	893	988	1388	967	183
Millet yields (kg/ha)							
Control	458	174	330	252	465	335	
2t/ha manure	780	332	529	544	707	578	73
4t/ha manure	890	361	689	531	673	628	83
2t/ha compost	1248	765	1250	762	1020	1009	200
4t/ha compost	1055	611	1038	1054	1404	1032	208

Source: Adapted from Westley (1997)

populations in the compost. For this reason, rainy season compost is now most often recommended. This also does not require the labour of turning. However, if there is excessive rain, a significant amount of nutrients is lost.

Now composting is moving into the cities of many African countries. City composting is more complicated and has different constraints than does village composting. In the cities, conditions are not good for making pits, so metal containers are often used. Water, a major input in compost, may have to be carried long distances to the sites in the city, and it is more difficult to add water to containers than to a pit. In addition, a lot of screening needs to be done with city wastes before they are composted. Certainly, in both cities and villages, composting can make a major contribution to improving food production. It is encouraging to see city garbage now being turned into compost that ends up on urban gardens instead of in city dumps.

PROSPECTS

To maximize the sustainable potential of the vulnerable soils of Senegal, their productive capacity must be restored. This requires a combination of soil conservation and regeneration measures. From the conceptual model presented here and from experience in Senegal, we know that degraded soils can be rehabilitated by nurturing soil biological communities, and that soil fertility can be maintained and managed through investments of organic matter and soil moisture conservation.

To summarize, every management step is focused primarily upon improving the soil ecosystem in favour of retention and availability of moisture and nutrients, supporting the populations of micro-organisms living and working in the soil using biological processes to solve chemical and physical problems. This regenerative strategy represents a major shift in emphasis from a short-term, production-oriented strategy to a long-term rehabilitative one in which farmers invest in their soil as a first priority. The subsequent long-term benefits are increased crop yields and sustained production in a much healthier environment. Rather than focusing on 'feeding the plant' one feeds the soil, and this in turn will amply nurture the plant and animal life, as well as human life, that depends on it.

NOTES

1   The Rodale Institute works with people worldwide to achieve regenerative food systems that renew both environmental and human health, seeking agricultural solutions to hunger, malnutrition, disease and soil degradation that do not rely on the use of agrochemicals and inorganic fertilizers. Applied research, educational programmes, information exchange and networking are supported to share success stories from wherever they can be found with farmers, researchers, consumers, food industry leaders, policy-makers and young people.

Chapter 11

Combining Traditional and New Knowledge to Improve Food Security in the Sahelian Zone of Mali

Mamby Fofana

Mali is one of the largest countries in West Africa, with an area of 1.24 million square kilometres, but over half of this is Saharan desert. It is one of the poorest countries in the region, with diverse ecosystems ranging from local forests and grasslands in the south to scrub and scattered vegetation in the middle to eventually none in the north. Average annual rainfall varies from 1200mm in the south to 150mm in the north. The country's agricultural and pastoral potentials, though abundant, were severely affected by the repeated droughts that occurred from 1968 to 1985. In the Sahelian region in the centre of the country, repeated cycles of drought have caused erosion of soil and vegetative cover levels, reducing agricultural and pastoral productivity. This is where since 1987 the Unitarian Service Committee of Canada (USCC) has been operating a rural development programme in Mali, supporting initiatives at the grassroots in Douentza district whose results are reported here.

Activities under the programme are initiated by or with the populations themselves and are jointly carried out by them with partners such as local and/or foreign non-governmental organizations (NGOs). The programme is based on two main methodologies:

1 *Participatory rural appraisal* supporting research and community activities through people's involvement. This informal, systematic learning activity strengthens farmers' abilities for analysing their milieu to maintain initiative when faced with changes in their environment.
2 *A land planning and management approach* which assumes that all available land has to be utilized more rationally through the development of sustainable land use plans.

CONDITIONS IN DOUENTZA DISTRICT

This district, situated in the eastern part of Mopti region, covers an area of 23,312 square kilometres. With a population in 1993 of 156,695 inhabitants, according to estimates made by Save the Children-UK, the density is less than seven inhabitants per square kilometre. The population is very heterogeneous but consists mainly of people of the Fulani and Dogon cultures.

Being located in the Sahelian zone, Douentza's climate is characterized by two seasons: a short rainy season from June to September, and a long dry season from October to May. Rainfall is unevenly distributed over space and time, with the average yearly rainfall of less than 400mm varying over the last two decades from a high of 547mm to a low of 194mm; the number of rainy days in a year ranged from 23 to 38. Average temperature varies from 10–20°C during nights in December–January to 35–40°C during days in April–May.

From west to east, the district of Douentza is crossed by cliffs. Sandy plains stretch out across the south and north of this chain of mountains. In the west, the soils are either clay or lateritic. The economy of the district is based almost entirely on agriculture and animal husbandry. The agriculture is mostly dryland food crops, with the main crops being millet (85 per cent of the sown area), sorghum, rice, cowpeas and peanuts. Animal rearing includes cows, sheep and goats. Table 11.1 shows the irregularity of cereal production.

Table 11.1 *Yearly Cereal Production in Douentza, 1981–1998, in tons*

Years	Millet	Sorghum	Combined
1980–1981	28,955		28,955
1981–1982	25,176		25,176
1982–1983	2438		2438
1983–1984	1417		1417
1984–1985	5685		5685
1985–1986	7719		7719
1986–1987	6324		6324
1987–1988	3197		3197
1988–1989	19,153		19,153
1989–1990	3196		3196
1990–1991	12,168	1298	13,466
1991–1992	15,838	1178	17,016
1992–1993	10,151	1515	11,666
1993–1994	16,262	1127	17,389
1994–1995	26,880	2242	29,122
1995–1996	6000	350	6350
1996–1997	4950	575	5525
1997–1998	4950	570	5520

Source: Ministry of Agriculture, Agricultural Extension Department, Douentza office

Reporting an average annual production of millet and sorghum (11,628 tons) is not very meaningful because the range has been from 3197 to 29,122 tons, almost a ten-fold variation. The average annual production per inhabitant is calculated to be 75kg, only about one-third of what is required, but this too varies by a similar range.

The following calculations show that Douentza as a whole is prone to chronic and serious food shortages unless there can be dramatic increases in production. The cultivable area in the district amounts to only 46,000ha. Even with an average cereal yield of 600kg/ha, which is more than double the average over the past 20 years, annual production would reach only 27,600 tons. Given a population of 150,000, this amounts to only 175kg per person. Meeting the United Nations Food and Agriculture Organization (FAO) norm for annual cereal consumption (225kg per capita) would require about 35,250 tons per annum, three times more than the average production over the past two decades. Thus, even yields that are presently considered high would produce only about three-quarters of what is needed. This leads to seasonal and sometimes permanent outmigration in search of employment and income.

The difficulties that people face in Douentza can be summarized as follows:

- A steady shortage of rainfall as a result of unfavourable and changing climatic conditions.
- A cereal shortage of at least 12,000 tons per year, even in years with good rainfall.
- A resulting rural exodus of both men and women, affecting about 40 per cent of the economically active population.
- Soils that are highly eroded due to their nature, the torrential nature of rainfall and their toposequence.
- Population pressure on the natural resource base for agriculture and animal husbandry leading to frequent conflicts between different ethnic groups with different livelihood strategies, some pastoral and others agricultural.

To address these problems, a participatory programme of land planning and management was initiated with 18 communities in different parts of the district. This effort had two main concerns:

1   Developing people's ability to manage their soils in ways that restore and conserve them and increase yields in order to draw from them greater and sustainable socioeconomic benefit. (The strategy followed for soil improvement is in accord with the principles offered in Chapter 10.)
2   Improving the genetic resource base available to farmers to have a more reliable and abundant food supply, particularly by promoting conservation, utilization, enhancement and extension by farmers themselves of the biodiversity of their main agricultural crops – millet, sorghum, maize, cowpea and bambara beans – and of fodder and forest plants: *Acacia albida, Acacia nilotica, Tamarindus indica* and *Adansonia digitata*. Farmers are given technical support to conserve, enhance, store and utilize

traditional seeds drawing on indigenous knowledge in Mali and elsewhere in Africa.

The activities being implemented include: soil conservation and agroforestry; conservation of traditional seeds and biodiversity; market gardening and fruit tree planting; management of cereal banks and supplying credit for agricultural equipment; information, education and communication; functional literacy; income-generating activities for women; village water supply, hygiene and sanitation; and supervision and follow-up. All these components are aimed toward food security and diversification of income sources.

RESULTS TO DATE

As food security in Douentza depends on the development of traditional agriculture, our first activity was to investigate farmers' conditions in order to understand better the constraints on raising production. The results of these inquiries permitted the design of activities that farmers could use to improve their production techniques. Then three main sets of activities were identified, taking local conditions into account and understood in terms of the climate, soil and socioeconomic factors that had been identified. These are reported on below.

Soil Conservation and Agroforestry

The populations of 12 villages, with about 2000 adults, have undertaken a number of improved techniques of water and soil conservation, receiving training from the programme on how to raise water levels, choose the best sites for anti-erosion works and carry out these works. More than 3000m of dykes have been constructed, protecting about 4000ha of arable land. In each of the 12 villages, a committee has been set up to manage these activities and to take further initiatives for local conservation of water and soils, including agroforestry and afforestation.

Agroforestry is a traditional production practice that has been rediscovered and put into practice, consistent with the points made in Chapter 8. People are sensitized through education and information campaigns to the need to prevent any further loss of trees. The programme also supports the replanting of trees, which has easy-to-notice effects as trees enhance microclimates, reduce wind damage, protect soils against erosion and regenerate them. Moreover, trees provide products for food, health and household energy requirements, which makes them multiply attractive. Agroforestry efforts are supported by the creation of four village nurseries with 24 agents trained in techniques of plant production and grafting.

These various efforts have contributed to much higher cereal yields in these communities, rising in some cases from 300kg/ha to 1700kg/ha. This is about twice the level needed to meet basic food needs in Douentza, since given the arable area and population, average yields of 850kg/ha could produce 225kg of cereals per capita.

Seed Conservation and Biodiversity

Traditional seeds, the result of a long selection process partly natural and partly purposeful, are usually quite hardy. Though less productive in some cases than exotic species, they are more capable of withstanding the rigors and vicissitudes of the climate. In addition, they can adapt to variations of rainy season length because they are photoperiodic. As a result, they can help ensure the food security of small-scale producers in the Sahel, who are regularly under the threat of climatic default. Villagers have identified the loss of plant bio-diversity as a matter of utmost importance in Douentza. This loss has resulted from successive years of drought that affected the region, and from pressures that derive from human activities.

The genetic variation of most plant species that are found in the area is being constantly reduced, as plants become more susceptible to climatic changes and to damage caused by pests and insects. An investigation under-taken at the farmers' level showed that they have genuine problems in obtaining seeds for many species and varieties. This limits farmers' options in terms of the plants they can utilize. The programme is organizing farmers to produce and store seeds of good quality that have been derived from the local genetic reserve. The initial focus has been on the staple crops that are most essential for food security.

Twenty-five farmers who are knowledgeable about local production practices and varieties were engaged by the programme to carry out a multi-local assessment of 41 landraces of sorghum and 11 landraces of millet. The results showed that many local varieties are better adapted to climatic condi-tions compared with the varieties being promoted by the national agronomic research service. The average yield from the 41 sorghum landraces evaluated was 830kg/ha, much more than the 480kg/ha from the improved variety (CSM 63-E) used as a control crop. Similar advantages were not observed with millet varieties, however; the 11 landraces tested averaged 1.88t/ha compared with 2.12t/ha from the improved variety that was used for comparison (Toroniou), not a significant difference.

Accordingly, 28 varieties of sorghum and two millet varieties that performed better than the varieties recommended by government researchers are being promoted locally. These varieties achieve better productivity from the rudimentary means of production available to farmers, adapting more readily to climatic changes in the Sahel and to the cropping cycle of the farmer.

A local gene bank, the first one in Mali, has been established in the village of Badiari where the seeds of all the evaluated sorghum and millet landraces are being kept and stored, along with the seeds of 56 local tree species and nine other plant species grown locally. Traditional techniques for conserving and/or dressing seeds have been applied to different crops. Seeds are tested annually in terms of germination to make sure of their viability, and plots have been estab-lished to multiply seed of those varieties that have proved outstanding.

This operation enables the programme to provide local communities with seeds of good quality and in sufficient quantity necessary for plant production and food security. From the production obtained at Badiari, another seed bank has also been established in the village of Gono, where a botanical garden and nursery have been established in the yard of the primary school. This helps establish a clear link between conservation and production, and it prepares children to become citizens responsible for and conscious of the world's future.[1]

Market Gardening and Fruit Tree Planting

These horticultural activities are also crucial for food security in Douentza. The Sahelian climate with its short rainy season and a long dry season leads to underemployment during much of the year. This induces many able-bodied villagers to migrate in search of additional monetary income. In many cases, seasonal departures become permanent, depriving families of the labour necessary for sufficient agricultural production the rest of the year. Market gardening and fruit tree planting, two productive dry-season activities, however, encourage people to remain at home by giving them economic opportunities within the village.

Five villages with a combined population estimated at 3000 inhabitants have become involved in such activities thus far. Each village has established areas for gardening and for fruit production, by 1999 benefiting 489 farmers, of whom 190 are women. The production of fruits and vegetables both improves the nutritional situation in villages and provides cash income. The most recent estimate is that village incomes are enhanced by more than 3,000,000 CFA francs per year, which works out to be about US\$10 per capita, a noteworthy sum in this cash-starved environment.

The production attained in the village of Ibissa gives an idea of the potential for such methods. Participating villagers growing shallots, onions, cabbage, lettuce, tobacco, garlic, potatoes and cassava were able to produce 260 tons of vegetables from 15.5ha cultivated. In addition, on three hectares planted with papaya, banana and mango trees they have produced almost 98 tons of fruit. Fruits and vegetables are crops that require much water, but with good soil and water management, significant production is possible under Sahelian conditions where sunlight and warmth are abundant. Careful use of water can give good results.

CONCLUSIONS

From the experience and lessons accumulated in Douentza, we see that achieving food security needs to be integrated in a larger vision of solving poverty problems. We are convinced that no isolated initiative or innovation can succeed in producing sufficient food for all. It is necessary to collaborate with all parts of the community and district, and at all levels, in order to move production frontiers and reduce poverty and food insecurity seriously and forever. With some well conceived and long-term assistance, rural people with mostly local resources are able to reverse many generations of food insufficiency in the Sahel. The production potentials with better soil, water, plant and animal management are quite substantial even in an environment as adverse as Douentza.[2]

NOTES

1   This arboretum has enabled 197 schoolchildren (116 of them girls) to become acquainted with techniques of seedling production and of transplanting and caring for trees. Each tree is the responsibility of two children who care for it. The arboretum contains 104 trees of 52 different species, 25 of which are local species. In addition to being a training place for children to acquire scientific knowledge (scientific names, morphology, physiology), it is also a place where they acquire indigenous knowledge about trees and their uses.
2   This chapter is based on mostly unpublished sources, since the Douentza experience has not been reported in the literature yet. Data cited in this presentation come from the *Report on Poverty in Mali* published by the National Direction for Statistics and Computer Science in Bamako, and from the Douentza office of the Ministry of Agriculture's Agricultural Extension Department, as well as from reports on the land management and planning project in Douentza from 1995 to 1998. These reports are available from the USC-Canada Mali office, BP E180, Bamako, Mali (email address: usc@spider.toolnet.org).

Chapter 12

Opportunities for Raising Yields by Changing Management Practices: The System of Rice Intensification in Madagascar

Norman Uphoff

The large expenditures made in recent decades on the genetic improvement of plants as well as animals imply that the existing genetic potentials of these various species have been rather fully exploited and need to be modified. The system of rice intensification (SRI) developed and being used in Madagascar suggests, however, that considerable genetic potential exists that can be tapped by altering management practices, exploiting the power of biological processes and the dynamics of agroecological relationships. As seen also from other cases in this book, such strategies can increase production by multiples, not just increments. Specific practices may not be universally applicable, but the principles underlying them should be examined and extrapolated as widely as is productive.

SRI experience indicates that significantly more food can be produced from the germplasm presently available by managing plants (and animals) differently. Rather than redesigning them and standardizing them, handling them essentially like machines to be manipulated to meet human needs, we might better regard them more purposefully as living organisms with their own innate capacities to adapt and be productive. By understanding organisms better and how to improve the environments in which they grow, we can derive greater benefit from existing genetic potentials.[1]

SRI was developed in Madagascar by working closely with farmers, particularly with innovative ones, and by observing rice plants carefully. It alters four practices that have characterized irrigated rice production for centuries, even millennia: rice seedlings are transplanted very carefully when they are still fairly immature, singly rather than in clumps, and widely spaced; also they are grown in unflooded fields during their vegetative growth phase. (After flowering, during the plants' reproductive phase, fields are maintained with a thin layer of water on the surface.) These practices, accompanied by weeding with a hand tiller and by application of compost, have been increasing rice yields several-fold.[2]

With SRI methods, yields reach six, eight ten or more tons, sometimes even 15 to 20 tons with the most skilful use of the practices. The highest yields make SRI controversial because they are above what is considered to be the biological ceiling for rice. This limit, however, has been calculated for rice grown in saturated soil. Such estimates of yield ceiling should therefore be considered as conditional rather than absolute.

Even if these top yields are not obtainable everywhere because of production constraints such as not having sufficient water control, substantial increases should be possible with adaptations of this strategy, drawing on SRI principles for growing healthier, more vigorous plants. Even achieving half as much increase in production obtained by farmers in Madagascar using this methodology would make a large contribution to world food supply. Recent trials with SRI methods in countries as diverse as Bangladesh, China, Cambodia, Indonesia, The Philippines, Sri Lanka and The Gambia have confirmed the yield-enhancing value of these methods. Perhaps as important, SRI may provide some new ways for thinking about and tapping the genetic potentials of other crops when raised under different, more favourable soil and water conditions.

FACTORING THE ENVIRONMENT INTO UNDERSTANDING PLANT PERFORMANCE

We have seen that rice plants have already the potential to produce 10–15t/ha, and even more. This can be achieved with better and more careful management, and with fewer external inputs than are now being used to raise production. A plant's genes are not a *blueprint* from which predetermined copies of an organism are made; they are more like a complicated *game plan* with many built-in, contingent courses of action (Lewontin, 2000). Depending on what the growing organism encounters in its environment, it has a large repertoire of responses and initiatives, some adding to grain production and others reducing this.

Plant breeders have long known that the organisms (phenotypes) resulting from a particular genetic inheritance (genotype) can vary considerably according to the environment in which they grow. Some phenotypical characteristics are directly attributable to the genetic make-up of the organism, but others are the result of environmental influences. Still other outcomes of the growth process result from *interactions* between genetics and environment.[3] Environmental influences at each point in time lead to particular *expressions* of genetic potential; subsequent interactions of the growing organism with its environment produce still different outcomes. Plant breeders are well aware of the importance of 'GxE' effects, the results of the genome interacting with its environment. In SRI, modifying soil conditions markedly changes the rice plant's physical structure, particularly the density and length of its root system and the number and grain production of its tillers.[4]

With SRI methods, increased production is accomplished not by adding nutrients to the soil through fertilizer, but rather by transplanting rice seedlings when they are very young (to maintain tillering and rooting potential) and then by growing them in well drained soil. The numbers of roots, tillers and grains per tiller are all increased by having more space between plants, which respond positively to their greater exposure to sunlight and circulating air. As with all management practices, the challenge is to determine what is *optimum* for growth and production. With SRI, the effects of changing management practices are synergistic rather than just additive.

The standard view of rice, which posits certain biological limits to plant growth and grain production, considers the plant and its immediate environment as a 'closed system' that encounters diminishing returns. This understanding of rice has led plant breeders and agronomists to focus on raising the *harvest index*, that proportion of plant biomass which can be consumed. It is believed that if the number of fertile tillers is increased, a corresponding *decrease* in the number of grains per tiller will result.[5] In this view, growing plants with more tillers is unwise because it will not lead to proportional increases in production because the number of grains per tiller will decline. Also, profuse root growth is considered 'a waste'. With SRI management practices, however, a *positive* relationship between the number of tillers per plant and the number of grains per tiller is observed. The increased root growth promoted by SRI methods makes the rice plant an 'open system'.

THINKING DIFFERENTLY ABOUT HOW TO INCREASE CROP PRODUCTION

Rather than go into much detail on SRI (for that, see Uphoff, 1999, and Stoop et al, 2002), the focus here is on broader issues that could be relevant to raising food production more generally.

Synergy

Each of the practices combined in SRI can by itself make a positive contribution to raising production.[6] The large increases in yield that result with SRI can only be explained, however, if each practice makes a bigger contribution to output when all are used together. This has now been shown through factorial trials.[7] The practices when combined, rather than being just additive, have multiplicative effects because of the way that root growth and tillering interact, each supporting the other. These dynamics contribute to greater grain filling – more grains and larger grains.

Why roots and tillers emerge in a synchronous, coordinated way is poorly understood, but we know that the number of tillers produced by a plant is constrained by the number of roots that the plant sends out, and this number reflects potentials sensed in its growing environment for supporting a larger canopy above ground, as follows:

- how much *warmth* is and will be available to support growth;
- whether the *moisture* in the root zone is and will be sufficient;
- whether there is too much water and the soil is saturated, so less *oxygen* is available to the root; and
- whether there are enough *nutrients* available and accessible to warrant putting out more roots, and so forth.

With vigorous root growth, a plant's canopy can grow fuller and taller as it has more access to nutrients and water for the growth of stalks, leaves and seeds. With more growth above ground to carry out photosynthesis, more energy is in turn available for root growth. Analytical methods that test only one variable at a time, looking for ceteris paribus effects, do not assess such synergistic processes. Seeking synergy is one of the basic principles of agroecology, where the whole becomes more than the sum of its parts, which we see with SRI.

Developmental Patterns

Growth differs from development in that the first is a matter of *scale*, representing an increase in the number and size of parts, while the second is a matter of *structure*, reflecting changes in the relationships among parts. Growth is essentially *quantitative* while development is fundamentally *qualitative*, although there is a dialectical relationship between the two such that development and growth support and facilitate each other.

SRI capitalizes upon an in-built pattern of physiological development in rice that was identified before World War II by the Japanese researcher T Katayama (1951). In studying the growth and development of cereal plants in the gramineae (grass) family to which rice and other grains belong, Katayama found that these plants did not simply grow more tillers en masse but put out their tillers in a regular sequential pattern. How much of its pattern of potential growth any particular plant realizes before it shifts from its vegetative growth phase into its reproductive phase, forming flowers and then grains, depends on how conducive are the soil and other conditions for its growth. This pattern of growth, analysed in terms of phyllochrons, has been discussed in Uphoff (1999), drawing on Laulanié (1993) and Nemoto et al (1995), and will not be elaborated on here.

Plants' developmental patterns are qualitative and structural; they are not captured by measuring simply aggregate, quantitative growth. Knowing the sequence of tillering is more important than knowing just the number of tillers, because one can better understand and promote increases in tillering by comprehending their pattern. Most rice research, however, has been satisfied with reporting and analysing rates and numbers of tiller growth, rather than how many phyllochrons have been completed. This has contributed to a misunderstanding that the rate of maximum tillering precedes, rather than being the same as, panicle initiation. With SRI these periods are identical.

Table 12.1 *Impact of Additional Weeding on Yield with SRI Practices,*
Ambatovaky, Madagascar, 1997–1998 Season

Weedings	(N)	Area (ha)	Harvest (kg)	Yield (tons/ha)
None*	2	0.11	657	5.97
One	8	0.62	3741	7.72
Two	27	3.54	26,102	7.37
Three	24	5.21	47,516	9.12
Four	15	5.92	69,693	11.77

* No weedings with mechanical weeder; only manual weeding
Source: Individual farmer data collected by Association Tefy Saina field staff

Management

As discussed already, management practices modify the environment in which genetic potentials are expressed and realized. That better practices can increase production has been known for thousands of years, so this is not new knowledge. But the extent to which alternative management practices for rice can make a really *large* difference in productivity, as seen from SRI, should lead to renewed interest in varying and evaluating management practices.

The effects of weeding as part of the SRI methodology illustrate this. Farmers using SRI methods around Ambatovaky in Madagascar have got one to two-and-a-half more tons of rice per hectare from additional weedings of their rice fields with a simple mechanical hand weeder. This 'rotating hoe' churns up the soil when pushed up and down between the plants, which have been set out, widely spaced, in a square pattern rather than in rows. The same benefit is not seen from weeding by hand (see Table 12.1).[8]

The apparent effects of soil aeration are quite remarkable, especially considering that farmers for thousands of years, and researchers for decades, have tried to get higher yields by growing rice under *flooded* conditions. This, however, cuts the root zone off from the atmosphere and makes it anaerobic and the roots hypoxic. Rice plants, whether bred for upland (unflooded) or for irrigated (flooded) conditions, develop air pockets in their roots called *aerenchyma* when grown under submerged conditions (Puard et al, 1986; 1989). This has been thought to prove that rice is an aquatic species of plant. However, little research has been devoted to whether or how much anaerobic soil conditions inhibit root growth and functioning.[9]

Not only can one expect well oxygenated roots to grow better than those that are oxygen-limited (Drew, 1997), but rice roots *degenerate* under flooded conditions. Experiments done several decades ago (Kar et al, 1974) showed that this degeneration increases over time when plants are grown under saturated soil conditions, with 78 per cent of roots degenerated by the flowering stage; on the other hand, the roots of rice grown under unsaturated conditions rarely degenerated. This research, perhaps because it was published in Italy, has had little impact on rice scientists' thinking about plant performance, though it has been recognized that such degeneration occurs. The die-back of roots under flooded conditions has been considered a natural, and thus unavoidable process, when in fact it is man-made.

We observe much higher yields when rice is grown in soil to which water is applied intermittently rather than continuously, provided that other SRI methods are also used. These alternate wetting and drying of the field, occasionally letting the soil dry even to the cracking point. This is anathema to rice farmers almost everywhere, who believe that their crop benefits from continuous inundation. SRI gets better results with water management that modifies the growing environment of rice and gets more oxygen into the root zone.[10]

The water management practices used with SRI also facilitate biological nitrogen fixation (BNF). Contrary to popular belief, nitrogen fixation is not limited to leguminous plant species. Under conducive soil conditions, complex associations of free-living micro-organisms around, on and in the roots can convert nitrogen into forms available for uptake by plants, including rice (Döbereiner, 1987; Baldani et al, 1997). It is known that 80 per cent of the bacteria found in and around rice roots have nitrogen-fixing capacity (Watanabe et al, 1981). Research by Magdoff and Bouldin (1970) found that disturbing soil to mix up aerobic and anaerobic conditions enhanced BNF by soil bacteria. This is another line of evaluation to be investigated for explaining SRI performance. As with other SRI management practices, there appear to be several benefits for the growing plant, not just a single benefit, from the recommended method of water application.[11]

Root Development

As suggested in Chapter 2, probably less than 5 per cent of plant research has focused on roots and their functions below the surface, though this ratio has begun to change in favour of more work on roots. One key to SRI results is apparently the much greater rice root systems when SRI methods of cultivation are used. The first measured comparison of root growth found that more than *five times* as much force was required per plant to pull up SRI rice compared to plants grown under standard conditions.[12] Especially because the soils in Madagascar are so deficient in nutrients, having a dense and extended root system will give the plant advantages, such as tapping better what few micronutrients are available.

The results of techniques that increase root growth with SRI should encourage both researchers and farmers to look more carefully at how they can help plants develop more ample and amplified root systems. Most research thus far has focused on what can be easily studied above ground, such as processes of photosynthesis, harvest index and translocation of nutrients. Yet, obviously, what meets the eye depends on the size and functioning of the roots, which are not visible. The *surface area* of root systems, a critical consideration, will increase more than proportionally when roots are greater in number, length and diameter. Enhancing root growth, which is mostly a matter of management, thus should be a special concern of anyone who wants to increase yields and in sustainable ways.[13]

Plant Nutrition

Everyone agrees that the supply of nutrients is crucial for plant growth. The favoured solution when soil nutrients are deficient has been to add chemical fertilizers to the soil to compensate for any deficiencies identified. There will continue to be a need to use inorganic fertilizers in many locations since soils often lack necessary nutrients, and fertilizers will often be the most cost-effective and assured way to provide what is needed, as argued in Chapter 8. But SRI has been successful without fertilizer on some of the poorest soils in the world, according to soil analyses done for North Carolina State University (Johnson, 1994). In the area around Ranomafana National Park, irrigated rice yields have been around 2t/ha. Poor soil quality has been seen as the main constraint on yield. Applying fertilizer amendments and using high-yielding varieties, a small number of farmers around Ranomafana who cooperated with North Carolina researchers were able to raise their average rice yields to 3t/ha, and in a few cases, as high as five tons (del Castillo and Peters, 1994).

During four seasons, 1994–1995 to 1998–1999, farmers around Ranomafana working with Tefy Saina averaged 8.2t/ha, most of them using compost rather than fertilizer, and some using neither. Those growers who are most skillful with the management practices got yields over 15 tons. In the first year only 38 farmers were willing to try the method; but by the fourth year, 275 farmers were using it, though most still on only part of their holdings. The method appears risky because the four practices it alters have been used by irrigated rice farmers from time immemorial with the expectation of reducing the chances of crop failure. However, we have not seen any greater risk than with standard methods if transplanting is done carefully and good management is maintained.

The extent planted with SRI methods around Ranomafana has now reached over 50ha, a substantial area and in diverse locations with elevations ranging between 400m and 1200m. In other parts of Madagascar, SRI methods have been used under a wide variety of growing conditions (altitude, temperature, rainfall), with yields in the same range as around Ranomafana. Over a five-year period, 1994–1995 to 1998–1999, farmers using SRI methods on the high plateau around Antsirabe and Ambositra averaged 8.2t/ha, with the area under these methods going from 36–542ha (Hirsch, 2000). Since 1999, researchers or non-governmental organizations (NGOs) working with farmers in a dozen African, Asian and Latin American countries who have tried SRI methods have got similar increases in yield.

We are ourselves still uncertain how such high yields are possible on such poor soils as those around Ranomafana National Park. Given the parent rock from which these soils were formed,

> *there are no significant areas of naturally fertile soils within tens of kilometres of the park boundary. The pH values in water range from 3.9 and 5.0, with most values between 4.2 and 4.6 ... The levels of exchangeable bases (Ca, Mg and K) are low to extremely low in all horizons. The subsoil horizons contain virtually no*

> *exchangeable bases. Phosphorous levels for all horizons are below 3.5 parts per million (ppm), far below the 10ppm level, which is generally considered to be the threshold at which large crop-yield reductions begin to occur* (Johnson, 1994, pp6–7).

Plant nutrition and growth are very problematic under such conditions. Yet SRI has been successful in spite of these limitations. Yields increase when compost is added to the soil, though not all farmers follow this recommendation. Compost is made from any biomass they can accumulate (rice straw, shrubs, banana leaves, weeds, etc). It has been the prevailing view among soil scientists that compost made from plants where the soil is already deficient in nutrients cannot remedy those deficiencies.[14] Our experience is not consistent with this view as we have found that a low-input technology can produce about *three times* the average and maximum yields that were achieved by using fertilizers. Findings from several lines of research could help to explain the high yields observed on the poor soils around Ranomafana.

Biological Nitrogen Fixation

Diazotrophic (nitrogen-fixing) bacteria are found universally in root zones in association with plants in the graminae family, which includes rice (Döbereiner, 1987). It was suggested above that BNF could be providing critically-needed nitrogen to SRI rice plants through their root systems. Brazilian researchers have studied BNF with non-leguminous plants, particularly sugar cane. Populations of bacteria and other micro-organisms in the root zone of sugar cane cultivars that have not had nitrogen fertilizer applied for several generations can fix 150–200kg of nitrogen per hectare. This makes the application of fertilizer unnecessary, given its cost.

Nitrogen-fixing capabilities were possibly easier to discover in Brazil with sugar cane because for historical reasons this crop had often been grown with little or no application of nitrogen fertilizer. For complex reasons, the symbiotic nitrogen-fixing capacity of diazotropic bacteria that inhabit the root zone, some living on the root surfaces and others even within the plant roots and other tissues, is suppressed when they live in a nitrogen-rich environment such as results from fertilizer application. This suppresses microbial production of nitrogenase, the enzyme needed for BNF (Van Berkum and Sloger, 1983; Döbereiner, 1987).

Organic Sources of Phosphorous

As noted above, the soils around Ranomafana are terribly deficient in measured amounts of phosphorous, less than half the amount usually considered necessary for acceptable production, let alone really good yields. It has been known for some time that soil micro-organisms can supply organic phosphorous to plants, but this has not been investigated very thoroughly. Some recent research has found that when lowland soils in England and Wales were rewetted after drying (the water management practice associated with SRI), the total amounts of water-soluble organic phosphorous in the soils increased by 185 to 1900 per cent, with somewhat more rapid solubilization at higher temperatures (30°C, compared with 15°C). This effect was attributed to the direct release of phosphorous from soil microbial biomass when microbes are killed by the osmotic shock and cell rupture of rapid rehydration following a period of drying. The researchers suggested that this effect could apply also to other soil nutrients and could partially explain a similar phenomenon observed for nitrogen mineralization in tropical soils (Turner and Haygarth, 2001). It is interesting to note that plants can respond to the lack of inorganic phosphorous in their rooting zone by producing more phosphatases which are exuded into the rhizosphere, where they make phosphorous available (personal communication, Malcolm C Drew, 13 June 2001).

Lesser Quantitative Requirements with Continuous Nutrient Supply[15]

Primavesi (1984) reports research, also done in Brazil, which suggests that plants can grow satisfactorily with much lower concentrations of nutrients than have previously been thought necessary *provided that* the limited supply is constantly available over time rather than given at a few points in time (p49). Plants appear to get significant benefit from quite small amounts of nutrients if these are available continuously, as discussed on page 82 above. Compost, used with SRI, furnishes nutrients in a more steady, even if lower, flow than is provided by fertilizer applications.[16]

With SRI there is much greater root growth, due in part to supplying water in limited amounts and only intermittently. (The greater root growth is attributable also to wider spacing of plants and to planting very young seedlings that have more tillering and rooting potential.) When roots grow in saturated soil and have chemical nutrients supplied directly to their root zone, they do not have to grow very much to satisfy their basic needs. To use simple language, they can be 'lazy', whereas plants under intermittent water stress need to expand into a larger volume of soil to survive. This gives them more access to whatever nutrients are available. If these come from compost, which releases them slowly, the conditions that Primavesi described from her laboratory experiments to measure nutrient uptake and plant growth are roughly approximated.

Moreover, roots accessing a larger volume of soil can take up a wider variety of micronutrients critical for plant growth and health, such as zinc, copper and boron. SRI results suggest that there has been too much attention given to nitrogen, and that having the equivalent of a 'balanced diet' is important for plant performance. That plant roots 'down-regulate' their absorption of nitrogen when it is available in the root zone but not needed (Kirk and Bouldin, 1991) indicates that the plant itself is not preoccupied with this nutrient.

Working with Plants as Active Organisms

It is hard to understand how SRI achieves such high yields on such poor soils, but there must be some tangible reasons. Possibly our standard concepts of plant/soil nutrition requirements may not be complete. Plants are not passive receptacles into which water and nutrients get poured; they are not like test tubes in a laboratory in which ingredients are mixed to get certain results. Rather they have potentials to be active agents, seeking their own wellbeing, and thereby contributing to ours. Unwarranted anthropomorphic or teleological thinking should be avoided, but on the other hand, the adaptive and flexible capabilities of plants should not be underestimated, as occurs with reductionist analysis. By learning more about these capabilities and by capitalizing on them, it should be possible to make agriculture more productive and sustainable than understood heretofore.[17]

Bangladeshi, Cambodian, Malagasy, Philippine, Sierra Leone and Sri Lankan farmers have reported fewer pest and disease problems when using SRI methods (personal communications). Since few of them can afford pesticides or fungicides, it is fortunate that SRI-grown plants appear less affected by pests and pathogens.[18] These farmer observations are consistent with integrated pest management (IPM) experience in Asia, which Peter Kenmore reported on to the Bellagio conference. This shows that the best way to protect plants from pest and disease attacks is to grow healthy, vigorous plants.

This sounds tautological, but it is not. Plants with well developed root systems, accessing sufficient water, nutrients and air, and given adequate spacing, are better able to protect themselves – not completely but reasonably effectively – against the predations of pests and the infections of pathogens. Plants have more active capacities than is usually recognized. An agroecological approach to agriculture seeks to understand and benefit from these.[19]

THE ORIGINS OF SRI

The system of rice intensification was developed inductively through the persistent and observant efforts of Father Henri de Laulanié, of the Society of Jesus, who went to Madagascar from France in 1961. He spent the next (and last) 34 years of his life working with farmers there, learning and teaching them how to improve their production of rice. He had already assembled a number of practices that could raise yields when in 1983, during a drought that shortened the growing season, he and students at his agricultural school near Antsirabe serendipitously transplanted some young rice seedlings that were only 15 days old, along with seedlings that were the more common age, 30 days.

By the end of the season, the spindly little plants had become large and loaded with grain, despite the poor climatic conditions. Early transplanting was tried again the next year, also with good results, so Laulanié and his students set out even younger seedlings, only twelve, then ten and finally eight days old. They continued to have good results, leading to the crystallization of SRI as a set of principles and methods for raising rice yields, not as fixed technology or recipe (Laulanié, 1993). More background on SRI's development and on the results from SRI methods have been reported in Uphoff (1999), so they will not be detailed here.

IMPLICATIONS FOR SUSTAINABLE AGRICULTURE

SRI is still being evaluated as an approach to raising rice production, requiring only changes in plant, soil, water and nutrient management. Its principles always need to be tested in and adapted to varying environments as there is no set formula for achieving the higher yields SRI can produce. Rather, the *logic* of the different components should be assessed and applied in order to see how the genetic potential that exists in all varieties of rice can be better evoked. The aim is to manage plants, soil, water and nutrients so as to achieve greater root growth and greatly increased tillering, to obtain major increases in grain filling and yield. Soil aeration and biological activity in the soil appear more important than the gross amounts of nutrients in the soil, as biological processes subsume chemical components.

How generalizable SRI will be remains to be seen, but it has been used successfully in a wide variety of conditions in Madagascar and increasingly now in other countries, showing that the yields in Madagascar are not unique. The principles on which SRI rests are not ones that should apply only in certain environments. They are:

- transplanting seedlings very early and very carefully to avoid compromising their potential for maximum tillering and root growth;
- keeping the soil well drained and aerated to oxygenate the root zone;
- spacing the rice plants so that their roots are not in competition and have room to spread, and so that tillering is stimulated by ample sunlight, air and room for the canopy to expand;
- subjecting the plants periodically to some water stress so that their roots are induced to grow more deeply and densely; and
- providing nutrients in organic form to have slow release and to make the soil's structure more amenable to root growth.

Moreover, these and other principles may be applied to other crops, such as wheat.[20]

If the principles of SRI are correct, current efforts to genetically engineer a 'super-rice' may be unnecessary. This new kind of rice is planned to have relatively few tillers – ten to twelve if transplanted, four to five if direct-seeded – all of which are expected to be fertile with 200 to 250 grains of rice (Khush, 1996; Conway, 1997). The results from the experiments have apparently not been very successful to date (Khush et al, 1998).

The assumption behind the 'super-rice' research is that rice – including the improved varieties bred to date – confronts a biological ceiling limiting yields, so a different rice plant structure is necessary to obtain higher production. The observed ceiling, however, may not be a function of inherent genetic potential, but rather of the *conditions* under which rice has been grown, in particular in flooded fields.[21] With the full set of SRI management practices, rice plants have 50 to 80 tillers, and sometimes even over 100, producing 150 to 200 grains per fertile tiller, and even 250 to 300 if the soil, water and nutrients are well managed.

Water Requirements

The main technical limitation on SRI adoption that we see at present is the requirement of having enough water control to be able to keep rice field soils moist but never saturated during the growth phase, and then to maintain a thin layer of water on fields during the reproductive period. Most rice paddies in the world have been constructed to retain as much water as possible, rather than to be well drained. Except in very low-lying areas where standing water accumulates naturally, it should be possible to modify fields' construction to permit draining off excess water. If SRI methods can achieve substantial increases in production, the cost of making modifications to control and drain off water should be both justifiable and affordable with greater income.

Where fields are watered in a field-to-field cascade system rather than from canals, it will be necessary to coordinate operations; ensuring sufficiency of water is harder to manage than providing a continuous surplus. Giving up some land area to construct field channels that give water control for individual fields should more than pay for itself through the higher yields possible on the remaining area. In countries like China, raising yields is economically less important than reducing the water consumed by rice production.

Rice seedlings that are carefully transplanted – allowing only 15–30 minutes between uprooting from the nursery and setting into the field, with due attention to the positioning of the infant root, so that downward growth will be quickly resumed – establish root systems that can withstand more water stress than will plants transplanted by more typical methods. The vulnerability of rice seedlings in the first weeks after transplanting appears to be more a matter of how they are handled than of inherent fragility. But there does need to be enough water control to assure that seedlings will not be either drowned or desiccated in their first weeks.

Labour Requirements

As its name implies, SRI requires more labour than standard management, although the amount of labour required is reduced considerably once farmers are acquainted and comfortable with the methods of handling the plants, soil and water. For households so labour-constrained that they do not have enough workforce to cultivate their whole landholding with SRI methods, it would make sense for them to cultivate just part of their land using SRI methods since these give higher returns to both land and labour. This will yield more rice and leave some land available for other cropping when labour is available.[22]

So far, adoption of SRI by farmers, except where Tefy Saina or other promoters can work intensively with them, has been slow to spread, and around Ranomafana there was some disadoption when technical assistance was abruptly withdrawn (due to a changeover in US Agency for International Development projects). The lack of rapid spread is currently the strongest argument against taking SRI seriously. Its methods are counterintuitive and go against generations of practice, especially important in Madagascar where 'the ways of the ancestors' are revered in traditional religion. Still, farmers who have made SRI work for them are enthusiastic, and we expect it is a matter of time, and continuing work with farmers, before SRI increases national rice yields in a major way.[23] SRI is not presented as a package technology but rather as a set of principles that farmers are encouraged to experiment with to find the best combinations of practices for local conditions.

The research strategy of Father de Laulanié was to observe rice plants closely, seeking to learn from them under what conditions they will thrive, rather than trying to force or change them to do what he wanted. This approach will sound strange to researchers and practitioners who seek knowledge to better control and manipulate plants (or animals). Laulanié considered it more fruitful to seek to gain knowledge *from* plants than *about* them (Laulanié, 1993a).

The evaluation of SRI can and should be thoroughly empirical, systematizing through careful testing whatever is learned from observation and experience. This methodology did not derive from hypothesis formulation and testing, however, but from talking with farmers, testing different practices singly and in combination, assisted by some serendipitous events (not unknown in conventional science). Plant and soil scientists are invited to do more formal hypothesis-testing and evaluation so that whatever in SRI is broadly valid can be incorporated into the body of scientific knowledge that can assist and guide us in utilizing further the potentials of nature for human benefit. As we undertake to double global food production in the next 30 to 50 years, such advances in knowledge will be essential.[24]

NOTES

1   Knowledge that rice roots 'down-regulate' their uptake of nitrogen when plant tissues have sufficient nitrogen (Kirk and Bouldin 1991) has led, for example, to a suggestion by Ladha et al (1998, pp48–49) that this capacity be modified genetically to induce plants to take up more nitrogen, acknowledging that such interference with plants' functioning might have some adverse effects. Knowing that plants have sophisticated mechanisms regulating their nutrient acquisition could suggest instead that these capabilities be worked with, rather than blocked through genetic changes. SRI creates more favourable conditions for the uptake of nitrogen and other nutrients by plant roots. This is not to argue against continuing research to enhance genetic potentials, however, since some 'improved' varieties give the best SRI results.

2   The system was originally developed using chemical fertilizer to increase nutrient availability in the soil, and it can give good results with fertilizer. But usually the best yields with SRI have been obtained using compost or manure.

3   For example, E O Wilson (1998, p137) points out that the arrowleaf plant, whose leaves resemble arrowheads when it grows terrestrially, will have leaves shaped more like lily pads when it grows in shallow water. When it grows under submerged conditions, the leaves are still different, like elongated ribbons. The same genome thus produces very different biophysical results depending on the conditions under which the organism grows.

4   Tillers are the plant's upright stalks, some or many of which will flower and produce panicles filled with grains.

5   Ying et al (1998) write of 'a strong compensation mechanism' between the number of panicles (grain-bearing tillers) per plant and the size of panicles (number of grains per panicle); there is a 'strong negative relationship between the two components' that determine yield, they say (p72). Figure 2 in their article illustrates this assumed constraint by showing an inverse relationship between tillering and grain-filling. SRI methods give opposite results.

6   This is acknowledged in a memorandum from the International Rice Research Institute (IRRI), 29 August 1998, responding to papers that we had provided to IRRI on SRI. However, the issue of synergy that we raised in our papers was not addressed.

7   These trials were conducted on the west coast of Madagascar near Morondava by Jean de Dieu Rajaonarison from the Faculty of Agriculture of the University of Antananarivo during the minor season of 2000. Detailed analysis is being done for his thesis, but analysis of the basic data shows that SRI practices (young seedlings, one per hill, in fields with careful water management and using compost) raised yield 2.4 times with the high-yielding variety (2798) compared with standard practice (more mature seedlings, planted three per hill, with fields kept flooded, and using nitrogen-phosphorous-potassium fertilizer); the differential with a local variety (riz rouge) was 2.8 times.

There were 96 combinations of practices with three replications (total trials = 288). Because the spacing values were very similar (25x25cm and 30x30cm) – both are within the recommended SRI range and yielded no differences; both sets of 144 trials averaged 3.28t/ha – combining spacing trials gives *six* replications of each combination analysed below. This increases the reliability of the conclusions. Using *any one* of four SRI practices (spacing and weeding were not evaluated) raised yield an average of 0.5t/ha over standard practice. Using *any two* SRI practices added, on average, another 0.75t, as did moving up to *any three* SRI practices. However, using *all four* SRI practices together raised the average yield by almost two tons compared with using any three of the four practices, a demonstration of synergy (Rajaonarison, 2000).

Absolute yield levels reflect local growing conditions and varieties, so relative yield differences are what is important for such analysis. The trials showed consistent synergistic patterns, as did measurements of other plant growth factors (tillers per plant, length of panicle, grains per panicle, root depth and root density). Factorial trials were repeated in a different location, the high plateau rather than on the western coast, during 2001. They showed similar cumulative effects of SRI practices though not as strong synergistic effects in the somewhat better growing conditions. The full set of SRI practices produced an average yield of 10.3 t/ha (Andriakaja, 2001).

8   Ambatovaky, where 76 farmers used SRI during 1997–1998, is on the western side of Ranomafana National Park, where Tefy Saina and CIIFAD were working with farmers to raise lowland rice production so as to reduce slash-and-burn upland rice cultivation that encroaches on rain forest ecosystems.

9   Kirk and Bouldin (1991) note that the literature contains 'surprisingly little recorded information' on these questions. The key articles they cite (p197) on the details of rice root structure and functioning are from 1943 and 1953! They suggest that the disintegration of the root's cortex to form these *aerenchyma* – which they say is 'often almost total ... must surely impair the ability of the older part of the root to take up nutrients and convey them to the stele'. Yet they then refer to rice as 'well adapted' for growing in flooded soils, not considering whether these are the best growth conditions for rice.

10 Research by Ramasamy et al (1997) comparing rice yields in identical soils that were either flooded or well drained found yields across a number of different trials to be 10 to 25 per cent higher under the latter conditions. Since these trials were done with mature seedlings (25 days old) and close spacing (10cm x 15cm or 20cm), still greater differences could be obtained, we think, from the synergistic interactions among SRI practices. Ramasamy et al suggested that well-drained soils could have less effect of potassium or phosphorous deficiencies and/or that the improved root conditions, observed with well drained soils, could prolong the synthesis and improve the transport of cytokinins in the roots, in turn enhancing photosynthetic activity and leading to more translocation and deposition of carbo-hydrates in the grains.

11 Another benefit from not having standing water on rice fields is that less solar radiation is reflected, so that more is absorbed in the soil, raising soil temperature. This may not be important at low elevations in the tropics where ambient temper-atures are already high, but it could be important at higher altitudes and in temperate zones. Another benefit from SRI can be reduction in the production of methane gas from flooded rice fields, which add substantially to methane emissions on a global scale (Conway, 1997). Research at the University of Florida has found that periodic draining of soil for rice crops reduces methane emissions drastically (press release from USDA Agricultural Research Service, 5 May 2000), so SRI could also help reduce greenhouse gases in the atmosphere, though there would be offsetting increases in nitrous oxide (less with SRI practices that do not add nitro-gen fertilizer to fields).

12 A simple summary (proxy) measure of root development developed at IRRI was used: the amount of force required to pull a plant out of the ground under compa-rable soil conditions (Ekanayake et al, 1986). This represents the amount of surface contact and friction that a root system has with its surrounding soil, reflecting the total number, length and size of roots, many of which are tiny and hairlike. Joelibarison (1998) found that, on average, 28kg of force was needed to uproot a clump of *three* rice plants grown under typical submerged conditions, whereas it took an average of 53kg of force to pull up *single* rice plants grown with SRI methods. Such a large differential makes possible measurement errors or soil differ-ences not very significant.

13 Recall the research by Kar et al (1974) cited above that the roots of rice plants grown under flooded conditions begin degenerating after their first growth stage if they remained inundated.

14 Johnson (1994, p8) reflects this prevailing view among soil scientists when he states: 'The two principal soil fertility constraints [around Ranomafana] are low nutrient levels and soil acidity. *These constraints cannot be realistically managed by low-input technologies such as composting or even manuring* [emphasis added]. The nutrient-poor soils give rise to nutrient-poor plant residues and manure … The only viable strategies for producing sufficient agricultural yields are to use man-made fertilizers or to continue slash-and-burn practices.'

15 This proposition is presented in greater detail in Bunch (2001). Bunch was led to explore these possibilities by Primavesi's research, cited here, his own experience in the tropics and his acquaintance with SRI results.

16 When we analysed the agricultural practices and yields for 76 farmers using SRI methods around Ambatovaky during the 1997–1998 season (see Footnote 8 and Table 12.1), there were no statistically significant regression coefficients for the application of chemical fertilizer or manure relative to rice yields beyond the use of compost.

17   Research in the UK on the growth of ivy found that plants produced two-and-a-half to four times more biomass from the same total amount of soil nutrients when they were distributed 'patchily', ie unevenly within the growing box, because plants put down more roots where nutrient supply was richer. Plants were able to sense, and respond to, gradients and spatial distribution of nutrients with unexpected, and beneficial, growth strategies, contradicting the conventional wisdom that homogenizing soil nutrients produces the best results (Gail, 2000; Wijesinghe and Hutchings, 1999; Hodge et al, 1999). I thank Gordon Conway for calling this research to my attention.

18   SRI rice is, of course, no less vulnerable than other rice to gross pest attacks, such as from hordes of locusts. But SRI methods have helped withstand even these attacks. Tefy Saina reports one farmer working with them in Mahabo, near Morandava, who had her newly-transplanted SRI crop eaten down to the ground by locusts in the 1997–1998 season, when locusts were an island-wide plague. Since new seedlings can be grown in just 12–15 days for SRI, she replanted her field, but locusts demolished the crop once more. She gave up hope of any harvest that year. Yet a month later she found a normal crop growing there; a whole stand of rice had restarted from the surviving roots that she had carefully laid into the ground. Similar experience is reported from the Marovoay region in Madagascar (Patrick Vallois, personal communication, 29 May 1999).

19   For example, Wightman (1998) cites research on protection against the Mexican bean beetle which showed how soybean plants work cooperatively to resist insect attack by producing antibiotic insect-resistance chemicals that are triggered by reaction to a substance in the saliva of the insect attacking them. Plants can communicate chemically with each other (a) that they have been attacked, and (b) that neighbouring plants should activate their resistance factors. 'Injured tomato plants release a chemical called methyl jasmonate that switches on the genes that produce the chemicals that repel caterpillars. It is volatile and, when released from one plant, quickly prepares all the neighbouring plants for attack.'

  Wightman also reports a fascinating biological response, known since 1990, whereby chemicals released from plants that have been attacked by the beet armyworm (*Spodoptera exigua*) attract beneficial parasites (*Cotesis marginiventris*) to come and infest the attacking pest. Another example cited by Wightman of plants being active in their own self-defence and maintenance of health is aphids' sending out a 'wound signal' that stimulates the production of a chemical that attracts predators. A demonstration that roots can operate as active agents rather than as passive recipients is the identification by plant and soil scientists of molecules, called siderophores, that are exuded from plant roots to facilitate the acquisition of iron ions from the soil.

20   A raised-bed system for growing wheat developed and now widely used by farmers in the Yaqui valley of Sonora, Mexico, has similarities with SRI, including wide spacing and reduced water application, using furrow rather than flood irrigation. Reduced costs for labour, agrochemicals and seeds mean that along with a 14 per cent increase in yields, profit per hectare is increased by 117 per cent compared with standard modern methods (Sayre and Moreño, 1997). With this system, seeding rates of 15–25kg/ha produce yields as good as seeding rates of 200kg/ha (p6). This is consistent with the SRI observation that fewer plants having more space to obtain light, air and nutrients – and more room to grow – outperform larger numbers of plants. One study supported by the French development agency found that a low SRI seeding rate of 7kg/ha, compared with 107kg/ha used in

conventional cultivation on the Madagascar high plateau, gave twice as much yield when combined with other SRI practices (MADR/ATS, 1996).

21   Some of the practices being used to develop the 'super-rice' are the same as those used with SRI. The trials are being conducted with 14-day-old seedlings, planted singly, and spaced 25cm by 25cm, though still usually with flooded soil and heavy fertilizer applications (Khush and Peng 1996, p39). One can speculate how much of any increased yield achieved with this 'super-rice' should be attributed to management practices rather than to genetic changes. With such young seedlings, we would expect greater productivity than has been observed in IRRI's trials to date. However, plant density is still very high – the seeding rate of 120kg/ha is 10–20 times greater than with SRI – and the plots are kept flooded during the growth phase. In our experience, these two practices would inhibit much or most of the productive potential created by early transplanting.

22   Bonlieu (1998) in his study area on the west coast of Madagascar found that SRI methods required about 500 hours of additional labour per hectare. At prevailing agricultural wage rates, this added cost of production could be recouped by selling just 500 additional kg/ha at harvest time, or just 250kg/ha three months later when the market price was higher. Since SRI raised yields 2t/ha on average for farmers in his study, additional labour investments were well repaid.

23   CIIFAD has supported research on adoption and disadoption of SRI. A study of Christine Moser found that labour constraints were considered very significant by farmers who had not adopted SRI or had given it up, with water control a less important consideration. The poorest farmers found it most difficult to adopt SRI because they needed to undertake income-earning activities at the start of the planting season (end of the 'hunger' season). Moser found disadoption in the community near Antsirabe which she surveyed. However, in that region, within French-assisted small-scale irrigation systems, use of SRI methods between 1994–1995 and 1998–1999 went from 12ha to 476ha (Hirsch, 2000).

24   The analysis here is based on the work of Association Tefy Saina, an NGO in Madagascar, together with two universities and now the government's agency for agricultural research, FoFiFa. Tefy Saina has been promoting SRI in a number of locations since 1990, maintaining records on the practices and production of those farmers who experiment with SRI and adapt its methods to their own conditions. Most knowledge about SRI comes from Tefy Saina, particularly its president Sebastien Rafaralahy, its secretary Justin Rabenandrasana, and staff including Ludovic Naivohanitrinianina and Edmond Rataminjanahary. The École Supérieure des Sciences Agronomiques (ESSA) at the University of Antananarivo has supported field studies to evaluate and analyse SRI. Special thanks go to Professor Robert Randriamiharisoa, ESSA director of research, to Joelibarison and William Rakotomalala, who did thesis research on the agronomics and economics of SRI during 1997–1998, and to Jean de Dieu Rajaonarison and Andry Heritiana Andriakaja, who did the factorial trials reported in note 7. The Cornell International Institute for Food, Agriculture and Development (CIIFAD) has been working with Tefy Saina since 1994 to introduce SRI around Ranomafana National Park where an integrated conservation and development project funded by the US Agency for International Development has sought to protect biodiversity, and now in the region around Moramanga. CIIFAD's agricultural advisor in Madagascar, Glenn Lines, has played a key role in coordinating these efforts and helping to develop insights into SRI. Erick Fernandes at Cornell has helped to link SRI insights to the agronomic literature.

Chapter 13

Increasing Productivity through Agroecological Approaches in Central America: Experiences from Hillside Agriculture

Roland Bunch

The experiences of two agricultural development programmes, one in Guatemala and another in Honduras, indicate the potential of agroecological approaches in Central America. Each programme did a baseline survey at the time of its initiation, with annual evaluations done each year during their eight or nine years of operation, including the measurement of crop yield levels. When an evaluation was done in 1994 to assess the programmes' impact five to fifteen years after their termination, these studies showed many sustainable improvements in basic grain yields and in other agronomic and economic indicators, as well as in such factors as educational levels, local organization and leadership abilities within the communities involved. Most important, we saw a continuous process of experimentation and innovation that changed farming systems.

Yields of maize in the four villages evaluated in the San Martin Jilotepeque area of Guatemala increased during the programme's tenure from 400kg/ha in 1972 to nearly 2.5t/ha in 1979, and nearly doubled again, to 4.5t/ha by 1994. In the three villages assessed in Guinope, Honduras, the average yields were 600kg/ha at programme initiation in 1981, 2.4t/ha at programme termination in 1989, and 2.7t/ha in 1994. Thus, between the initiation of these programmes and their evaluation in 1994, yields increased by a factor of 11 in San Martin and a factor of four in Guinope, all within 22 years in the first case and 13 years in the second (Bunch and López, 1995).

In the case of beans, the yields started at even lower levels and increased as dramatically. In San Martin, in those villages where beans were cultivated both before and after the programme, bean yields rose from 170kg/ha to more than 1.53t/ha, a nine-fold increase in 22 years, while in Guinope they increased eight times, from 100kg/ha to 800kg/ha over 13 years.

Contradicting the argument that low-external-input agriculture cannot generate significant increases in yields among small farmers, these yields were all produced with no or very low levels of chemical inputs. No herbicides are used in either case, and the use of insecticides and fungicides on maize and beans is virtually unknown in the two townships. Small-scale farmers in San Martin use very modest amounts of chemical fertilizer (the alternatives to chemical fertilizer that can now be recommended were not known in the early 1970s), while those in Guinope use virtually no chemical fertilizer on either maize or beans.

These programmes achieved a number of additional impacts worth noting. First of all, the increases in productivity after programme termination cannot be explained by the technologies introduced by these programmes. Most of the after-intervention gains are attributable to a variety of technologies that the villagers developed by themselves *after* programme termination. Thus, the sustainability of impacts does not (and cannot ever) come from just choosing a good technology. Rather, it comes from the creation of a social process in which villagers are the protagonists of their own development.

This dynamic process includes, among other things, farmers analysing their major problems, searching for solutions, experimenting with the most promising of these, and then adopting, modifying and disseminating those solutions that prove to be the most useful. Sustainability depends not on any specific technological package, but rather on a process of innovation that results from engaging farmers as active agents for development.

Other impacts of the programmes include a 90 per cent reduction in temporary emigration from San Martin. In Guinope, which suffered from a heavy net outmigration before the programme started, there was a net in-flow of people returning from city slums or from other rural districts by the end of the programme. There were also increased educational levels, more organization within the communities, and greater innovation in agriculture, including the development of entirely new systems of production such as intensive cattle-raising and the use of fruit trees to shade coffee, technologies which were not part of the original programme.

The most dramatic result of these programmes, however, especially in San Martin, is the number and the quality of farmer leaders trained. Some 23 villagers from San Martin alone, all with less than six years of elementary-level education when they were children, have been hired to fill a total of 63 different positions (as agricultural extensionist or higher) by 31 different development agencies working in five countries. Twenty-two of these positions have been as programme directors, three as national directors of agriculture for non-governmental organizations (NGOs), and three others as international consultants.

In each case, these organizations hired these persons as staff because of the quality of their work and, in most cases, because of a desire to learn either the technology or the extension methodology that these people had mastered in San Martin. In many cases, additional villager-extensionists were employed after the NGO or government agency had seen the results achieved by the first villager it hired from San Martin. This evidence of leadership ability and technical know-how achieved among villagers trained in a such programme shows that such development programmes need to go beyond the usual concept of 'participation' and to embrace the concepts of villager protagonism and community empowerment.

These programmes provide evidence that agroecological techniques, even those available ten years ago are capable of greatly increasing the supply of food. With newer, more promising agroecological technologies such as green manures and cover crops (GMCCs), microharvesting and recycling of water, natural pest control and dispersed tree systems, the potential is even greater.

THE SAN MARTIN PROGRAMME IN GUATEMALA

The programme in the San Martin Jilotepeque township was initiated in 1972 by the NGO World Neighbours with funding from OXFAM/UK; it was closed, according to plan, in 1979 (Bunch, 1977; Bunch and López, 1999). This was an integrated rural development programme, with the agricultural component being one of its two largest focuses. A major earthquake in 1976 interrupted its planned activities for a year and a half. Thus the agricultural work with farmers lasted only for six to seven years.

San Martin is an area with extreme disparity in land tenure, with a few very large farms that are quite underutilized, surrounded by a huge number of extremely small, intensively cultivated farms. The average farm size among those with whom the programme worked was under 0.5ha. The programme worked mostly in some 45 villages in the southern half of the San Martin Jilotepeque township, an area with a population of about 28,000 people. The elevation of the area varies from about 800–2000m above sea level. In most years, the rainfall supported a good crop of maize, although the soils, of volcanic origin and naturally fairly fertile, were heavily eroded and deteriorated. Slopes generally varied from 15 to 35 per cent. Maize yields in 1971 averaged just 400kg/ha, according to the baseline survey done.

The local population – predominantly Cakchiquel Indians – was unable to produce more than a fraction of the maize and beans needed for subsistence, so most of the people were forced to spend two or three months each year on the malaria-infested South Coast, working on huge coffee, sugar and cotton plantations.

The programme in San Martin began its work with an emphasis on just two simple innovations: the use of urea as a nitrogen side-dressing on maize, and the construction of contour ditches and contour napier grass hedgerows to control erosion. These two innovations were chosen because soil fertility – especially nitrogen availability – was an obvious limiting factor for productivity in the area. Previous experience in and near San Martin had shown that these two innovations, by themselves, could at least double maize yields. Actually, it was the side-dressing that doubled productivity, while the contour barriers held in place the nitrogen and the small amount of remaining topsoil on the hillsides. Subsequent technologies of major importance were increasing plant populations on cultivated fields, natural and chemical control of insects and diseases, the use of various sources of organic matter to increase soil fertility, and a rotation of maize, beans and wheat.

THE GUINOPE PROGRAMME IN HONDURAS

This programme, based in the department of El Paraiso, Honduras, began work in 1981 and ended in 1989 (Bunch, 1988). It was funded and managed by World Neighbours, together with another NGO and the Honduran Ministry of Natural Resources. The programme, which worked in 41 villages in the contiguous townships of Guinope, San Lucas and San Antonio de Flores, concentrated on agriculture and health. Its leading agricultural technologies were the installation of drainage ditches (constructed at a 0.5 per cent slope) with contour live barriers, and the application of commercial chicken manure. Later innovations included improved bean production, the application of other sources of organic matter including GMCCs, cattle manure and compost and commercial vegetable growing and marketing.

The Guinope area climate is fairly similar to that of San Martin, although the slopes, which average approximately 15 per cent, are somewhat less severe, and the landholdings, averaging about 2.5ha per household, are larger. Nevertheless, an impenetrable subsoil underlies Guinope's topsoil at a depth of 15 to 50cms. When this thin layer of topsoil erodes away, traditional agriculture becomes impossible. Before 1981, emigration from the Guinope area was heavy. Some residents referred to it as a 'dying town'.

PROGRAMME RESULTS

Both programmes did baseline surveys before beginning, gathering data on agricultural productivity and crops produced, followed by annual checks on productivity each year of their operation. According to data collected at the end of the San Martin programme, some 4000 families (representing approximately 20,000 people) had made contour ditches and hedgerows, and an equal number had improved their fertilizer applications during the programme's operation. The total cost of the agricultural side of the programme came to less than US$50 (in 1979 dollars) for each family that had at least tripled its basic grain production.

In Guinope, a similar analysis showed that over 1500 families (approximately 9000 people) had at least tripled their maize and bean productivity during the life of the programme. The total cost per family, again counting only those who had tripled their productivity, was US$300 per family, at 1990 prices.

In 1994, the Honduran NGO COSECHA did a detailed study, financed by the International Institute for Environment and Development (IIED) in the UK, covering four villages in San Martin and four in the Guinope area. This study sought to determine whether the impact of the programmes had been sustainable, and which of the technologies were still in use. The study was conducted five years after the conclusion of the Guinope programme and 15 years after the San Martin programme terminated (Bunch and López, 1995). Although the 1994 study documented considerable beneficial long-term impacts on a whole series of variables, from land prices and daily wages to return migration, organizational development and educational levels of the farm population, the discussion here addresses observed trends in the performance of agricultural technologies used and the resulting levels of food productivity.

In neither evaluation were the studied villages selected at random. In San Martin, the villages were chosen because they were where the four people carrying out the study (employees of COSECHA) were born and could, we expected, get the most cooperation and most reliable answers. This could cause some bias in favour of more successful villages, inasmuch as those villages with the best leaders could be ones where the programme had better results, but no criterion related to programme success in each village was used for selection.

In the case of Guinope, the 41 villages that had been worked in were divided into three equal-sized groups: high programme impact, medium impact and low impact. COSECHA selected one village from the first group, two from the second and one from the third to study in depth. Within these groupings, the villages chosen were ones where little or no development work had been done by any other organizations between 1989 and 1994, and ones that had the most geographic distance between them and thus maximum variation in altitude and access to markets.

Productivity of Maize

In Central America, maize is the basic staple grown by the vast majority of poorer farmers. Its widespread importance as a subsistence crop and its almost universal presence on small farms make it a good measure of food productivity in general. Its sensitivity to soil fertility and moisture levels makes it also a good indicator of conditions important for the sustainability of agricultural productivity.

The average harvests listed below are rounded off to the nearest 100kg/ha. This is done because they are figures based on the number of bags of maize that farmers carried to their homes after harvest. They do not include grain lost to thievery, ears harvested earlier for eating on the cob or the occasional ears given to labourers as partial compensation for their work. The data in Table 13.1 represent, in each case, production levels at the time of programme initiation taken from the baseline survey; when the programme terminated; and then in 1994, when the follow-up evaluation study was done.

Major increases in yield were obviously achieved, both during and after the programmes' operation. In Guinope there were no after-programme increases in two villages, perhaps because of the shorter time period between termination and the 1994 study. There were no cases of village-wide post-programme decreases in productivity.

Table 13.1 *Changes in Yields of Maize (100kgs/ha)*

| | Guatemala: San Martin Jilotepeque | | | | |
	San Antonio Cornejo	Las Venturas	Xesuj	Pacoj	Average for 4 villages
Initiation (1972)	4	3	3	5	4
Termination (1979)	<24	<28	<20	<28	<25
Evaluation (1994)	48	52	32	48	45

| | | | | | Both countries |
| | | Honduras: Guinope area | | Average for 3 villages | Average for 7 villages |
	Pacayas	Manzaragua	Lavanderos		
Initiation (1981)	6	6	6	6	4.9
Termination (1989)	32	20	20	24	<24.6
Evaluation (1994)	42	20	20	27.3	37.4

Productivity of Beans

Beans (*Phaseolus vulgaris*), the second most important subsistence crop in Central America, were not grown in all the villages studied. Production data obtained from four villages where beans are produced are shown in Table 13.2.

In neither case were there any significant agricultural development programmes operating in these villages during the years between programme termination and 1994. In San Martin, the Guatemalan civil war had gone on during those years. The violence was so intense that many people in the studied villages were killed, and most of the others emigrated to nearby inaccessible mountainous areas for several years, unable to plant any crops at all on their own lands. In Honduras the villages studied were to some extent selected in order to avoid any 'contamination' caused by contact with other programmes, although one other agricultural programme worked some in Pacayas during the ensuing years.[1]

Even though the yields from maize and beans increased dramatically, for both crops by over seven times per hectare, the total production of these crops did not increase by more than two to three times (except in the case of beans in Guinope, where they were a good cash crop in the early 1990s because of crop failures elsewhere in Honduras). Once farmers could produce enough

Table 13.2 *Changes in Yields of Beans (100kgs/ha)*

	San Antonio Cornejo	Las Venturas	Xesuj	Lavanderos	Average for 4 villages
Initiation (1972)	2	1	2	1 (1981)	1.5
Termination (1979)	12	2	12	8 (1989)	8.5
Evaluation (1994)	18	8	20	8	13.5

staples to meet the consumption needs of their own households, they reduced the amount of land dedicated to these crops in favour of more economically advantageous commercial crops, mostly vegetables, coffee and fruit.

The Technologies Used

Our study differed from many others in that it did not examine the impact of one or two distinct technologies or types of technology on farm productivity and economics. It is, rather, a study of overall increases of productivity of thousands of farm families associated with their increasing use of a whole range of low-external-input agricultural technologies. While the two programmes studied have been characterized as concerned with soil conservation, they were integrated development programmes for which soil conservation was only one aspect. Of the dozens of agricultural technologies introduced by these programmes, most were not for soil conservation. Many of these technologies were adopted, but still more technologies were developed and adopted after programme termination.

Since the results of these programmes are not attributable to any one technology or one kind of technology, it is not possible or relevant to analyse the specific economic costs and benefits of any one technology or group of technologies. Rather, one should consider whole sets of new practices or technologies. Our concern here is not primarily with the amount of benefit that accrued to producers, but rather the increased amount of food becoming available for others. The key consideration is not the economic benefits for producers, though most technologies that give increased yields also bring increased economic returns. If this does not occur, farmers, being economically rational, will not use these technologies.

The specific technologies used by the end of the two programmes varied considerably between them, and also among the different villages within each programme area. In addition to the practices and products promoted by the project, even more came from suggestions of programme personnel that were picked up and elaborated by farmers or from farmers' own technological innovation. Some idea of the variety of such technologies has already been given in Chapter 4 (page 54). Large numbers of farmers in San Martin at various times raised broccoli, cauliflower, potatoes, tomatoes and other vegetables, coffee, half a dozen species of fruit and cattle. To grow these crops and animals, they employed technologies ranging from strip cropping, the abandonment of agricultural burning and fallowing and crop rotations to low-input pest control and increased applications of all sorts of organic matter. In Guinope, activities that did not come from the programme included the cultivation of carrots, peaches, sorghum and corn-on-the-cob. Innovations that facilitated this new production included the use of green manures and cover crops, crop rotations, half a dozen new plant species used to stabilize contour barriers that conserved soil, several technologies for natural insect and pest control and an irrigation project costing US$150,000 (Bunch and López, 1995).

In fact, in San Martin, where the post-programme period was longer, whole new systems of production have been developed, including an economically very profitable system for growing coffee in the shade of fruit trees, an intensive cattle-raising system with as many as five animals/ha and a sustainable forestry production system.

Most of the few programme-promoted technologies that did last through 1994 were changed almost beyond recognition. In Guinope farmers who had very few animals pulled up the grass that formed their vegetative contour barriers and planted in its place sugarcane, fruit trees and medicinal plants, nevertheless making sure the arrangements of the new species would continue to control water erosion (López et al, nd).

Most of the technologies promoted by the programmes tended over time either to be abandoned – as markets changed, new technologies made older ones obsolete or conditions changed – or to be modified. Thus, the sustainability of programme impact was not to be found in the technologies introduced. Rather, what was lasting and what allowed villagers to go on raising yields and incomes after the programmes ended was *the social process of continuing innovation*, of analysing problems and experimenting with possible solutions. The sustainability of agricultural development is not to be found in any particular technologies promoted, but rather in the processes, motivations and capabilities that are created and nurtured.

This book project was undertaken to consider whether and how food productivity can be increased sufficiently to meet future generations' needs, and particularly whether such a increase could be achieved with low levels of agricultural inputs. In answer to the first question, experience in Central America suggests definitely 'yes'. Increases of over seven times in food productivity would more than meet population increases in most places in the world. With such increases, producers can feed themselves and their families and have a surplus available to market.

Would such increases be possible in other situations? Obviously, the potential for increasing food production depends to a great extent on present levels of productivity. In highly productive irrigated areas, even doubling present yields will, in many cases, be virtually impossible, with or without a higher levels of inputs. But in areas of rainfed farming, where present productivity of staple grains is low, often below 1t/ha, a tripling of production is within the possibilities of poorer farmers in a large majority of cases, even in environments where soil acidity or water stress, rather than nutrient levels, are the limiting factors, as discussed below. Chapter 11 gives an example of this.

Answering the second question – can such yield increases be achieved with a low level of external inputs? – is more complicated. The technologies that were used to improve maize and bean productivity in the two cases here focused on how to increase soil fertility over time. In San Martin, small to moderate amounts of chemical fertilizers were being used even before the programme's inception in 1972, and they are still being used now. Nevertheless, they are now supplemented by large amounts of organic matter from a variety of sources. Some farmers use large quantities of commercial chicken manure, while virtually all use crop residues and some use compost or animal manure obtained through barter. A thriving barter market has developed in some villages in which abundant napier grass from programme-introduced contour barriers is provided to cattle ranchers in exchange for the manure of the animals for which the fodder is being provided. Organic inputs to enhance soil fertility are well understood and sought after in San Martin.

In Guinope, virtually no farmers in the programme area used chemical fertilizer until recently. The most common source of fertility enhancement has been commercial chicken manure, followed by crop residues, GMCCs and on-farm sources of animal manure. However, since 1994 heavy demand for chicken manure has forced its price to rise considerably, causing some farmers to begin using chemical fertilizer, while others are finding ways to utilize GMCCs with less cost.

In one comparison made during the 1994 study, farmers in one San Martin village were harvesting an average of 4.4t/ha of maize, while farmers in a nearby village outside the programme area were harvesting an average of less than 2t/ha – while using three times more chemical fertilizer per hectare as in the programme village. Many variables always impinge on productivity, so more needs to be considered than just the amounts of fertilizer used. But the comparison does indicate that conventional amounts of fertilizer may well be associated with lower yields than are obtained from so-called 'low-input' systems. In virtually all villages in the two programme areas, the amount of chemical fertilizer used by farmers participating in the programmes, even to achieve maize yields of over 4.5t/ha, was less than 400kg/ha per year, well below usually recommended levels.

The use of pesticides also varies considerably from one area to another and between villages. None of the farmers studied in either programme area reported using herbicides. This may be partly due to the relatively cooler climate in both areas; but in Guinope it is also partly attributable to the fact that farmers know that the use of GMCCs, together with normal hand tillage, can control weeds adequately and at less cost than by using herbicides. Certain GMCC species are being used in some villages in Guinope to control nematodes.

Because the programmes found and promoted alternative, non-chemical methods for controlling the major pest problems in maize (white grubs and corn borer), no insecticides or fungicides are used for this crop. However, quite a lot of insecticides and fungicides are used with commercial vegetable production in both programme areas. Many farmers see this as a serious problem, especially in Guinope, and some farmers in both areas continue to search actively for less toxic solutions to these problems, although the search advances slowly, on an insect-by-insect and disease-by-disease basis.

Thus, the remarkable increases in both maize and bean production in both areas have been accomplished with little use of chemical inputs. It can be expected that productivity will continue to increase in both townships. I think, based on some scattered evidence, that over the next 10 to 20 years there will be greater increases in Guinope than in the five years prior to 1994, where increases during that period were rather low compared with San Martin. In the Guatemalan programme, farmers may be approaching levels of diminishing returns in productivity, so that future increases will come more slowly. In neither case is there any apparent reason to expect that productivity will be significantly reduced in the future, barring major natural catastrophe. Even Hurricane Mitch in 1999, the worse storm in decades, has not caused any lasting decrease in production in Guinope.

Ecological Impact

The ecological impact of most of the technologies introduced or developed has quite evidently been positive. Overall, soil erosion has been greatly curtailed; run-off has been greatly reduced; resistance to damage from drought and storms has increased; soil fertility and organic matter content have been enhanced dramatically; and the application of chemicals has been held to a minimum. Three factors that have historically increased forest destruction in these areas – shifting agriculture, the need for fallowing and the use of large areas to produce crops with very small yields – have largely been eliminated, allowing for the survival of significant forested areas in both townships.

An exception to this benign pattern is the increasing spread of commercial vegetable production in a few Guinope villages that have good access to roads and markets. This production has resulted in increased deforestation and in the abundant use of chemical pesticides in these communities. We have no idea how the deforestation there can be reduced, but we can hope that as farmers experiment more with non-toxic methods of controlling insects, the use of chemicals will subside. Except where a few relatively large-scale vegetable producers are operating in Guinope, the vitality of ecosystems in both townships has improved dramatically, and in many different ways, while food security and overall food productivity have also increased substantially.

Farmer Involvement

The methodology used in these programmes has been described previously (Bunch, 1982) and was summarized in Chapter 4. As suggested in the preceding discussion, this process involves villagers actively in the generation and dissemination of agricultural technologies – as farmer-experimenters and as villager-extensionists – to the point where they become the main protagonists in their own agricultural development process. The entire process is carried out in such a way that, gradually, over a period of six to ten years, the villagers become empowered and as the programme fades away, those activities that are necessary to carry on the process are taken over by a movement of, by and eminently for the villager-farmers themselves.

CONCLUSIONS

These cases provide evidence that food productivity can be increased greatly, by several hundred per cent. These large increases in productivity have been achieved with only meagre applications of chemical fertilizers, very few pesticides (except on commercial vegetables) and no herbicides. Similar yield increases should be possible among most traditional small-scale farmers around the developing world if they manage their soil, water, plants and animals in a more integrated and synergistic manner.

Larger farmers and 'modern' agriculture will have a role to play in the future, but they are likely to encounter serious energy and water constraints in the decades ahead. There is substantial potential in the people and the agro-ecosystems of so-called marginal areas to meet food needs in both rural and urban areas. As world markets expand, small farmers with supportive policies, services and organization should be able to match the agricultural expansion of the past 30 years provided that their prices are remunerative.

NOTE

1   This was an agricultural extension programme operated by the PanAmerican School of Agriculture at Zamorano.

Chapter 14

Raising Smallholder Crop and Livestock Production in Andean Mountain Regions

Edward D Ruddell

The high mountain regions of Peru, Bolivia and Ecuador represent some of the most difficult areas in the world for growing crops, due to adverse temperature, soil and topographic conditions. The situation is made more difficult by a lack of infrastructure and high rates of illiteracy that reflect generations of discrimination against the indigenous population in highland villages. These people are economically some of the poorest to be found anywhere, though they are richly endowed with cultural resources and traditions.

PERU

Half of Peru's 12 million inhabitants live in extreme poverty. Most people in the very mountainous departments of Apurimac, Ayacucho and Huancavelica in the southeastern part of the country live at elevations ranging from 2000–3800m, on hillsides with very limited productive capacity. The farming practices employed result in major losses of top soil and a reduction in the organic matter in what soil remains. The climate is variable and quite irregular, with most rainfall occurring between October and April.

Violence in the countryside during the 1980s, which continued into the 1990s, forced many rural people to migrate to urban areas. Some of these have now returned but migrate on a seasonal basis during the dry season. Indigenous forms of peasant organization called *ayni* and *minka* continue to operate in many communities, raising food on communal land. The most important crops are potatoes, barley and corn.

Initial Improvements in Pasture and Livestock Production

In 1970, World Neighbours, a non-governmental organization (NGO) based in the United States, was invited to assist the staff of an international development organization in the department of Ayacucho to disseminate technologies that its ten-year research programme had shown in experimental trials could increase local pasture production by up to 500 per cent. World Neighbours was chosen because its approach was to help indigenous local leaders promote their own people-centred development.

During the next several years, the community leadership training efforts began to flourish. As pasture forage doubled, tripled and quadrupled, the communal livestock herd began to increase in size for the first time. This increased the quantity of meat, the quality of wool and the supply of animal manure, which in turn was used to improve the production of potatoes, the principal food staple of the area.

Next, improved breeds of sheep were introduced, which led to a 50 to 100 per cent increase in animal size. The local people were pleased, and local field-days conducted in the Quechua language soon drew hundreds of men and women farmers from a dozen villages surrounding the pilot centre. Indicative of economic benefits from the new production methods was the fact that 80 per cent of the families in the first community to participate in the programme were able to purchase land on the outskirts of the city of Ayacucho. They began to construct their own adobe homes there, and close proximity to town allowed them to send their children to secondary school. The development organization marvelled that the annual budget for this programme was only US$15,000, whereas its budget for the country was US$800,000 or more per year.

Identifying Barriers to Broader Dissemination

Four years after the start of this programme, World Neighbours sought to expand these benefits beyond the 400 families in 12 communities. We invited Roland Bunch from World Neighbours' Central American programme to assist our Andean staff to find effective ways to accelerate the dissemination process. Bunch's first observation was that the technology used for improving pastures required imported seed. This was seldom available in local stores and generally cost more than most farmers could afford, unless there was a subsidy from the international agency.

A second barrier to dissemination of these practices was the need for tractors. The operational costs of these machines was being subsidized by the international agency since isolated indigenous communities could not otherwise pay for the use of tractors. As the demand for subsidized tractor services grew, the agency was not able to purchase and operate enough tractors to provide timely services to small farmers distributed throughout this mountainous area.

A third barrier identified by Bunch was the difficulty of teaching farmers new pasture management methods. Overgrazing was considered normal, so it was not uncommon to see new improved pasture destroyed within the first year. Once the first demonstration trials failed to survive, what was disseminated was the observation of failure.

Simple Means for Overcoming Limiting Factors

The barriers to dissemination appeared so formidable that programme staff nearly lost hope that the technology could be replicated on a larger scale. Then one day while travelling in a neighbouring department where the development agency had not yet worked, Bunch observed improved pastures. However, the clover was not uniformly distributed; it was found in clumps. Bunch stopped the car so he could ask a local farmer how these improved pasture practices were introduced.

This farmer had observed the good forage on improved pastures planted by large farmers elsewhere. However, he was unable to acquire the expensive seed, and he had no tractor to prepare the seed bed for planting it. So he decided to graze his cows along the road that passed by the improved pastures when the clover plants began to flower and produce seed. When he brought his cows back to his own unimproved pastures, the seed passed through their digestive tracts and were deposited on the ground, well saturated and fertilized. A year later, clumps of clover pasture could be found throughout his small corrals. A broad smile broke across the weathered face of this Quechua farmer as he reported that the process was automatic. It eliminated the need for expensive tractors and reduced the destructive grazing of new plants, thanks to the presence of other more mature plants.

When programme staff began to share this idea with small farmers in the department of Huancavelica, they discovered that clover plants did not produce seed in many areas of this department, however, because of its higher elevation. This problem was eventually overcome by planting clover nurseries at lower elevations. When the plants were mature, children from interested families transplanted them into unimproved pastures by digging holes and sticking them in the ground while they were grazing their animals.

Efforts to replicate the results of this programme led to closer collaboration with the Ministry of Agriculture, which selected 500 leaders from the region to attend a field-day organized by the programme. There, farmer-promoters enthusiastically shared the stock of information they had gained about how to improve pasture and livestock production during the previous five years. At the end of the event, participants all received a packet of seed to take back for planting in their communities.

Improving Potato Production through Farmer Experimentation

After identifying, testing and disseminating appropriate technologies for improving livestock production, indigenous leaders expressed interest in evaluating new potato varieties that might improve their tuber production. Aware that the spark behind the success of this programme was local enthusiasm for testing and disseminating new ideas, the programme coordinator agreed to help acquire several new varieties developed by plant breeders. Farmers promised to plant them alongside their native varieties in demonstration trials and then to organize field-days at harvest time to share the results with their neighbours.

These field-days occurred several months later when the potatoes were harvested. At this first set of demonstration trials, the potatoes harvested were ranked according to their yields. Everyone was impressed to find that some of the new varieties surpassed the production of native varieties by as much as 300 per cent.

However, the next day another field-day was conducted by another Quechua farmer at his demonstration plots, reached by a 45-minute drive up the mountain road. These trials included the same four varieties shown on the first day. However, the varietal ranking was exactly the reverse in this location. It looked like something had gone dramatically wrong with the trials until the programme coordinator explained that the difference in performance was quite likely caused by the change in elevation. This confirmed the merit, indeed the necessity, of doing site-specific experimentation, especially in such varied conditions as in mountains.

At this time, the Ministry of Agriculture was only recommending the use of new varieties of potatoes in the area. These varieties were indeed well suited for lower elevations. However, some of the best soils and climates for potato production were at higher elevations, where using 'improved' varieties could actually diminish production. No wonder indigenous farmers distrusted the technical recommendations of the ministry.

Combating Barley Rust

Four years after the programme had begun promoting the improvement of pasture, livestock and potato production, an international seminar was organized to share the methodology and results with interested agencies and individuals. The seminar attracted 20 participants from six programmes working in four countries. It started with the participants visiting the pilot centre to enquire from peasants there what were their most pressing needs for improving their agricultural production. Both the men's and women's groups reported that their greatest problem was no longer pasture and livestock production, but instead combating the rust disease that was destroying their barley. While the stomachs of their livestock were now protected from hunger and starvation, they said that they themselves were in grave danger, because barley was the only crop they could store for more than one year. If drought, severe frost or a hailstorm destroyed their annual potato crop, they would suffer famine if their barley crop also failed because of rust.

Initially, the programme coordinator was disinclined the change the focus of the programme, having been trained for improving pasture and livestock production. The only knowledge he possessed about barley production had been learned in a university classroom. We assured him, however, that farmers were less likely to abandon a programme that was trying, even unsuccessfully, to solve an urgent problem than one that did not seek a solution to what they considered their most critical need. Fortunately the coordinator had heard that the national brewery in Lima, Peru, was working with plant breeders at the Universidad de la Molina to develop a new rust-resistant variety of barley called Zapata. After much effort he secured a sample of this variety for several demonstration trials.

The results of these trials were dramatic. Zapata variety was not only rust-resistant; it doubled production with traditional production practices and matured one month earlier than the native variety, reducing the danger of crop loss due to drought or hailstorms. Before long, small farmers were walking as much as 16 hours round-trip to get assistance in barley production. When the programme did not have enough funds to hire another agronomist and purchase a second vehicle, farmer-promoters offered to help.

With a small stipend for doing part-time work, these energetic locals travelled in rickety trucks to each of the weekly markets to display pictures and offer the improved seed for sale. The programme loaned one kilo of Zapata seed to any community whose authorities requested assistance on two conditions: they had to promise in writing to cooperate in testing this new variety in a communal demonstration plot, and if the seed improved their production, they should repay the loan by sharing a similar amount of seed the following season with some interested neighbouring community. The rest of the seed produced was to be distributed among the people who had planted and harvested the trial.

The national brewery could hardly believe that the programme had sold or loaned eight tons of Zapata seed to farmers by the end of the first planting season. It was also impressive that efforts by the Shining Path to stop expansion of the programme had failed totally.[1] Asking communities to pass on Zapata seeds to neighbouring communities interested in repeating these tests was an unbeatable strategy for rapid replication.

The outreach of these farmers was so effective that rural teachers became interested in collaborating with the programme.[2] This led to cooperation agreements with the Ministry of Education in two departments. During the next two years, the teachers from 200 rural schools worked with their students and the parents in those communities to plant demonstration trials of Zapata barley. School children enjoyed these activities, and parent–teacher relations improved dramatically (Bebbington, 1991).

The total programme budget for the three-year period of expansion did not exceed US$100,000. This included the salaries for a local agronomist and eight part-time farmer-promoters, money for their transportation and for the field-days they organized, operational expenses for a jeep and a revolving fund used to purchase barley seed sold at market days. The cost-effectiveness of this programme was practically unparalleled in the region. A survey conducted by the Ministry of Agriculture four years later in 400 communities revealed that for all practical purposes, rust disease had been wiped out of the area. The value of the barley produced during that period was estimated to be worth close to a million dollars.

Lessons from the Peru Experience

These experiences demonstrated the value of training a large number of farmers, not just a few, to conduct their own site-specific demonstration trials to ensure that recommendations are appropriate to diverse microclimates such as found in the Andean region. Twenty years after programme activities were terminated in the area, World Neighbours asked a former programme participant, Maximo Beingolea, to assess the long-term impact of the project. He had helped prepare audio-visual materials for the programme while still in high school and was now the director of an NGO that he had formed to develop the same type of programme in a neighbouring department. He was received warmly by farmer-promoters at the former pilot centre who informed him that they continued to reap benefits from improved pasture and livestock management. Every time they entered their sheep in the annual livestock fair in Ayacucho, they won the grand prize for the best quality. Other farmers were travelling as far as eight hours to purchase locally improved sheep from them. No new problems had been experienced with rust on their barley crop. Thanks to the community solidarity developed during the programme's operation, they were tithing 10 per cent of their potato harvest each year to support their local pastor and the 14 women who had been widowed by the tragic violence during the 1980s.

They went on to explain that one of the most valuable skills they had gained from the leadership training provided by the programme was the self-confidence and ability to communicate their needs and concerns to government and NGOs. For example, they had been able to negotiate fair compensation for the loss of some communal grazing grounds that were submerged by a lake created by the dam for a large government irrigation project. This gave them funds to construct a communal stable for milk production and for building a simple restaurant to feed tourists who came to fish on the lake. These negotiation skills also enabled the villagers to acquire funds and technical assistance for building a pure drinking water system for the community. Such capabilities were making their local development sustainable.

BOLIVIA

Five years after the World Neighbours programme was initiated in Peru, a similar programme was started in Northern Potosi, Bolivia. Increasing potato production in this area in a sustainable way was far more challenging because of the acute erosion of its sandy loam soils. Population density had forced farmers to cultivate hillsides with slopes of up to 45 degrees, at elevations ranging from 2800–4000m above sea level, with average rainfall of only 500mm per year, average annual temperatures of only 9 to 15°C, and occasional hail that could destroy crops. High illiteracy rates, the predominance of local indigenous languages and a lack of agricultural research or extension services posed additional challenges.

Villagers' suspicion ran high when the programme was initiated. Visitors were so rare that locals were sure that any strangers who came wanted to take away their land. In fact, five years after programme activities had been initiated, villagers would still not reveal the number of families living in each community. Powerful people in the area, benefiting from the subordinated position of the Indian population, sought to dampen any enthusiasm for the programme by threatening to kill any members of the communities who participated in training seminars conducted by our 'communist' organization.

Even so, some community members began to cooperate in efforts to improve their food security. After initial trials to improve livestock and corn production failed due to lack of pasture and insufficient rainfall, some local farmers asked programme staff to assist them in improving potato production. Their primary interest was in testing new varieties that might help improve output with traditional agricultural practices. This led to a variety of demonstration trials that helped farmers increase their yields by 50 to 100 per cent. After eight years, enthusiasm for the programme was such that farmer-promoters were walking up to 24-hours round-trip to conduct training programmes in outlying villages where demand was great.

Then disaster struck. The area suffered serious drought, frost and hailstorms from 1983 to 1985. These climatic conditions led to the production of potato tubers the size of a finger. The dire shortage of this major food staple forced thousands of villagers to abandon their animals and homes in rural areas and migrate to cities. It was tragic to watch the plight of Quechua women and their children squatting on urban sidewalks with hands outstretched for money to purchase food.

Up to this point, the results of farmer-managed demonstration plots had not been documented in a systematic manner. Farmers simply reported the number of sacks of potatoes that they harvested for every *arroba* (5.2kg) of seed planted. The length and width of the experimental plots, the elevation of the sites, the rainfall received, the dates of planting and harvesting, exact weight of the harvests, a classification of the potatoes harvested and the number of farmers trained were never recorded because of high illiteracy rates. Population and environmental problems were growing faster than farmers' ability to improve production under these conditions.

Increasing Potato Yields through Systematic Experimentation

At this point, a former director of the College of Agronomy at the University of Huamanga in Ayacucho, Peru, approached World Neighbours about joining its staff. Here was an opportunity. Julio Beingolea had a deep respect for the indigenous peoples of the Andean area, spoke Quechua, had completed a Masters degree in agriculture in the United States, and had done research on Andean crops at his university, so the programme hired him to try to find ways to improve the food security in the area (Ruddell, 1995).

During an annual planning meeting, Beingolea explained to project staff how information recorded from small sub-plots, planted on a random basis, could be used to project results on a larger scale with a reasonable degree of confidence and accuracy. This would cost less than previously used methods, which were generally to plant irregular-sized plots three to ten times as large.

The staff decided to hold three-day seminars on this methodology in each of their five programme areas so that interested farmer-promoters could learn these more systematic methods.

After Beingolea joined the programme, farmer-promoters began designing, conducting and evaluating small experimental field trials, 10m by 12m. Since small farmers in the area had developed a thirst for new knowledge from previous site-specific trials, they were fascinated with the idea of conducting field trials in the same way as scientists. This would demonstrate to city residents, most of whom looked down upon the indigenous rural people, that they too could practise methodologies used by researchers.

Each farmer was free to select the treatments (variables) used in his or her own experiment. Frequently they chose to compare the productivity of native potato varieties with that of new varieties provided by research stations. In other cases they chose to compare the results of different fertilization practices. A six-page handout summarized the procedures to be followed when conducting these experiments.

After participating in training sessions, farmers helped plant two field trials designed by a promoter in the community where their seminar was held. Beingolea emphasized that three random replications of each treatment within the trial, planting different seed varieties set out within a randomized block design, would enable them to carry out valid trials on small plots (3m x 3.2m). In order to improve production under the worst climatic conditions, Beingolea asked farmers to plant their trials on land prepared with deep cultivation (deep tilling), to use uniform spacing, to select seed of appropriate size and to apply adequate fertilization (a minimum of 10t of manure per hectare). He felt confident that the synergy produced by using these four practices would make it possible to improve yields under even the worst environmental conditions.

Monthly two-day courses followed these seminars. During these events, farmer-promoters visited the field trials they had helped their neighbours plant, promoting farmer-to-farmer exchange. Sound agronomic practices for controlling weeds, pests and mounding up the soil around the plants were also reviewed.

Farmers began to appreciate the value of including a control plot to compare new and traditional practices. Farmers had previously resisted this practice because they felt it wasted land and valuable resources. But now poorer farmers who visited these sites and had previously thought that higher yields were impossible on their own land could see that the yields on control plots were similar to those from their own land.

One month before harvest, farmers were invited to a second seminar. Beingolea prepared another six-page handout describing the process of harvesting yields, recording the data, and carrying out systematic data analysis. As part of the training process, farmer-promoters assisted in harvesting the initial plots planted at the first seminar, recording the results, and noting any major variations in randomized replications of the same treatment that could result from differences in the slope of the land, from the presence of a large tree near the experimental plot, etc. Then, the amounts harvested from the plots receiving the respective treatments were gathered together in piles, to be ranked by yield. This visual comparison was extremely important for the semi-literate farmer-promoters. Then the weights of these groups were recorded for the sake of more precise evaluation.

Farmers quickly came to appreciate the impact of combining deep cultivation, appropriate seed size, better spacing and higher rates of fertilization. Beingolea taught them to classify and weigh the potatoes in each group. This was important because larger potatoes are worth much more in the market than small potatoes. Farmers could get more income from growing more large potatoes within any given harvest. Finally, the participants were taught how to calculate the average yield of each sub-plot and to use statistical formulas and tables to evaluate the results of each experiment.

The results of variety trials carried out in two communities, Qayastiya and Limaya, demonstrated dramatically the importance of site-specific trials since yields there ranged from 2777kg/ha to 44,444kg/ha. Differences in rainfall played a big role in these differences, but appropriate varieties and adequate fertilization were demonstrably important.

Several months later a government agency began working in Northern Potosi to promote the adoption of four new potato varieties at elevations above 4000m. Its staff were impressed when the programme coordinator showed them the results of site-specific experimental field trials with these varieties, which were disappointing. Although two of the varieties developed in The Netherlands at low elevations produced well at the research station in Cochabamba (2900m), they had never been tested above 3900m. The staff thanked the coordinator for this information and gathered up these two varieties, averting another unwise technical recommendation by technical personnel.

A third important finding emerged from an experiment organized by Filemon Colque to determine the optimum seed size for potatoes. He discovered that seed potatoes of at least 60g planted with deep tillage and 10t/h of animal manure almost doubled yields on average from 8333 to 14,352t/ha. Often bad climate and short harvests led Andean farmers to eat their best potatoes to survive. What they have left to plant is often too small to re-sprout if drought or frost strikes early in the next growing season. Colque showed the value of keeping larger potatoes to use as seed if farmers expect a good harvest to follow.

Dissemination of Farmer Learning

The information acquired from these new methods and technologies soon motivated farmer-promoters to request mathematics classes at their monthly seminars. The following year we discovered that many farmer-promoters had resumed their high school education via radio correspondence courses to complete their degrees.

Between 1991 and 1993, farmer-promoters elected by their local community assemblies planned and conducted 342 experimental field trials in a wide variety of elevations for 89 varieties of eight crops, including potatoes, corn, wheat, barley, quinoa and legumes. Three fertilization regimens were also tested repeatedly for each crop (Ruddell et al, 1997).

Systematic documentation of field trials conducted by farmers themselves began influencing other NGO programmes, government agencies and universities. The results of field trials presented at biannual staff meetings had more credibility and motivated programme leaders in adjacent areas to replicate the tests. The government of Bolivia also made a grant through its social investment fund to help the programme to train leaders in 120 communities in two more provinces on how to use the technologies that had been developed by local farmers during the previous ten-year period. These one-day seminars on how to plan and conduct experimental field trials involved 1936 men and 794 women over an 18-month period. The grant also provided funding to produce a series of pamphlets that summarized the results of farmers' experiments over a three-year period. The UNICEF programme PROANDES subsequently reproduced them for use in the literacy classes it was conducting in 600 communities in the area.

In addition, the Ministry of Education assigned ten teachers to work with the programme full-time. This opened the door to including training on how to conduct experimental field trials in the new curricula for rural schools. Once universities began to see the results from systematic documentation of site-specific, farmer-promoter field trials, they asked if their students could do their theses with local programmes. The programme coordinator agreed, provided the students spoke one of the indigenous languages. To date, seven theses have been completed. This has generated a small pool of professionals, both male and female, who have continued to work in the area.

It is important to note the succession of potato varieties used in the area after systematic evaluation began. Initially, a variety called Sanimilla identified through the selection process in Cochabamba gave the best yields. Waycha and Qorosongo became more popular about two years later. Now Colombiana has gained popularity because it raises yield potential up to 44,444kg/ha.

Farmer-promoters have become conscious of the numerous factors that affect production. One recently concluded that scientists must have developed a way to test two or more treatments within a single experiment. This pleasantly surprised Beingolea, who offered to teach farmer-promoters how to do this if they wished. This represented another advance in involving farmers as active participants in the process of improving agricultural capabilities.

Improving Yields by Planting Lupine as a Green Manure[3]

Once the site-specific field trials conducted by farmer-promoters had documented that a broad selection of native and new varieties developed by plant breeders had solid potential for improving production, the most important limiting factors were soil fertility and water. With adequate moisture, deep tillage, appropriate seed size and moderate soil fertility, family food security could be virtually guaranteed. When 10t/ha of manure are applied to the soil, productivity can be improved up to four times.

With regard to soil and water conservation, the World Neighbours programme in Central America was ahead of that in the Andes, having started work on this in the 1970s. In 1985, farmer-promoters from Central America were brought to Bolivia to teach their Andean counterparts how to initiate this process. The initial efforts met with no success, however, in part because the natural resource base in the Andes is much poorer, with greater frequency of drought (Ruddell, 1996).

Also, the primary food staple in Central America is corn rather than potatoes, the most important food crop in the Andean region. The legume *mucuna*, when grown in association with corn, could double yields (Flores, 1995). However, this plant would not grow at the elevations above 2800m in Northern Potosi. Furthermore, *mucuna* would have been destroyed by the process of mounding up soil around the potato plants which is done towards harvest time.

Still, perseverance eventually paid off. During a subsequent visit to Honduras by 13 farmer-promoters from Peru and Bolivia, we learned that a suitable alternative to mucuna for the Andean area might be lupine (Larson et al, 1989). Andean farmer-promoters were initially incredulous, since this leguminous plant was already grown in parts of Northern Potosi. Could it possibly provide 200kg of nitrogen per hectare per year when turned under as a cover crop? Fortunately their experience with experimentation at the field level persuaded them that they should test the idea rather than discard it.

Thomas Vilca made the first lupine trial in the community of Luqu, planting it as a cover crop on sloping land in October 1990 and turning it under four months later. His results were promising enough that the following year, 25 small farmers tried this same method. The growing season had normal rainfall, although there was one hailstorm in March 1992. On 8 April 1992, the potatoes were harvested. Table 14.1 shows that where lupine had been planted as a cover crop the previous year compared with no nutrient addition, potato yields increased from 1.78t/ha to 8.57t/ha. This represented a huge increase in yield over the average for the area. By itself, fertilizer could give yields greater than just lupine, but the greater cost of fertilizer made lupine attractive to farmers. If they had the time and money to use both lupine and fertilizer together, they could get a yield almost twice that obtainable from either alone.

Table 14.1 *Potato Yields from Fertilization Trials in Luqu, Northern Potosi, 1991–1992*

Fertilizer treatment	Yield (t/ha)
Lupine[a] + commercial fertilizer[b]	16.44
Lupine[a] + sheep manure[c]	13.19
Commercial fertilizer[b]	10.19
Lupine[a]	8.57
Sheep manure[c]	5.56
Farmers' practice without nutrient amendments	1.76

Notes: a) Lupine incorporated as green manure. b) Commercial fertilizer applied at the equivalent of 80kg/ha of nitrogen and 120kg/ha of phosphate. c) Sheep manure applied at the equivalent of 10t/ha; nutrient content not determined

Table 14.2 *Potato Yields from Fertilization Trials in Vitora, Northern Potosi, 1994–1995*

Fertilizer treatment	Yield (t/ha)
Lupine[a] + commercial fertilizer[b] + sheep manure[c]	14.86
Lupine[a] + sheep manure[c]	12.50
Lupine[a] + commercial fertilizer[b]	10.69
Lupine[a]	7.36
Commercial fertilizer[b] + sheep manure[c]	6.94
Sheep manure[c]	4.58
Commercial fertilizer[b]	2.50
Farmers' practice without nutrient amendments	1.39

Notes: a) Lupine incorporated as green manure. b) Commercial fertilizer applied at the equivalent of 80kg/ha of nitrogen and 120kg/ha of phosphate. c) Sheep manure applied at the equivalent of 10t/ha; nutrient content not determined

Pablo Choque conducted a second trial in the community of Vitora. Here lupine was planted as a cover crop on a slope at approximately 3500m elevation. Sixteen farmers helped Choque plant the site with the Waycha potato variety on 24 October 1994. These results are shown in Table 14.2. An analysis of variance of potato yields from eight different fertilizer treatments found that the variation between replications was not statistically significant, but the effect of fertilization treatments was highly significant. There was quite a difference in the productivity of fertilizer between the two locations as seen from these two tables.

The comparative costs per hectare of using lupine vs commercial fertilizers to improve yields are presented in Table 14.3. The total cost for using lupine was only about three-fourths of the cost of fertilizer. But more important than the total comparative cost was the fact that *cash outlay* for the use of lupine was only US$18 per hectare as compared with US$167 for chemical fertilizers. For farmers earning an average of US$300 cash income per year, the latter cost is prohibitive. In addition, transportation of lupine seed into remote areas is far easier than hauling in fertilizer (Ruddell, 1996).

These positive results prompted farmers to organize 18 more demonstration trials the following year; 276 farmers observed the harvests, and 209 families then replicated this new technology on their own farms. Under the best conditions of production, an investment of US$18 in lupine seed plus labour produced an additional three to four tons of potatoes per hectare, worth US$1200.

The trials that followed demonstrated that other varieties of green manures are better for improving soil productivity at elevations below 3000m, with the most promising of these species appearing to be vetch. These tests demonstrated once again that continued growth and development in agriculture requires continual experimentation, adaptation and innovation.

The spread of lupine in the area eventually led to the appearance of a fungal disease that affected this legume. Something similar happened with

Table 14.3 *Variable Costs of Using Lupines and Commercial Fertilizers for Potato Production on Peasant Farms in Northern Potosi, Bolivia*

	Cost/ha (US$)
Use of lupine	
Seed (60kg)	18
Oxen (5 oxen-days)	17
Labour (10 person-days)	21
Food and coca for labour	28
Oxen for turning lupine under (15 oxen-days)	51
Total cost for use of lupine	135
Commercial fertilizer alternative	
Cost of fertilizer[a]	167.38
Transporting fertilizer	13.35
Labour for spreading fertilizer (2 person-days)	4.24
Food and coca for labour	1.40
Total cost for use of commercial fertilizer	186.37

Notes: a) Fertilizer applications were equal to 4 bags of ammonium phosphate, 2 bags of urea, and 1 bag plus 10kg of triple super-phosphate, totalling 80kg of nitrogen and 120kg of phosphate/ha

leucaena, a fast-growing leguminous tree when it was used in Asia as a monoculture to improve soil fertility and forage production. Research programmes should accordingly identify not just one, but a series of suitable alternatives for improving soil and water conservation.

Lessons from the Bolivia Experience

Important things that the Bolivian experience added to our understanding of rural development include the following:

1. Follow democratic and participatory processes in rural development programmes.
2. Always seek a variety of solutions to a problem.
3. Adopt an integrated approach.
4. Coordinate efforts among multiple institutions.

We were pleased that the democratic and participatory process followed in helping people test variables of their own choosing eventually motivated them to play a more active role in the political process, for example, in the election of their local representatives. Two of the teachers actively involved in these programmes were elected as mayors in the two most prominent rural towns in the area. A third was selected to be the sub-prefect of San Pedro de Buena Vista, and a fourth was elected to the national House of Representatives.

COMPARATIVE OBSERVATIONS

The evolution of farming practices in certain African locations between 1945 and 1995 was documented in Chapter 7. The problem today in Latin America and other parts of the underdeveloped world is that we cannot wait 50 more years for poor rural farmers to develop more productive systems when 800 million people are hungry and malnourished.

Preceding chapters have demonstrated that such increases can be achieved more quickly when a participatory, people-centred approach is used to increase food production in food-deficit areas. This approach recognizes that the delivery of another 'improved variety' or 'better technology' will not in itself end their hunger or enable them to produce surplus food for others. Instead, people must be seen as the key resource to develop.

By consulting with them to determine what are their ideas and priorities for raising agricultural production, in the process improving their skills to do their own experiments, document the results, and share them with their neighbours, we build their dignity as well as capacity. This reduces language and cultural differences and geographic isolation as barriers to their development.

Each new problem solved generates stronger group solidarity, more capacity to organize themselves for the task at hand and greater conviction that their future depends not only upon agricultural technology but on their capacity to negotiate with government and other external entities to improve the roads, health and educational infrastructure in these areas. These are important factors in any sustainable development.

NOTES

1   The Shining Path was a group of approximately 10,000 Peruvians who sought to overthrow the political and economic regime of Peru. This and other terrorist groups in Peru promoted violence during the 1980s and early 1990s that led to the destruction of $US30 billion worth of property and the loss of 25,000 lives.

2   World Neighbours has observed that whenever crop yields of a major food staple are doubled, rural teachers ask for an opportunity to participate. This led to major expansion of the programmes in Peru, Bolivia and Ecuador. Parents in those communities were rural teachers did not desire to participate began to request the transfer of more progressive teachers to their towns. These teachers insisted that community leaders join the farmer-extension programme.

3   This section is based on the account and documentation in Beingolea (1993).

Chapter 15

The Spread and Benefits of No-till Agriculture in Paraná State, Brazil

Ademir Calegari

Soil degradation is to be expected whenever there is not good management of soil and crop residues. The challenge facing farmers is to manage organic inputs so that the release of nutrients is synchronized with plant growth demands. The practice of growing winter cover crops, mainly legumes, which has spread widely in Brazil not only protects the soil surface from water erosion and enhances soil organic matter, but also contributes to nutrient recycling and/or nitrogen fixation that benefits the following crop. Understanding how crop residues influence nutrient cycling and soil chemical properties, and then integrating residue management strategies into different cropping systems, are essential for soil fertility management. The no-tillage system using cover crops and crop rotations can minimize soil degradation and diminish the chemical inputs needed. It increases the stability of crop performance and also enhances cash crop yields. Especially in tropical conditions, improved residue management and reduced tillage should be encouraged because they contribute to sustainable production.

Concern with preserving soil and water was not a priority in Brazil until the 1970s, when annual-crop monoculture spread and tractor mechanization had almost doubled in Paraná state, with practically no conservation methods used. More than half of the area that was cultivated during the summer was not planted in the winter, leaving the soil bare. This accelerated erosion and decreased organic matter and nutrients in the soil, reducing fertility and productivity.

Indiscriminate and short-run exploitation of natural resources has been the mode of operation in Brazil since the beginning of colonization. Seldom were the natural carrying capacities of areas respected. Happily, this situation has begun to change, mainly through relevant results in research and experimentation, together with rural extension work that helps growers adopt new practices to manage and conserve their soil.

In Brazil, the no-tillage system began in the state of Paraná at the start of the 1970s with the work of a farmer, Herbert Bartz, in Rolandia in the northern region of the state. The initial aim was to control water erosion in areas where soybean and wheat were intensively tilled. Soon, corn also began to be cultivated under the new system. The state of Paraná is important in Brazil because although it covers about 200,000km², only 2.3 per cent of Brazil's territory, it contributes almost 20 per cent of the gross national product (GNP) and 24 per cent of national grain production. It has been a trend-setter for agriculture in whatever it does.

Annual precipitation in Paraná ranges between 1200mm and 2000mm. There is a wide array of soil types, often with low phosphorous content, excessive exchangeable aluminium and medium to high potassium levels. Of the more than 6 million hectares cultivated with summer crops, three-quarters are under soybean or maize. Other crops include beans, cotton, irrigated rice, rainfed rice, sugarcane, cassava, coffee and fruit.

The absence of adequate planning and regulation for use of natural resources has dramatically decreased forests, from 87 per cent to less than 10 per cent of the total land area, with agriculture covering both fertile and marginal areas. In the 1980s, technical data from the Agronomic Research Institute for Paraná (IAPAR) showed that the no-till system being experimented with should be more than an alternative soil management method; rather, it should evolve into a system fully integrated with different cropping practices soil (Darolt, 1998). Already, by 1984, there were around 300,000 hectares under no-till agriculture in Paraná – 5 per cent of the cultivated area; by the next year, this system covered 800,000 hectares throughout all of Southern Brazil (Derpsch et al, 1991).

The no-till system involves growing different species of green manure, incorporating dry matter into the soil, and rotating crops as basic strategies for sustainable management of annual production. These practices have been systematized and spread through extension work in hydrological micro-basins so that these systems now occupy more than 3.5 million hectares in Paraná. By the start of 2001, no-till systems cover close to 13.5 million hectares in Brazil.

MAINTAINING SOIL FERTILITY IN THE TROPICS, WITH SPECIAL REFERENCE TO BRAZIL

The major challenge in no-till agriculture is to balance and organize the various cropping and soil management components appropriately: one must keep weeds under control and choose, if necessary, the right herbicides to be applied; apply them correctly while using fertilizers together with green manure and cover crops; mulch fields rationally; and rotate crops. This system does not reject the use of agrochemicals, and indeed often requires them. But it tries to minimize their use, for both economic and ecological reasons.

With good management of other practices, reliance on agrochemicals can be greatly reduced and possibly even ended (Petersen et al, 2000). In neighbouring Paraguay, it has been demonstrated that the use of herbicides can be eliminated with a three-year, no-till cycle that alternates cash crops and cover crops (Kliewer et al, 1999). A roller-cutter is used to manage suitable, quick-growing cover crops planted between crops that are grown for cash: soybean, wheat and maize. A rotation of sunflower-black oat-soybean-wheat-soybean-white lupine-maize, for example, gives a farmer four harvests of cash crops within three years, yielding revenue well above costs of production. A productive no-till system does require continuous observation of the field; farmers need to watch plant development closely and to monitor continually changes in the physical, chemical and biological characteristics of the soil. A grower therefore needs good training and management skills.

The costs of careless use of natural resources are becoming clearer to everyone. Already ten years ago, the Brazilian National Agency for Agricultural Research (EMBRAPA) estimated that the use of technologies available for soil conservation in the southern and central regions of Brazil could avoid losses of up to US$110 billion, calculated just in terms of the fertilizer equivalent of soil nutrients lost unnecessarily through erosion (Vergara et al, 1991).

For productive, competitive and sustainable agriculture, we must establish and maintain agroecosystems that promote, through biodiversity in time and in space, effective nutrient recycling and recovery and conservation of the physical, chemical and biological properties of the soil. Integrating these practices into farming systems helps to improve not only agricultural production, but also the socioeconomic conditions of rural growers, as no-till cultivation leads to better employment of labour throughout the year (Calegari, 1995a).

The reduction of erosion attainable by reduced or eliminated tillage can be seen from the results of a four-year experiment reported in Table 15.1. In the no-till system soil losses are insignificant, whereas the more that soil is ploughed, the greater are soil losses. In Paraná, average soil losses of 10 to 40 tons of fertile soil per hectare per year have been reported from traditional soil tillage systems (Sorrenson and Montoya, 1984). Erosion is so severe in many areas that fertility has already been irreversibly lost to agricultural production.

Because most soils in Paraná State have a high clay content, they have been considered relatively erosion-resistant. But the high rainfall intensity creates erosion hazard whenever the soil is tilled but not protected by a leaf

Table 15.1 *Soil Losses in Different Tillage Systems with Animal Traction in Álic Cambisol, Ponta Grossa, Paraná (average of four years of evaluation)*

Treatments	Soil losses (t/ha/yr)	Losses relative to mouldboard plough (%)
Ploughed soil uncovered	113.78	1307
Mouldboard plough	8.70	100
Chisel plough	4.35	50
No-till	0.84	10

Note: Annual average rainfall during the four years of evaluation: 967.5mm
Source: Araújo et al (1991)

canopy or plant residue. Even though the soils of Paraná have a high pore volume and show high infiltration rates in the laboratory, surface sealing impedes the infiltration of precipitation and makes the soil very susceptible to surface erosion (Roth, 1985). When the land is covered with grass, crop plants or plant residues (mulch), the impact of raindrops is absorbed by the biomass. The water reaches the soil's surface gently, without the force to detach soil particles, and it then infiltrates through an undisturbed pore system. Infiltration studies with a rainfall simulator in Paraná have shown that independent of the tillage system used, there was 100 per cent water infiltration when the soil was covered with plant residues, while only 20 to 25 per cent infiltration was observed under bare soil conditions (Derpsch, 1986).

Research in the United States has shown that conservation tillage which covers even 30 per cent of the surface with crop residues can reduce soil erosion by approximately 50 per cent compared with a soil surface with no residues (Allmaras and Dowdy, 1985). With the advent of commercial fertilizers after World War II, utilization of crop residues as a source of nutrients and consequent soil protection diminished drastically in many countries, however.

TILLAGE PRACTICES

Covering the soil with growing plants or plant residues has been shown by research and by farmer practices in many places around the world to provide important erosion protection (Meyer et al, 1970; Lal, 1975; Roose, 1977; Derpsch et al, 1984, 2000; Sanchez et al, 1989; Hargrove, 1991; Calegari, 2000). Conventional methods of tilling and planting protect the soil only by the leaf canopy of crops in advanced growing stages. In Europe, this tillage system has not caused major problems, due to low rain intensity as well as favourable topography and soil characteristics. In contrast, these methods of cultivation on sloping land in the tropics, where rainfall occurs with high intensity, have led to widespread erosion and unproductive soils.

Farmers' practices in Paraná and other parts of the world, as well as research results from the UK and elsewhere, have shown that tilling the soil is not necessary to produce a good crop (Russell, 1961). This started becoming accepted in the late 1950s when British industry developed a new generation of herbicides, and techniques emerged for drilling seeds without disturbing the soil, referred to variously as mulch tillage, no-tillage or direct drilling.

Conventional Tillage

Standard soil tillage in Southern Brazil presently requires one ploughing and two diskings with a light harrow to level the ground and prepare the seedbed. When such tillage is performed properly and does not exceed two diskings, it produces less erosion damage than traditional tillage, where a heavy disk harrow penetrates 10–15cm into the soil profile and leaves a finely disaggregated soil surface layer.

No-till Alternatives

No-till agriculture – also referred to as zero-tillage, direct planting, direct drilling or direct seeding – involves planting crops in previously unprepared soil by opening a narrow trench or band wide and deep enough to obtain seed coverage (Phillips and Young, 1973). Some no-till machines now on the market bury the seed at regular intervals without opening any trench. No other soil preparation is done. Herbicides can be used to control any unwanted weeds and grasses. In the last decade the importance of such techniques for developing countries has been increasingly recognized. Soil management can be improved, for example, by replacing conventional tillage with the planting in winter of a cover crop that is cut in the spring, with crop seeds then planted into its residue (Morgan, 1992).

No-till aims to keep the soil covered with plant residues or growing plants throughout the entire year. In long-term studies conducted in Indiana, Kladivko et al (1986) found that on sloping, well-drained soils, higher maize yields were obtainable with no-till methods than from ploughing. Conversely, on poorly drained soils with high organic matter, no-till was not as good for continuous maize cultivation – though a maize/soybean rotation produced only slightly less than did ploughed systems. This is a static evaluation, however, not taking into account changes in soil structure and fertility with no-till. The researchers concluded that on poorly-drained soils with high organic matter, although maize yields would be lower in the first several years, they would increase over time when cultivation practices are changed. With a maize–soybean rotation, no-till yields would exceed yields from ploughing after just a few years of soil structural improvement.

The great challenge in a no-till system is to identify the best management practices for the given soil, climate and other conditions so that all components are integrated properly. Defining the best system requires continuous observation in the field and a monitoring of soil characteristics.

COVER CROPS AND GREEN MANURES

Cover crops are planted primarily to protect the soil from the direct impact of raindrops. Protection is given by the growing plant's canopy and root system, and by litter or residues on the ground. Total cover of the soil with plant residues improves the infiltration of rainfall. However, cover crops have the potential also to improve soil fertility as green manure (Muzilli et al, 1980; Derpsch and Calegari, 1992; Calegari et al, 1993). Leguminous cover crops in particular can add biologically fixed nitrogen to the soil.

In natural systems, nutrient release from litter and growing plants; uptake of nutrients generally occur in synchrony, resulting in the efficient use of nutrients. Where the two processes of release and uptake are separated in time, low nutrient-use efficiency results. This is particularly acute with nitrogen, as any amounts not used by plant roots are liable to be lost by leaching, denitrification and ammonia volatilization.

In agricultural systems that incorporate cover crops into the soil, these decompose gradually and release nitrogen to the following crop with less nitrogen loss. A low mineralization rate, for instance in wheat and sorghum residues, is beneficial for the subsequent crop (Wagger et al, 1985). In general, the losses of nitrogen due to ammonia volatilization from litter or crop residue on the soil surface are greater than for residues that have been incorporated into the soil. The trade-off is that incorporating residues leads to less surface soil cover.

The role of soil cover in enhancing rainfall infiltration rates and reducing erosion has been documented by many researchers, eg Mannering and Meyer (1963); Greenland (1981); Lal (1990); and Sidiras and Roth (1984). Results obtained by Roose (1977) after 20 years of erosion research in Africa showed that even with high precipitation, covering the soil with 4–6t of mulch/plant residues per hectare can protect the soil from erosion as much as would a secondary forest about 30m in height.

Not ploughing the soil but incorporating plants as green manure and/or maintaining crop residues on the soil surface preserves and promotes soil organic carbon equilibrium. The use of green manure species is now widespread in Southern Brazil to improve soils; they are used as mulch for no-till cash crops, for perennial crop intercropping (coffee, rubber, citrus and other perennial fruit crops) and also for animal fodder. Many research studies and results obtained by farmers with different green manures in no-till systems, conducted under various agroecological conditions in Paraná, have shown the efficiency of these systems for improving soil properties and promoting more productive equilibrium in plant–soil–water systems. Also, the practice of rotation, including more species in cropping cycles, has contributed to increased biodiversity and as a consequence has reduced pest problems in these systems (Derpsch et al, 2000; Calegari, 2000c).

Many mixes of cover crops are in use.[1] In various ways, these contribute to the following:

- Soil physical effects – an increase in soil aggregate stability and enhanced soil water infiltration rates.
- Soil chemical effects – higher levels of nitrogen, phosphorous, potassium, calcium, magnesium and organic matter in the soil surface, improved nutrient recycling and/or nitrogen fixation, and decreases in aluminium toxicity (Miyazawa, Pavan and Calegari, 1994).
- Soil biological effects – increased soil microbial populations and reduced populations of nematodes; also reduced impact of weeds due to changes in weed species and weed biomass.

Many studies in Southern Brazil have shown that the use of winter cover crops, by improving soil properties, can increase the yields of crops such as maize, beans, soybeans, wheat and cassava substantially if the proper cover crop is included in a rotation system (Muzilli et al, 1980; Derpsch, 1986; Santos et al, 1990; Bairrão et al, 1988; Medeiros et al, 1989; Paiva, 1990; Calegari et al, 1993, 1995, 2000). Almeida and Rodrigues (1985) have shown that cover crops like black oats, oil seed, radish and hairy vetch can substantially reduce weed populations in a no-till system and consequently reduce the amount of herbicide needed. There is a direct correlation between the amount of biomass produced by cover crops and the suppression of weeds. Weed reduction may be due not only to competition for light but also to the allelopathic effects of plant exudates (Altieri and Doll, 1978).

Sorrenson and Montoya (1984) have found that herbicides are the greatest single cost in no-till cultivation as it has developed, amounting to about 25 per cent of the total outlay on soybeans and maize production. This gives farmers an economic incentive to identify and include cover crops in their rotation system that can reduce weed problems.

CROP ROTATION

After evaluating agronomic results from many parts of the world, Cook and Ellis (1987) have drawn three conclusions:

1   Rotation of crops is better than continuous cropping even when the crops are not leguminous or are depleting in nature, eg maize–oats is better than continuous maize.
2   To be highly effective, a rotation system must include a legume.
3   Rotation alone is not sufficient to maintain productivity over time – the addition of some plant nutrients will be needed to compensate for nutrients exports and losses.

The value of a legume in a rotation is not only that it adds nitrogen to the system but that soil structure is improved by the bacteria, fungi and other soil micro-organisms that decompose leguminous organic matter. The utility of legume green manures for maintaining or building up soil fertility has long been recognized, and legumes have been traditionally used for this purpose in many regions (Sturdy, 1939).

Most data show that rotational cropping improves soil properties (physical, biological and chemical), enhances crop stability, reduces weed infestation, diminishes total production costs and over several years gives greater yields than does continuous cropping with the same species. Green manures and cover crops as well as crop stubble residues can have different fates: they can remain on the surface as mulch, be used as animal feed, incorporated (tilled) into the soil or burned. Mulch is the use that best preserves and/or recovers the productive potential of farming soils with the passage of time.

Plant residue on and in the soil causes important changes in soil properties, due to the effects of biomass on the surface or in the soil. It increases the stability of aggregates in water (the cementing action of organic matter, polysaccharides and fungus hypha, discussed in Calegari and Pavan, 1995) thereby increasing water retention capacity and water infiltration ratios; it promotes greater soil porosity and aeration due to larger root systems and reduced evaporation due to the effect of mulch on the surface; and it decreases soil density due to the effects of organic matter.

The population density, diversity and activity of micro-organisms in the soil is directly related to the volume of organic matter available (Chapter 10). This constitutes one of the main sources of energy for these organisms. Ploughing operations and the lack of mulch under conventional cultivation cause greater fluctuations in temperature as well as in moisture. These conditions contribute to a decrease in the populations of soil organisms, both micro-organisms and macrofauna such as earthworms, which are important contributors to soil fertility. A no-till system thus increases biological activity in the soil, which is important for both availability and uptake of the nutrients that plants need.

Some plant species can control weeds, either through their mulch effects or due to the physical and chemical (allelopathic) effects of their roots. Examples include sorghum, pearl millet, mucuna, crotalaria and pigeon pea, which have been shown to affect the quality and quantity of certain weed species.[2] The rotational use and management of green manures is thus important when there is the need to decrease populations of some weeds. With no-till and good weed management it is possible to keep these species from fructifying and completing their cycles, thereby decreasing the degree of soil infestation over time.

The use of green manure and crop rotation can also provide an economic method to control different species of nematodes. Several results from research and grower practices have also proven the efficacy of crop rotation to decrease the risk of attack by stem canker (*Diaphorte phaseolorun f.* sp. Meridonalis), an important disease that can occur in sorghum.

The various plant species that could be used as a green manure must be carefully tested, assessed and validated by growers on their farms to be sure of best results. The experiences of many farmers in Southern Brazil, Paraguay, Argentina and other South America countries have shown that after several years of no-till cultivation and management, the need for fertilizers decreases significantly, while grain yields increase compared with the conventional system of one ploughing and two turns with disk harrows. Improvement of soil characteristics not only permits reduced fertilizer use but also diminishes labour requirements. These changes together with increased crop productivity lead to greater profits from farm operations with no-till cultivation, which explains its widespread appeal.

NO-TILL ON SMALL FARMS

Generally smallholders occupy areas with lower soil fertility and use family labour intensively as well as animal traction. They find in the no-till system an important way to save labour, improve their soil, and decrease their production costs. However, few efforts have been made to improve the mechanical implements available to small farmers for no-till cultivation. Starting in 1985, IAPAR began developing a machine it called *Gralha-Azul* that could make no-till work easier. After using the machine for five seasons to grow beans and corn, however, it became clear that mechanical improvement by itself was not enough. There was need to improve soil fertility by introducing appropriate green manure species into the crop rotation. Mechanical and biological processes thus needed to be integrated in a complementary fashion.

New experiments beginning at 1990 evaluated the effects of no-till where black oats were mixed with common vetch, and rye grass with ornithopus and velvet bean, as winter cover crops and green manure (Araújo et al, 1991). The results prompted the following observations:

- After two cuttings, the cover crops provided good biomass production for the no-till system.
- The cover crops could be managed by using a roller-cutter and herbicides (hand-sprayed).
- The no-till machine performed well in cutting the straw and also in distributing seeds and fertilizer on the soil, promoting homogeneous plant development.
- Compared with the conventional system of ploughing where two weedings by hand were needed during the beans cycle, only one control operation was needed in the no-till system because of allelopathic straw effects.

In general, the no-till treatments showed better plant development and also earlier soil cover than the conventional system.

In these studies, farmers' participation and evaluation was integrated into the testing and validation process. The no-till system is a dynamic one and leads to changes not only in soil management but also in farmers' behaviour. Traditional customs and common practices of ploughing the soil every year need to be changed for this new system. One of the main requirements for no-till success is continual monitoring of soil properties and other variables like weed infestation, soil compaction, soil fertility changes and effectiveness of cover crops and crop rotations. Rotations need to be adjusted according to soil, climatic, technical and economic factors, seeking to maintain soil equilibrium, decrease insects and disease, and diminish weed infestation (Darolt, 1998, Derpsch et al, 2000; Calegari, 2000).

When tobacco is the main crop, residual fertilizer can be utilized by planting two rows of beans or one of corn along with each row of tobacco. (This can be facilitated by using a hand machine known as *saraquá*.) Some farmers intercrop corn with *mucuna*, which is sown at the corn's flowering stage; after the corn is harvested, *mucuna* continues growing, sometimes together with a grass (*Brachiaria plantaginea*) until the next spring when it is cut down by a roller-cutter if not killed by frost in the winter; no-till tobacco is then planted in small holes. This improves the yield of both tobacco and corn and also decreases production costs, enhancing net farm income.

There are various green manure options that can be used in the summer season to increase the nitrogen content of the soil, eg *mucuna*, crotalaria, pigeon pea, canavalia and cowpea. Some non-leguminous species like *Penissetum americanum* can be very useful as mulch, covering the soil and increasing soil humidity while also being effective in weed control. Another option for small farmers in Paraná State is the use of spontaneous vegetation growth, mainly of *Brachiaria plantaginea*, which develops during the corn-growing season. When killed by herbicides afterwards, this creates a valuable layer of mulch (4–7t of dry matter) which is useful for no-till cultivation of beans, corn, cotton and soybeans.

According to the United Nations Food and Agriculture Organization (FAO), approximately 4 million farmers in Brazil presently use animal traction combined with human labour. These farms are characterized by limited natural resources inadequate for intensive agriculture, steep slopes, shallow soils, and low natural fertility. Small farmers in Paraná State have made improvements in their animal-drawn no-till implements, giving better distribution of labour requirements throughout the year as well as better maize and bean production. Yield increases vary between 78 and 106 per cent, respectively, compared with conventional systems, with positive net income and system adoption by large numbers of farmers. Such results show that the no-till system is technically, economically and socially viable for small farmers, and an important contribution towards small farm sustainability.

MEASUREMENT OF NO-TILL EFFECTS ON SOIL FERTILITY, CROP YIELD AND PROFITABILITY

Various studies conducted in Paraná have showed significant reductions in soil acidity, increased cation exchange capacity and available soil nutrients, as well as lowered aluminium saturation near the soil surface under no-till compared with the conventional system (Sidiras and Pavan, 1985; Sá, 1993; Calegari, 1995; Calegari et al, 1995). Such evaluation is complicated as the effects of residue incorporation on soil productivity are difficult to separate from tillage effects because incorporation requires some type of tillage operation. Also, soil water content, soil temperature and porosity are influenced by the incorporation of residues. Differences in climate, soil type and residue quality all affect changes in the rate and degree of organic matter accumulation associated with surface residues.

Carbon and nitrogen availability within crop residues, along with lignin content, greatly influence decomposition rates and the resulting availability of nitrogen to plants (Hargrove et al, 1991). Decomposition of residues with low nitrogen content such as black oats (*Avena strigosa Schreb.*) may result in the immobilization of soil and fertilizer nitrogen by microbial activity, thereby reducing the nitrogen available to plants. Normally, residues with nitrogen concentrations below 1.5 per cent, or carbon to nitrogen ratios greater than 25 to 30, are expected to immobilize inorganic nitrogen. Despite this, residues with very similar carbon to nitrogen ratios can have different decomposition rates because of variations in chemical concentrations (Stott and Martin, 1989).

Many results have shown that faster decomposition rates occur with buried residues than with surface residues, due to greater soil-residue contact, a more favourable and stable microbiological environment for decomposition and increased availability of exogenous nitrogen for decomposition by micro-

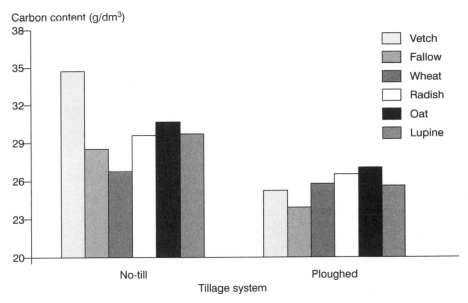

Figure 15.1 *Effect of Different Winter Cover Crops in No-till Systems on Carbon Content in the Soil Profile (0–5cm)*

organisms (Unger and Parker, 1968; Cogle et al, 1987). Results obtained by Calegari (1998) in Pato Branco in the southern region of Paraná, working over 11 years with cover crops and crop rotation in no-till systems showed that soil fertility is improved, as are crop yields, when these soil management practices are followed. Experiments evaluating four cover crops plus fallow and wheat, comparing no-till with plough tillage, showed that accumulating organic material on the soil surface together with a no-till regime (practised over 11 years) maintains higher levels of organic carbon and phosphorous in the top horizon of soil (0–5cm) than does the conventional tillage system (see Figures 15.1 and 15.2). Organic matter accumulation causes changes in nutrient dynamics that have a positive impact on soil fertility.

The experiments also showed significant differences for the various cover crop regimes with no-till for calcium, magnesium and potassium at the surface (0–5cm), though in the next lower horizon (5–20cm) no significant difference was observed for phosphorous or organic carbon compared to the ploughed system. This contributed to higher base saturation and probably together with the higher organic carbon also to complexing of the aluminium, decreasing its toxic effects. At the same time this could contribute to higher pH values and enhance phosphorous release. In the lower layer (5–20cm), the ploughed system presented higher values for calcium and magnesium, compared with no-till, probably induced by residue incorporation and also the effects of adding lime (calcium carbonate) to the soil.

Maize yields obtained in these experiments showed significant differences according to tillage regime and also for nitrogen application rate (see Figure 15.3). The no-till system even without any addition of nitrogen gave higher

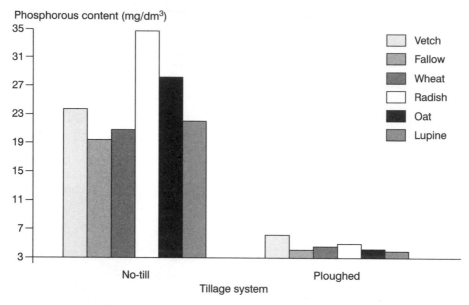

Figure 15.2 *Effect of Different Winter Cover Crops in No-till Systems on Phosphorous Content in the Soil Profile (0–5cm)*

maize yields for four treatments than did conventional tillage with nitrogen (90kg/ha). The exceptions were wheat and radish as winter cover crops. Probably nitrogen immobilization in the wheat straw contributed to a decrease in this yield with no-till.

Also when nitrogen was applied with no-till methods, greater maize production was attained for all trials than with conventional tillage. Maize yields are affected not only by what was done for that year's crop but also by the effects of crop residues accumulated in previous years. Because with no-till there was no response to nitrogen application in the legume cover crop treatments (lupine and vetch) – actually, yield was lowered – and also with oats planted as a cover crop, these results suggest that there was higher availability of nitrogen with these species than with the wheat, radish and fallow treatments. Probably oat straw accumulation during several previous years and the organic matter layer generated on the soil surface contributed to this.

These experiments showed that it is possible to obtain from no-till practices with legume cover crops the equivalent of more than 90kg/ha of nitrogen through biological processes which is not produced with conventional tillage means. The maize yield increase was correlated with soil phosphorous availability and with phosphorous content in maize plant leaves. The greater phosphorous uptake is attributed to higher moisture content below the mulch and consequently to increased root growth and a higher phosphorous diffusion rate as compared with non-mulched or tilled plots.

The data show that crop residues together with tillage regime caused significant alteration and redistribution of nutrients within the soil profile. There are also likely to be effects in nutrient cycling and certain soil physical and

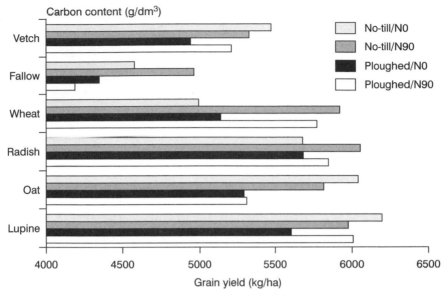

Figure 15.3 *Effects of Tillage Practices and Nitrogen Application on Maize Yields*

biological properties that were not evaluated in this study (Calegari, 1998). Even though the no-till system caused nutrient concentration in the upper soil layer, this was not a disadvantage for maize production. The no-till system promoted soil conditions that consequently led to phosphorous and nitrogen becoming more available and taken up for the maize crop.

To match the specific soil and climatic conditions in different regions, several crop rotation systems have been developed in Paraná State. For example, maize can be rotated with oat-beans or beans-oat, or with vicia-maize-radish-beans or just beans; soybeans can rotate with oat-wheat-field pea-maize, or with pearl millet-maize-pigeon pea. In a clay soil in the Itapua Department of Paraguay, working under on-farm conditions, Herter (1998) obtained consistent results that varied according to the number of years which the different systems were used (Table 15.2).

Table 15.2 *Organic Matter at Different Depths Comparing No-till and Conventional Tillage Systems, Paraguay, 1998*

Years	Soil organic matter (%) at different depths				Organic matter average (%)
	0 to 5cm	*5 to 10cm*	*10 to 15 cm*	*15 to 20cm*	
Conventional system	2.89	2.63	2.43	2.14	2.52
No-till (4 years)	3.11	2.89	2.74	2.11	2.71
No-till (7 years)	3.25	2.99	2.94	2.48	2.92
No-till (10 years)	3.71	3.16	3.01	2.59	3.12

Source: Herter (1998)

Table 15.3 *Soybean Yield (kg/ha) and Organic Matter Content Comparing No-till and Conventional Tillage Systems, Paraguay, 1998*

Soil management	Soybean yield (kg/ha)	Yield (%)	Organic matter (%) at 10–20cm
Conventional (10 years)	2050	100	2.52
No-till (4 years)	2956	144	2.71
No-till (7 years)	3199	156	2.92
No-till (10 years)	3188	155	3.12

Source: Herter (1998)

The organic matter increase at different soil depths is directly related to the number of years that no-till has been used and the rotation in effect, compared with the conventional system of ploughing and harrowing the soil. The addition of residues each year on the soil surface and not disturbing the soil contributed to increased soil organic matter content and also enhanced soybean yields (Table 15.3).

The results attained by these farmers in Paraguay show that over time, the no-till system enhances soybean yield as it increases soil organic matter. Further, since the costs of herbicides and fertilizers were lower with no-till, this led to higher net income than did conventional practice. A further example of how soil organic matter content is increased by no-till cultivation with suitable crop rotations over a six-year period is given in Table 15.4.

The savanna region is characterized by well defined dry and rainy seasons and also by soils with low clay content, high temperatures, and rapid organic matter mineralization. When the soils are disturbed to plant annual crops, there is a decrease in soil organic matter content. The data after six years of soil management and planting show the following:

Table 15.4 *Organic Matter Content with Different Tillage Systems and Crop Rotation in Savanna Soils, North Central Brazil, 1986–1992*

Soil management and crop rotation	Soil depth (cm)	Soil organic matter (%)
Heavy disks on monocropped soybean	0–10	1
	10–20	1
	20–30	1
Disk ploughing on soybean–corn rotation	0–10	1.5
	10–20	1.3
	20–30	1.3
No-till on soybean–corn rotation	0–10	3.8
	10–20	3.4
	20–30	2

Source: Seguy et al (1995)

- With the crop rotation (soybean–corn) in a no-till system, soil organic matter content was higher at all depths than with other treatments (plough or heavy disks).
- When the soil was ploughed with disks, soil organic matter was mineralized, showing that soybean-corn was not sufficient to maintain and/or increase soil organic matter content.
- The use of heavy disks in a soybean monocrop, year by year, accelerated soil organic matter decomposition rates, showing that this soil management system is not sustainable.

Controlled studies conducted on the St Antonio farm in Floresta, North Paraná, comparing both tillage systems on a 1.6ha cultivated area over a six-year period, found that no-till systems yielded 34.4 per cent more soybeans and 13.7 per cent more wheat than did conventional tillage. Growing these crops in rotation with cover crops rather than as monoculture added 19.2 and 5.8 per cent, respectively, to soybean and wheat yields (Calegari et al, 1995). A separate study on a 50ha experimental site in North Paraná gives further evidence that a well designed no-till system with soybeans in crop rotation can generate important net income gains compared with conventional systems (Table 15.5).

Table 15.5 *Economic Evaluation of Soybean Production with No-till Rotation Systems Compared with Conventional Tillage in Northern Paraná*

No-till and crop rotation system advantages	Value of yield increments (US$)	Reduction of production costs (US$)
Crop yield improvement	3960	
Cheaper machine maintenance		1145
Less fuel used		731
Less labour required		2880
Less fertilizer applied		186
Benefits	3960	4942
Total net benefit on 50ha		8902

Source: Calegari et al (1999). The price for soybeans was US$166/t at the time of evaluation

CONCLUSIONS

The understanding and use of new technologies of soil management and conservation has prompted a substantial increase in land under no-till systems in Brazil, utilizing cover crops, green manure and crop rotations. This has contributed to higher production efficiency and increased net farm income. No-till systems contribute to a better distribution of agricultural labour throughout the year, practically eliminating ploughing, harrowing and mechanical weed control. This leads to more available time to arrange, plan and manage different activities for better farm diversification.

The system of cultivation substantially reduces soil losses, improves soil fertility, contributes to higher biodiversity, diminishes weed infestation, pests and crop diseases, increases crop yields and enhances production stability, making permanent land use possible. At the beginning of this new century, sustainable land management represents one of the biggest challenges for farmers and researchers alike. In Brazil, we have found no-till agriculture with its associated practices to be a fundamental component for strategies that improve soil management to better meet economic and social needs and maintain our natural resource base.

NOTES

1   Examples include: avena and vicia, avena and raphanus, avena and lupine, or avena and radish and vicia, avena and lupine and vicia, and pennisetum and crotalaria. The main cover crop species used in Paraná are: *Avena strigosa, Lupinus sp., Vicia sp., Secale cereale, Raphanus sativus, Pisum sativum, Mucuna pruriens, Crotalaria juncea, Cajanus cajan, Vigna unguiculata, Canavalia ensiformis, Canavalia brasiliensis, Penissetum americanum,* and *Calopogonium mucunoides* (Calegari et al, 1993).
2   Recall also the example given in Chapter 1 of intercropping controlling stem borer and striga in Kenya (Khan et al, 2000).

Chapter 16

Diversifying Rice-based Systems and Empowering Farmers in Bangladesh Using the Farmer Field-school Approach

Marco Barzman and Sylvie Desilles

Over the past eight years, CARE-Bangladesh has been developing a set of sustainable agriculture projects to improve production in rice-based farming systems in this country. These efforts have contributed to an emergent programme to increase food security for poor and marginal farmers. The programme raises agricultural productivity by diversifying agroecosystems and optimizing yields, reducing costs of production and creating new income streams.

Although the programme is associated with a number of agricultural techniques that are readily adopted and disseminated, it strives to achieve more than the simple transfer of sustainable agriculture technologies. It follows a philosophy of adult education that emphasizes personal experience and builds up people's confidence and ability to manage their future development (Hagmann et al, 1999). The main difference between this and other agricultural development programmes has been its emphasis on knowledge and experimentation, rather than technologies and material inputs. The programme enables farmers to take initiative in experimenting with and understanding ecological processes in their fields.

These principles and methods have been developed as part of a regional programme to refine and spread IPM practices within Asia, supported since 1982 by the UN Food and Agriculture Organization (FAO) (Röling and Van de Fliert, 1994). This chapter reports on experience in Bangladesh, focusing on the programme's philosophy and operation as well as its results. In Chapter 17 a parallel programme in Sri Lanka is reported on, with more detailed data on yield and income effects as well as on the issue of agrochemical use.

PROGRAMME OVERVIEW

The rice programme in Bangladesh has had two aims: to increase rice field productivity and to empower farmers in terms of their decision-making and management capacities. The rice field is the main source of revenue for the category of small farmers targeted by the projects. Appreciating this, the programme seeks to enhance the capacity of households and communities to independently generate, access and disseminate knowledge. This is achieved through the facilitation of a team of skilled and motivated staff who are given the task of creating an effective learning environment in which farmers test a variety of sustainable agricultural practices.

The dual objective of exposing farmers to agricultural innovation and of empowering them is achieved through the creation of farmer field-schools (FFS) adapted from experience in other Asian countries (Kenmore, 1997). Important elements of the approach adopted by CARE-Bangladesh include the development of a broad curriculum that focuses on the entire rice cropping system and provides farmers with a 'menu of topics' to choose from, and an experimental approach in which farmers are considered researchers (Kamp and Scarborough, 1996).

The agricultural techniques in the programme have been:

- sustainable agriculture practices in rice;
- vegetable production on rice field dykes;
- fish production inside the rice field;
- production of fish fingerlings; and
- tree planting on rice field dykes.

These practices increase rice-field productivity through optimized use of the natural resources available. The way in which these practices are learned is intended to result in a certain level of farmer empowerment.

An innovative, field-tested model for adult education has been developed through experience with three related projects supported by the UK's Department for International Development (DFID) and the European Commission. The programme during its initial phase (through 2000) involved around 150,000 farmers. The second phase, now started, disseminates this model to other organizations, non-governmental and governmental. It expects to reach more than 1 million farmers within the next six years. The activities of the expanded rice programme will include more emphasis on homestead gardening and vegetable growing.

IMPACT ON FARMERS

The core agricultural practices of the programme – sustainable rice production, rice–fish farming, vegetable dyke crops and trees on dykes – aim at sustainable and optimal landuse, a fundamental issue in rural Bangladesh.

Households with limited access to land need farming systems that generate more income and food in an efficient, reliable and sustainable manner.

Rice Production

Between 1995 and 1998, small-scale rice growers in the project areas obtained 4.7 and 3.2t/ha yields in the dry and monsoon season respectively (n=2927).[1] Data from 1995 to 1997 show that project participants generate yields that are on average 4 per cent and 11 per cent higher than non-participants in the dry and monsoon seasons respectively (n=1640). This was achieved mainly through changes in fertilizer and pesticide use, transplanting time, spacing, variety and seed and seedling quality. While these increases are modest, they combine with reduced production costs to raise household incomes.

Most of the reduction in production costs is from reductions in insecticide use. In 1999, for example, the percentage of participants using insecticide on their rice at least once dropped from 86 per cent (n=6045) in 1995 to 11 per cent (n=7700) in the dry season, and from 76 per cent (n=6045) to 19 per cent (n=7700) in the monsoon. We estimate there have been savings of 486 takas per 0.11ha plot (n=360), which amounts to 4418 takas per hectare per year (US$92).[2]

After having obtained yield increases, farmers become willing and able to spend more on quality seed of improved or local varieties or to produce and preserve their own seed. In the highest external input use district, the percentage of participants producing their own rice seed went from 40 per cent to 83 per cent (n=200 respondents from 21 FFS) before and after project intervention. Irrigation costs do not change, while labour costs increase slightly as farmers give more importance to certain crop management practices such as planting in rows and weeding. Overall, the costs of production for dry season rice decreased by 12 per cent, from 1.70 takas per kilo of rice produced (baseline) to 1.40 takas after two years. For monsoon rice, the per-hectare costs of production declined 30 per cent, from 1.30 takas to 0.90 takas per kilo of rice produced.[3]

For farmers, the other main consideration in rice production is reducing year-to-year yield fluctuations. Participating farmers have developed the capacity to maintain a minimum threshold of production from their rice fields while reducing their costs of production. This is essential to food security. Farmers working with the project experienced variation in yield averaging plus or minus 5 per cent, while control farmers experienced average variances in yield of 9.5 per cent.[4] The data from pairs of farmers matched by rice variety and irrigation source during seven seasons between 1996 and 1999 show that the total crop failure rate of participants was 0.3 per cent, while that of control farmers was 0.54 per cent (n=1673 pairs of farmers).

Other Production Within the Rice Farming System

The programme is also promoting the integrated cultivation of rice, fish, dyke crops and timber trees. As rice is the staple food in Bangladesh, and also probably due to the influence of the Green Revolution, farmers invest considerable time and effort in rice while neglecting other crops. By introducing other crops within the same field and managing them through an ecological approach, farmers are able to lower their dependency on rice both as a crop and in their diet.

Throughout the year, farmers participating in the programme harvest vegetables and fish in addition to rice from their fields. They sell and consume these products in varying proportions depending on their needs and their production. The cost of production of these complementary crops produced in the rice field is minimal, and rice production remains the same or possibly increases (Barzman and Das, 2000).

Rice–Fish Production

The number of farmers who produce fish in their rice fields varies from area to area depending on the local physical characteristics of the soil and the water management. By the end of 1998, 30 per cent of participating farmers (n=11,383) were experimenting with this technique, and data taken two-and-a-half years after project intervention show that their numbers actually increase after the initial trial phase. Farmers practising this technique dig a small ditch in the corner of the field to provide shelter to fish that they introduce as fingerlings at the beginning of the season. The fish find their own food within the rice field, making their maintenance cost negligible. In the case of the higher valued native fish species, the main cost of production is the purchase of fingerlings. For common carp on the other hand, farmers typically produce their own fingerlings.

Most of the labour in this fish production system concerns the raising of dykes and the digging of the corner ditch. The cost of water can be considered a rice production cost. Usually, family members take care of these complementary production activities. In 1997, the average cost of production of fish per rice–fish plot (n=212) was 791 takas per year, which is equivalent to 5271 takas per hectare (US$110). All rice–fish farmers consume a part of the harvest, and most also sell the remainder.

In 1997, after production failures were excluded, the average annual potential net return per farmer was 6241 takas (n=212), which is equivalent to 53,290 takas per hectare (US$1110).[5] Some rice–fish growers make extra income and reduce their fish production costs by producing and selling their own common carp fingerlings during the irrigated rice season. By the end of 1998, 33 per cent of participating farmers (n=11,383) were doing this and obtained on average 609 takas per farmer (US$13) from their sales of fingerlings.

Rice and Dyke Crops

Vegetables can be grown on the dykes surrounding the rice field if these are raised and widened. With this technique, project participants have been producing country bean, yard-long bean and a number of squash species that are planted in small pits filled with compost. In 1998, the average potential net return from dyke crops is 1470 takas per farmer (n=145).[6] This is equivalent to 9800 takas (US$204) per hectare of rice land over the year. The inputs required are mainly seed, organic compost for pit preparation and labour. Farmers usually just use a portion of the fertilizer that they have purchased for their rice production on their vegetable dyke crops and the organic compost is produced at no cost.

Vegetable seed is increasingly produced locally at no financial cost. One study conducted in the highest external input use district showed that the number of women participants producing their own vegetable seed went from 26 per cent to 69 per cent (n=200 respondents from 21 FFS).[7] This helps to make the activity reasonably remunerative.

Trees on Dykes

There are cultural practices, such as periodical pruning of roots and branches, which make it possible to grow trees on the dykes without affecting the rice crop. This type of pruning is not favourable to fruit trees but is perfectly appropriate to trees producing timber, cooking fuel and fodder. In December 1998, 28 per cent of project farmers (n=19,450 farmers) were experimenting with this technique. We have observed that two years after the project had left, farmers continue to tend their trees. It is too early to evaluate the economic benefits of this technique, but we are already observing that the number of farmers planting trees on their dykes is growing and that many have initiated small-scale tree nursery businesses to supply their community with the required tree saplings.

Rice–Fish–Dyke Crops

Since the integrated system does not decrease the yields of any one component, rice farmers who integrate fish and dyke crop production naturally compound their net returns. Already, 20 per cent of the poor and marginal farmers in the programme have adopted this combined practice.

Overall Economic Impacts

The exact economic impact of these practices on households is difficult to assess. The figures given above are based on a single 0.11ha plot whereas our participating households own about four plots averaging a total of 0.45ha on which it is impossible to say how many of the innovations are practised. Another difficulty is that households consume much of the increased production as food and donate a significant portion to relatives and neighbours, making an economic analysis nearly impossible. Even if the actual returns per household attributable to the innovations were known, we still would not know how much this would represent relative to the total income of the household and its needs. Nevertheless, the high adoption rates of the new practices make it clear that they make economic sense.

Non-economic Benefits

Measurable economic returns represent only a fraction of the benefits generated by the programme. Households and communities also benefit from improved nutrition, decreased environmental degradation and empowerment.

Nutrition

Our indicators of diet show that the project contributes significantly to improving project participants' nutrition. The project is helping to reduce over-reliance on rice by increasing vegetable and fish consumption. We find that after project intervention, the relative number of project households consuming vegetables every day doubled while those consuming vegetables only once a week decreased 12-fold. Similarly, the number of households consuming fish every day more than doubled while those consuming fish only once a week decreased nine-fold. And lastly, the amount of edible oil used by households – a good indicator of the total amount of food cooked – increased by 21 per cent on average.

Environment

It is well established that simply foregoing the use of insecticide dramatically increases the abundance of beneficial insects and will reduce farmers' and consumers' exposure to hazardous chemicals (Pingali and Gerpacio, 1997). The reduced reliance on insecticide is long lasting. Our post-project survey shows that even two-and-half years after project intervention, 77 per cent of participating farmers (n=1200 farmers) continue to grow rice during the dry season – the high input season – without using insecticides, a long-term change up from 14 per cent before project intervention (n=400 farmers). Another intervention with a probably significant impact on environmental sustainability is the planting of trees on dykes. Trees add a structural dimension to the rice field and provide refuge to many life forms. The data set just mentioned also shows that two-and-a-half years after phase-out, 41 per cent of FFS participants (n=1200 farmers) still have trees growing on some of their dykes, a technique totally unknown before project intervention (Barzman and Banu, 2000).

Empowerment

Through farmer involvement, the programme hopes to strengthen the capacity of farmers and their communities to continue to innovate and to respond to future challenges without requiring project intervention (Röling and de Jong, 1998). With a better understanding of the agroecosystem and familiarity with the experimental method, farmers build up their confidence in their own knowledge, learning capacity and decision-making capabilities. Field staff who work directly with farmers have no doubt that the programme does achieve this other objective. Yet this is a difficult entity to quantify.

There are signs that, as a result of project intervention, farmer innovation is taking place. Some participating farmers moved beyond the 'prototype techniques' and adapted them to suit their own needs. For example, there are farmers now experimenting with the cultivation of shrimp instead of, or together with, fish. They develop their own knowledge by studying shrimp production by larger producers, asking questions and using their neighbours and other members of the group as a source of knowledge. Whereas the programme focuses on the use of common carp – an exotic but easy-to-rear species – as the fish species of choice, many farmers are producing native species, including some that are endangered. The same is taking place in vegetable dyke crops where farmers are trying vegetable species not associated with the programme. Some farmers have taken the vegetable crops and planted them around the homestead or along roadsides and the same is true for trees originally intended for dykes.

Another indicator of empowerment is the level of organization and collective action. Some FFS, independently of project staff, have formed formal grassroots organizations. The potential impact of these organizations is great and the programme will study their development as a first step towards enhancing this process. Already, some farmer groups have taken collective action in marketing their products. Others are producing fish in large areas made up of a number of adjoining individual fields in which fish fingerlings are jointly procured and from which returns are shared among the group.

Within the FFS, farmers experiment and compare their results. They observe and monitor other farmers' fields where organizations have already installed demonstration plots. FFS participants are better able to evaluate new technical options presented by agricultural researchers or extension services. They are also more critical of proposed new technologies.

The project is also contributing to the empowerment of women, who in rural Bangladesh are confined to the household and kept away from sources of power. The project encourages women to work away from the homestead in the rice field. One survey showed that two-and-a-half years after the project's departure, 74 per cent of them were still tending vegetable dyke crops and all of them were involved in rice–fish culture (n=1200 responses). Farmer leaders, half of whom are women, receive additional training and subsequently serve as resource people in their communities. Such women often report a newly acquired sense of worth and enhanced social status. Some of these women have been elected to the local government.

LESSONS FROM THE RICE PROGRAMME IN BANGLADESH

After eight years of programme implementation, some valuable lessons have emerged. The programme, which can be considered successful on several levels, has received regional and even international recognition. Some of the operational reasons for this success include the following:

- Projects sought ways to place farmers in the centre of the learning process and enabled project staff and farmers to communicate productively in spite of major socioeconomic differences between them.
- The experiential approach, by treating the farmer as a researcher, created an environment conducive to learning and confidence building.
- Projects used sustainable agriculture techniques modified from native practices that had already been intensely tested and therefore presented minimum risks to subsistence farmers.
- Men, women and sometimes children from the same household were provided with training to ensure support from within the family.

- Working with groups on collective activities has meant a better use of resources, increased sharing of knowledge, better coordination within the community, and more acceptance of women's involvement.

The programme developed several strategies to promote wider acceptance of women working in rice fields. Staff had to be willing to learn and be flexible enough to regularly modify strategies, and a system was developed that enabled farmers to conduct their own monitoring and analysis of results.

DIFFICULTIES FACING THE PROGRAMME

The success that the programme has met with respect to its objective of increasing rice field productivity has overshadowed its other objective: empowering farmers. Both farmers and staff tend to focus on the relatively rapid agronomic results obtained. And since the agroecological conditions in Bangladesh are rather homogeneous, the same agricultural practices tend to yield similar results in different locations. Because of this, projects tend to become promoters of a technological package which, even if itself sustainable, is not conducive to sustainably increasing farmers' decision-making and management capacities.

The agricultural practices associated with the programme were designed as *entry points* to obtain quick results and from which projects could move on to long-term changes. But they sometimes become an end in themselves in a way that is reminiscent of the old transfer-of-technology extension model. Obviously, farmers' needs are connected to the opportunities that exist in their area in terms of land type, land tenure, irrigation, availability of agricultural inputs and resource people or organizations. Considering these factors, programme implementation, extension strategies and priority agricultural practices need to differ from area to area.

Another difficulty concerns access to services. Farmers generate knowledge through experimentation, observation and sharing, but they still need access to extension services, research organizations and markets to get the most out of their learning. There are a number of constraints that make these services inaccessible to farmers. Sometimes these are a matter of psychology. Poor and marginal farmers often lack confidence to meet 'professionals' and ask them questions. Most farmers are barely literate and rely on somebody else to read for them. Sometimes the constraints are geographical. Research centres are often distant from farmers not only in physical terms, but also in the way they present research materials, subjects and results. Such limitations need to be tackled and overcome.

While implementation of the rice programme is proceeding very well overall, there remain many other challenges. Establishing equal working and learning opportunities for men and women is one such major challenge in Bangladesh. Another challenge has to do with the way we measure success. Organizations are under pressure from government or donor agencies to demonstrate quantifiable outputs. This can impinge upon the quality of the service provided as well as on the principle of putting farmers at the centre.

For example, a programme may have contractual responsibilities with donors to train a certain number of farmers annually in order to justify budget allocations. If the number is overestimated, the rush to reach this number before the end of the year will decrease the quality of the work and the sustainability of the learning process. The organization needs to protect the quality of its interventions by avoiding an excessive preoccupation with target numbers.

CONCLUSIONS

The rice programme is still learning from on-going activities and innovations. It strives to be an education programme rather than just agricultural extension. Clearly, it has been quite successful in facilitating a process in which farmers can get more out of their rice fields. To what degree this is due to their increased capacity in decision-making and management is not yet clear. It is clear, however, that it represents an improvement over the training-lecture types of programmes whose limitations are well known in terms of their doubtful sustainability, dubious quality, and poor fit with farmers' needs.

The programme needs to continue to focus on giving farmers opportunities to take control of their own learning. The social, economic and cultural milieu in Bangladesh does not easily support this. To start working with poor and marginal farmers, it has been necessary to adopt specific and beneficial sustainable agriculture entry points that offer a high likelihood of success. But the programme must ensure that these remain entry points and not sole objectives. In any event, the high adoption rates and the increased returns from the innovations brought about by the programme attest to the appropriateness of the intervention. Longer-term impacts beyond the adoption of particular sustainable agriculture practices may be taking place but still remain to be measured.

NOTES

1   The growers surveyed owned, on average, 0.45ha of land. Note: Through the 1999 Bellagio conference, CARE/Bangladesh learned about the system of rice intensification (Chapter 12). Farmer field-school participants using SRI methods in Kishoreganj district in 2000 averaged 6.5t/ha. Farmers cooperating with the Department of Agricultural Extension in Kishoreganj averaged 7.5t/ha with SRI methods.

2   48 takas = US$1. One kilo of rice (paddy) is worth 5–7 takas (US$0.10–0.14).

3   These data are from the NOPEST mid-term review and monitoring report.

4   In compiling results, the areas considered were ones with no climatic disasters during the three seasons.

5   Since a large part of the fish production is consumed by the household, potential net return – defined as the value of the harvest minus the cost of producing it – is used rather than actual net return.

6   As with fish, most vegetables are consumed, so potential net return – the value of the harvest minus the cost of producing it – is also used here.

7   Caring for seed is traditionally a woman's responsibility.

Chapter 17

Integrated Pest and Crop Management in Sri Lanka

Keith A Jones

In Sri Lanka, approximately two-thirds of the population depend on the agricultural sector, and approximately 40 per cent of the population are engaged in farming. The rural poor account for about four-fifths of national poverty, with 24 per cent of the rural population living in absolute poverty, according to the World Bank (1995).[1] Agricultural production, excluding the plantation sector, is dominated by rice, with significant production of vegetables and other field crops (OFC), particularly chillies, grams and pulses, in some areas.

Agricultural productivity, which increased dramatically after independence in 1948, has stagnated in recent years. Taking rice as an example, the average paddy yield in 1950 was approximately 1.5t/ha; however, with the introduction of new, improved varieties and the use of increasing amounts of agrochemicals, yields had increased to about 3.5t/ha by the mid-1980s (Central Bank, 1998).[2] Since then, productivity has remained at or below this level. While rice yields are similar to those now obtained in certain Asian countries such as India and The Philippines, they are lower than those obtained elsewhere in Asia, for example Indonesia, which reached 4.35t/ha between 1990 and 1994 (FAO, 1995), though yields there have not advanced since then. These levels are well below potential yields for rice (see Chapter 12).

Overall, the increases in production since 1950 have not kept pace with increased demand, mostly as a result of population growth. Imports of rice, wheat and flour rose from about 730,000t in 1950 to almost 900,000t in 1994; although Sri Lanka's population growth is relatively low for a developing country – 0.9 per cent in 2001 – shortfalls in supply will continue and probably rise if production is not increased. Total output of rice can be raised by bringing more land area under production – which inevitably encroaches onto increasingly marginal and/or environmentally sensitive areas – or through increases in productivity.

Coupled to these macroproblems are the difficulties faced by individual farmers. The vast majority of small-scale farmers face low or even negative economic returns from farming, resulting from a combination of high-cost inputs and stagnating yields.[3] This has led to a loss of interest in farming, increasing rural unemployment and underemployment, and it ultimately threatens the livelihood security of the rural population.

A further problem associated with the excessive and often inappropriate use of agricultural inputs is environmental contamination and compromised human health, both of which affect livelihood security (see note 3). These problems have led to a growing realization that increasing agricultural production cannot rely on ever-greater inputs, or on the continued expansion of agricultural land. Ultimately these approaches are unsustainable. The challenge facing Sri Lankan farmers, and Sri Lanka as a whole, is to intensify production in a sustainable manner, one that is both profitable for farmers and based on environmentally sound principles.

It has been demonstrated in a number of countries in the Asian region that adoption of systems of *integrated pest management* (IPM) can both reduce the unnecessary use of chemical inputs, and also increase yields (Ruchijat and Sukmaraganda, 1992; Ooi, 1996; see also Chapter 16). IPM has been described as 'a set of practices that maintain pest problems at a level below that which causes economically significant losses; it emphasizes minimal intervention – particularly with synthetic biocides – and husbandry of natural regulating mechanisms be they biological or cultural.'

This chapter reports on the impact of an IPM extension project on yields and farmer incomes in Sri Lanka.[4] The programme has targeted 30,000 farmers across the whole of the country, covering a wide range of agroecological and socioeconomic conditions, including both irrigated and rainfed agricultural systems. Project locations included the lowland dry zone in the north and east, lowland wet zone areas in the south and the upland intermediate zone in the central highlands of the island. Project sites were located also in some of the conflicted areas of the island, including those held by the Liberation Tigers for Tamil Eelam (LTTE) in the north and east, where much of the infrastructure, including irrigation systems, is damaged. Access to extension services is extremely limited or non-existent there, and many other agricultural inputs are limited, unavailable and/or expensive – for example, urea-based fertilizers are not available.

Landholdings of project farmers ranged from less than one hectare to more than ten, with an average of approximately one hectare, which is close to the average for the country as a whole.[5] Although the main crop covered by the project was rice, vegetables and OFCs were also included. It was important to cover this wide range of conditions in the programme to demonstrate that the approach taken and the impact obtained are not dependent on a certain set of conditions, but are, with appropriate adaptation, broadly applicable within Sri Lanka.

The project was originally designed to reduce farmers' exposure to hazardous chemicals. However, it was soon realized that the most effective way to achieve this was to reduce their unnecessary and excessive use. It is axiomatic that farmers will not change their practices if this results in reduced yields and/or incomes, or if it increases the risk of crop failure. Farmers generally believe that reducing pesticide or fertilizer application will result in greater crop losses and lower yields. To address this belief, farmers need to understand the impact of their pest control and cultivation actions if they are to be confident enough to change their practices on a sustainable basis.

The project therefore adopted the promotion of IPM as its core activity. The project has been described previously in some detail (Jones, 1996; 1999a). The methodology adopted to promote IPM was the farmer field-school (FFS), an approach developed by the United Nations Food and Agriculture Organization (FAO) in Southeast Asia and explained briefly in the preceding chapter. This participatory training approach aims to enhance both farmer knowledge and empowerment. Farmer groups are formed, and through a process of experiential learning the members develop a thorough knowledge of their agroecosystems, as well as the effects of their interventions on these systems. As a result, farmers gain the confidence to assess their own problems and to formulate solutions.

The approach does not deal solely with pest control; rather it focuses on the need to grow a healthy crop as a key to reducing pest problems and maximizing yields. Thus, land preparation, water management and appropriate cultivation practices are included in the hands-on training along with pest management. It is more accurate to refer to this as the promotion of *integrated crop management* (ICM) rather than IPM. Farmers are encouraged to conduct their own experiments. For example, they test different fertilizer regimes, artificially apply different levels of damage to small areas of their crop to assess the effect of pest damage on yield, or observe the impact of different levels of natural predation on pest numbers. In all cases, the farmers by collecting the data themselves and analysing it become convinced that the results are genuine.

The FFS programme lasts a whole season, with farmers attending for one day or morning each week. Observations are made in one of the farmers' own fields. No subsidized inputs are provided as this would rely on unsustainable material incentives and encourage a dependency attitude. Further, crop yields are not underwritten by the project, so that farmers are conscious of the importance of their own decisions. The programme wants them to become persuaded that the IPM practices are worth pursuing for farmers' own benefit, without any outside payments.

PROJECT IMPACT

Over a four-year period covering seven growing seasons, over 130 FFS programmes were undertaken. The majority (93) were on rice, with the remainder on aubergines, beans, bitter gourd, cabbage, capsicum, chilli, cowpea, longbean, okra, onion, potato, snake gourd and tomato. The results, summarized below, clearly demonstrate that adoption of IPM results in significant increases in yield and farm income.

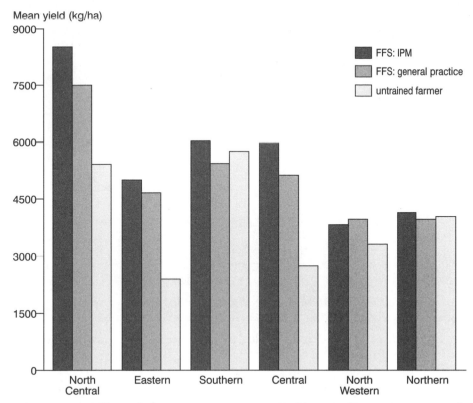

Figure 17.1 *Effect of IPM Practices on Rice Yields in Different Provinces of Sri Lanka: Results Obtained in the Main (Wet) Season, 1997–1998*

Rice yields were, on average, between 11.5 and 44 per cent higher for fields where IPM practices were used.[6] Yields differed significantly between different areas: for example, lowest overall yields were recorded in the LTTE-held areas of the north, and the highest yields in the irrigated areas of the North Central Province. This difference can be easily explained by the lack of water and available fertilizer in the northern areas, compared with adequate supplies in North Central Province. However, in all areas, whatever the conditions, average yields were higher in the fields where IPM practices were applied (Figure 17.1). Similarly, although season-to-season yields varied considerably, reaching a low between the 1996 and 1997 dry seasons as a result of a prolonged island-wide drought, yields were always highest in the IPM fields (Figure 17.2).

The same pattern was reflected in farmer incomes, with overall increases of 38 to 178 per cent in mean net income being achieved, from 8037 Sri Lankan rupees (R) per hectare to 22,346Rs/ha, equivalent to US$115–319. This rise in income was not only a result of increased yields but also of lower expenditure on inputs, in particular on synthetic chemical insecticides. On average, the number of insecticide applications to paddy fell six-fold, from 2.9 to less than 0.5 per season, representing a reduction in expenditure from 2762Rs/ha (US$39.50/ha) to 412Rs/ha (US$5.90/ha).

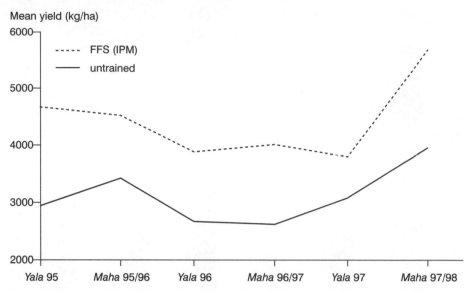

Figure 17.2 *Variation in Rice Yields over Six Growing Seasons*[7]

In contrast to rice, detailed field-level agroecology has not been extensively studied for many OFC and vegetable crops. Moreover, most alternative pest control techniques have not been fully tested or are unavailable. However, the act of observing the crop in the field and determining levels of pest attack, plus an understanding of the principles of agroecology, resulted in dramatic reductions in the amount of chemical pesticides used on these other crops, with associated cost savings.

The impact of adopting IPM practices on vegetable and OFC production was similar to that for rice. Table 17.1 gives examples for three crops: aubergine, chilli and potato. It reveals yield increases of between 7 and 44 per cent, with associated increases in income of 12 to 129 per cent being achieved. Again, this was coupled with a significant reduction in pesticide use, which fell on average by over two-thirds. The reduction in pesticide expenditure was from over 52,000Rs/ha to less than 15,000 – from US$743–214.

PROJECT SPREAD

Early in the programme, it became apparent that farmers in areas surrounding the field schools were following the programme with interest. Initially sceptical of IPM, as indeed were most FFS participants, neighbouring farmers were by the end of the season able to observe for themselves the positive impact of adopting IPM practices. They therefore sought information from the farmers participating in the FFS programme and started adopting IPM practices themselves.

Surveys were undertaken to measure this lateral spread (Jones, 1999). From the results of 20 surveys, an estimated average 'degree of spread' was observed to be 13-fold. Overall 73 per cent – between 30 and 100 per cent –

Table 17.1 *Impact of Adopting IPM Practices on Yield and Net Income for Vegetable and Other Field Crops*

| | Yield in kg/ha[8] | | Net income in Rs/ha (US$/ha)[9] | |
	IPM	General practice	IPM	General practice
Aubergine	23,694	22,008	540,325 (7719)	481,982 (6885)
Chilli	2076	1445	102,594 (1466)	44,800 (640)
Potato	10,216	9534	157,449 (2249)	98,096 (1401)

of farmers interviewed had adopted IPM.[10] Thus, from a total of 4287 farmers who attended FFS programmes during the project, it is estimated that approximately 55,000 farmers have adopted IPM (Jones, 1999b).

This project has demonstrated that adoption of IPM (or ICM) practices can increase the average yield of a number of crops. While not yet reaching the *full* potential yield, these practices have greatly increased yields under the conditions that farmers face. Thus, in the areas that are most suitable for rice cultivation in Sri Lanka, such as the North Central Province, yields as high as 8t/ha were achieved in some IPM plots.

Such high yields are not possible (at least at present) in some more marginal lands, or in the North Province where water and inputs are severely restricted. However, increases in yields obtained in any and all areas can significantly contribute to food supply. With rice, the average increases if achieved for all rice production in Sri Lanka would amount to between 189,000 and 555,000 extra tonnes of rice available per annum for consumption, using 1994 production figures as a base.

LEARNING FROM THIS EXPERIENCE

Similar increases in yield have been demonstrated in other countries where IPM has been adopted in rice (Ooi, 1996). There are a number of possible reasons for this increase. A significant amount can be attributed to improvement in general agronomic practices (water management, land preparation: see Footnote 6) and at least 10 per cent to a reduction in the use of chemical pesticides – the difference observed between the IPM and general practice plots (see Figure 17.1).

This could be the result of fewer negative effects of pesticide use that are reduced by IPM practices. The best documented is the destruction of beneficial insects, leading to the breakdown of natural control of pests and resulting in increased pest problems (eg Kenmore et al, 1984; Kenmore, 1986). This may be compounded by development of resistance to pesticides by certain pest species. In Sri Lanka, large outbreaks of brown planthopper occurred during 1997, and farmers observed that pest numbers were lower in IPM fields compared to general practice fields.

A number of other indirect as well as direct effects of pesticide on crop production can also occur:

- It is well known that crop damage can result from inappropriate choice, or use, of herbicides.
- It has also been documented that some fungicides and insecticides can suppress plant growth or lead to unproductive growth.[11]
- Pesticides can destroy beneficial soil micro-organisms that are essential for healthy plant growth (Brady, 1982).
- Some damage to plants by insects can have beneficial rather than detrimental effects; up to 25 per cent or more of the leaves can be removed from rice plants without loss of yield, and often a slight increase in yield has been observed (Jones, 1999).

Anecdotal evidence to support these effects was given by Sri Lankan farmers who reported that aubergines and fruits were larger in IPM fields, where pesticide applications were lower. Moreover, fruit cropping from IPM fields is lasting for a longer period than from non-IPM fields.

Inappropriate or excessive use of fertilizer can also result in increased pest attack and/or reduced yield, primarily through alterations in crop growth or nutritional level (Herzog and Funderburk, 1986). For example, high applications of nitrogen, phosphorous and potassium fertilizer increase the incidence of the rice leaf-folder insect (Crop Protection Compendium, 1997) and also promote unproductive leafy growth which can encourage pest attack. Excessive use of nitrogen (as well as high plant density) can promote a buildup of populations of brown planthopper (Uhm et al, 1985). In Sri Lanka, several farmers interviewed by the IPM project reported increased insect attack following excessive use of fertilizer, the latter being in response to a recently introduced government subsidy.

An understanding of the impact of inputs on crop growth and the wider ecology is essential if farmers are to optimize yields. Their lack of understanding of these factors is likely, ultimately, to result in negative rather than positive impacts. The results obtained in this programme have provided further support for the view that continued increases in agrochemical inputs do not result in commensurate increases in yield, and ultimately are detrimental to yields. This questions the wisdom of subsidies for agrochemical inputs. Support for other interventions such as widescale FFS programmes, or for marketing activities, is likely to have a more sustainable impact.[12]

The programme has demonstrated that adoption of basic IPM/ICM principles can result in improved crop productivity, increases in farmer income and an overall improvement in livelihood security of the rural poor. Undoubtedly, further improvements could be made through the further development of other appropriate technologies, eg biological pesticides, drought- or disease-resistant varieties and more selective, safer chemicals. However, this needs to be done in partnership with farmers, taking into account their needs and the practicalities of adopting new technologies in order to promote sustainability. The cost and availability of new technology are also important factors.

The key to sustainable uptake of improved cultivation (and pest management) practices is farmer education. The participatory approach adopted in the FFS has been shown to be an effective mechanism for human resource development. The approach builds farmers' confidence in experimentation and decision-making and results in greater empowerment. This can be a more significant long-term contribution to maintaining and improving crop production, as well as to increasing livelihood security, than IPM itself. The net result is that farmers are able to adapt practices to changing conditions and to seek appropriate technical support when required.

For any large-scale impact, mechanisms must be developed to provide suitable technologies, appropriate training and information to all farmers. In Sri Lanka, the task is made easier by the large degree of farmer-to-farmer spread already in evidence as well as the high literacy rate, currently almost 90 per cent. This means a wide range of training literature can be used and that training programmes can be accomplished in a single season. This can be contrasted to similar programmes in Bangladesh, where the literacy rate is much lower and, as a consequence, FFS programmes are planned to last several seasons. However, similar impacts on yields have been obtained, as seen in Chapter 16.

Poor spread has been observed between different communities in Sri Lanka. For example, in the Eastern Province, there is little interaction between Tamil and Muslim communities, and there has been therefore little spread from FFS-trained Tamil farmers to neighbouring Muslim villages. This problem was been locally mitigated by project-mediated cross-visits. However, mechanisms need to be in place for such actions to be repeated and extended elsewhere when similar problems arise. This will require commitment and cooperation among government institutions, community-based organizations and farmers. Continued and sustained support for the institutional framework to maintain these links will require continued support from all stakeholders, as well as recognition of the important role that is played by each group.

NOTES

1   The absolute poor are families/households that spend more than 90 per cent of their earnings on food. The figure of 24 per cent does not include data from the conflict areas of the north and east of Sri Lanka, where an estimated 16 per cent of the population live. Within the conflict areas, more than 800,000 people have been displaced and the socioeconomic infrastructure disrupted.
2   This doubling of yield was accompanied by a 13-fold increase in the application of inorganic fertilizers to rice between 1960 to 1990. During the 1970s, the total amount of pesticides applied to all crops increased 20-fold. From 1980 to 1986, the volume of insecticides used in Sri Lanka increased by 106 per cent, fungicides by 128 per cent and herbicides by 214 per cent.
3   A significant portion of production costs is for pest control. On rice, an estimated 50–60 per cent of farmers apply at least twice the recommended dosage of pesticide. This overuse and misuse of highly toxic chemical substances is not only wasteful and costly, but also causes environmental degradation which, in turn, affects the economic and physical wellbeing of farmers and the country as a whole. This is compounded by a lack of safety precautions taken by farmers when handling pesticides. Pesticide poisonings are notoriously high in Sri Lanka. From

1973–1978 an average of over 14,000 cases of pesticide poisoning were reported annually, of which almost 1000 were fatal. Many of these were deliberate suicide attempts. Unfortunately, comparable recent data are hard to come by. There are few reliable studies of occupational exposure. One study in two districts found that acute occupational poisoning among farmers was 7 and 22 per cent. It is also estimated that farmers spend approximately 10 per cent of their average net family income on pesticides.

4   This project, known as INTEGRATED, has been implemented by CARE International and was funded by the European Community, UK's Department for International Development (DFID) and CARE International. The views expressed in this chapter are those of the author and are not necessarily those of the funding or implementing organizations.

5   The average landholding in the non-plantation sector is 0.8ha, with 64 per cent of farmers cultivating less than this extent (Kudagamage et al, 1992). Farmers interviewed as part of the project baseline survey cultivated between 0.4 to 0.8ha on average, with a range of 0.06 to 12ha.

6   Three treatments were compared. Two, the IPM and general practice fields/plots, were part of the FFS programme and were located next to each other. The general practice plot was established to compare the effect of the farmers' normal practice with IPM practices. Land preparation and water management were the same for both plots. However, the pest control decisions of the farmers who managed the general practices plot and who were also part of the FFS programme soon mirrored those of the IPM plot. Therefore additional data were collected from farmers in the area surrounding the FFS site who were not part of the training programme; these are referred to in comparative evaluations as 'untrained farmers'. The small increase in mean yield resulting from IPM compared to 'general practice' is attributable to changes in pest control practices in the latter. The larger increase in comparison to 'untrained farmers' is a combination of changes in pest control practices, better land preparation, water management, fertilizer and pest control practices and, possibly, more careful harvesting from the smaller FFS sites.

7   *Maha* season is the main growing season, with rains and supplementary irrigation; *yala* is the dry season, when crops depend entirely on irrigation.

8   Undamaged fruit/pods/tubers, respectively.

9   Income per hectare is high, but farmers growing vegetables and OFC are cultivating small plots, often 0.1ha or less.

10 Defined as at least two IPM techniques, which included pest control operations based on regular crop observations, *in addition to* components such as proper land preparation, seed selection, proper water management, use of non-chemical control methods and/or proper use of pesticides.

11 Toscano et al (1982) have reported that use of methomyl reduced growth of lettuce seedlings by 70 per cent due to interference with photosynthesis; also, organophosphate insecticides can increase unproductive vegetative branches, delay plant maturity and lower yield in cotton (Lloyd and Krieg, 1987). Fungicides may also have a similar effect (Horn, 1988).

12 Farmer surveys carried out as part of the programme revealed that inadequate marketing was one of the major constraints faced by Sri Lankan farmers.

Chapter 18

Increasing the Scope for Food Crop Production on Sloping Lands in Asia: Contour Farming with Natural Vegetative Strips in the Philippines

Dennis Garrity

Alternative, agroecological or low-input agricultural production systems emphasize reliance on the resources present within the farm (Altieri, 1995; Pretty, 1995). Conversely, 'conventional' approaches to agricultural development are characterized as depending more heavily on external inputs. Our task in this collaborative effort has been to assess the extent to which so-called alternative agriculture can contribute to increased world food production in the future. This issue is significant for those making decisions on how to invest in agricultural research, and for those involved in assisting farmers to apply that research on the land.

There is reason to feel some discomfort with classifying agricultural practices into 'alternative' or 'conventional'. It can be more of an ideological divide than a logical one. The dichotomy is not particularly useful either scientifically or practically since most real-world systems defy such classification. A number of the cases in Part 2 have discussed practices or combinations of practices that might be identified as being either alternative or conventional depending upon one's point of view, or that have changed over time. The traditional rainfed systems analysed in Chapter 7 originally used local resources available to the household but now incorporate some external inputs to maintain soil fertility as population has grown, chemical fertilizer has become available and cash incomes have increased. On the other hand, some high-input conventional systems such as the irrigated rice systems considered in Chapter 16 are evolving towards alternative approaches through less pesticide use.

Our objective should be to find combinations of practices that best meet the complex needs and objectives of farm families and of their societies, and that also maintain the health of the production base. Farmers' most pressing objective is to maximize the productivity of every input they have available at reasonable cost. To achieve this objective, yield increases tend to be a dominant means of enhancing income and family food security. This requires a pragmatic, flexible approach that is open to all possibilities, without preconceptions as to whether external or internal inputs are 'good' or 'bad'.

Farming systems in many parts of the tropics are increasing in productivity through the use of commercial fertilizers. Some may consider it paradoxical that the application of external nutrients can increase the amount of organic matter available for enriching the soil. Meanwhile, conventional systems are being improved all the time by the application of ecological principles to the management of pests, weeds and nutrients, and their interactions.

As a systems agronomist who has worked on smallholder upland and lowland rainfed farming systems in Asia for 25 years, I have collaborated with farmers in researching crop rotations, intercropping, multiple cropping, green manures and other organic and inorganic nutrient management practices, conservation farming systems and agroforestry systems during that time. This experience has given me particular respect for the complexity of these systems, and the uncertainty that accompanies theory when applied in the real world. Solutions based either on agroecological or conventional approaches, when tested in the real world, often founder upon the shoals of one or more unforeseen factors.

This chapter explores these issues in the context of one particular case of changing smallholder agricultural systems in Southeast Asia. It argues that sharp distinctions between what is 'alternative' and what is 'conventional' miss the real point, raised in preceding chapters, that in the context of historically evolving farming systems, there is a compelling place for agroecologically-based practices alongside ones based on the best available chemical, genetic and engineering components. As proposed in Chapter 1, the most productive and sustainable farming systems will be some kind of 'hybrid'.

THE UPLANDS OF SOUTHEAST ASIA

In Asia, the land area having more than 8 per cent slope is estimated to be 900 million hectares, or 53 per cent of the total area (Magrath and Doolette, 1990). About 65 per cent of the region's rural population of 1.6 billion live in sloping areas. Approximately 19 per cent of this region is under closed forest, most of this being tropical rainforest, the reservoir of about 40 per cent of the biodiversity on the Earth. Degradation through overcutting and grazing is reducing productivity on much of the remaining stand (Doolette and Smyle, 1990). Forest cover is receding at a rate of about 1 per cent a year. The most recent estimates suggest that the rate of deforestation is not slowing, but rather accelerating. In much of the region, forest resources are integral to the agricultural system as sources of fodder and many other products.

The seriousness of soil erosion is not adequately known, but may be deduced from indirect evidence. The most striking indicator is the rate of sediment passing into the oceans from the major river systems of the world. Global data highlight Asia as being in a class by itself in this regard. Rates of sediment deposition in the oceans in this region are on the order of ten times higher than from areas of comparable size anywhere else in the world (Milliman and Meade, 1983).

Until recently, upland areas were not subjected to serious human settlement pressure. Currently, however, these ecosystems are undergoing major transformation in all countries due to greatly accelerated in-migration and rapid natural population increase.[1] Settlement is dominated by small-scale farms that produce cereal crops, mainly maize and upland rice to meet subsistence food needs. The soils typically are found on rolling to steeply hilly topography with high erosion potential, severe phosphorous deficiency, aluminium toxicity and low cation exchange capacity (CEC).[2] Large portions of these lands have been deforested and converted to short-fallow rotation systems or to permanent food crop cultivation, with an accelerating pace of ecosystem degradation (Garrity and Sajise, 1992).

The nations of the region are progressively opening their economies, and participation in global markets is accelerating. This is having profound changes on upland livelihood systems and on the upland environment. The economies of mainland Southeast Asia are interacting more vigorously as new roads, shipping and railroads facilitate cross-border trade. World market demand for perennial tree products produced in insular Southeast Asia is spurring smallholder expansion of production of rubber, oil palm, tree resins and various fruits, as well as timber production on-farm. These forces will continue to affect landuse change in complex ways well into the future.

EVOLVING FARMING SYSTEMS IN NORTHERN MINDANAO, THE PHILIPPINES

A research location in northern Mindanao, The Philippines – Claveria – was selected in 1984 by the International Centre for Research in Agroforestry (ICRAF) as representative of the problematic complex of acid upland environments in the region (Magbanua and Garrity, 1990). Intensive on-farm research has been done on sustainable upland farming systems for strongly acidic soils. The agroecosystems in the Northern Mindanao uplands were under dense humid *dipterocarp* forest until the early 20th century. Swidden agriculture was practised on a very limited portion of the land.

As substantial areas of the old-growth timber were harvested by logging companies, small-scale farmers from the central Philippine islands followed the logging operations. Dry-season burning, in association with swidden farming, converted large areas into grasslands. Farmers cultivated upland rice and maize in a grass-fallow rotation on the flatter areas. Coffee, coconuts and perennial fruit trees were planted on small areas during the 1950–1970 period. Their area increased from 4 per cent to 30 per cent of the land area between 1967 and 1988. Market tomatoes became an important crop in the 1970s.

The area of annual cropland doubled between 1967 and 1988, reaching 41 per cent of the total area (Garrity and Agustin, 1995). The previously scattered cultivated areas coalesced into large contiguous zones of tilled land. During a 40-year period, 1949–1988, the area under field crops increased five-fold. In the 1980s maize was the dominant crop, cultivated twice a year with local open-pollinated varieties and little or no application of fertilizer. Farm sizes averaged 3ha (Mandac et al, 1986).

The fallow rotation system evolved into continuous cultivation due to intense pressure for land. Although the clean-cultivated fields tilled with animal power extended to the steepest slopes (>40 per cent), there was little or no contour farming and no significant use of other conservation practices. Erosion rates were typically in the range of 60–200t/ha per year (Garrity et al, 1993).

Extensive farmer surveys established that farmers were clearly aware of the gravity of the situation. They were observing rapidly declining maize yields and were concerned about the consequences (Fujisaka and Garrity, 1989). It was evident that practical conservation farming options were needed for the range of slopes and farmer circumstances. There was much debate within the research team as to whether the serious degradation in land quality could be contained in time to prevent much of the farmland from being ruined beyond productive use.

New trends in landuse and farming practices during the past decade have supported a more favourable scenario. There has been widespread adoption by hundreds of farmers of contour farming based on natural vegetative strips (NVS), and the practice is rapidly spreading. Fertilizer use, which was practised on less than 10 per cent of the farms in 1984 (Mandac et al, 1986), had reached over 90 per cent of farms by 1998. Hybrid maize cultivars replaced local varieties to a similar extent, and maize yields that had ranged between 1–2t/ha in 1984 were, by 1998, between 2–3t/ha, depending on land quality and management practices.

Equally dramatic was an accelerated shift towards smallholder timber and fruit tree production systems. This was a market-driven phenomenon facilitated by strong productivity increases in maize and other annual crops. This enabled large parts of many farms to be released from food production to more profitable, and environmentally sustainable, tree-based systems with less soil erosion. The next section describes the development and adoption of the conservation buffer strip.

ADAPTATION AND ADOPTION OF CONTOUR BUFFER STRIPS FOR CONTINUOUS CROPPING

The main conservation farming practice prescribed for open-field intensive cultivation systems in Southeast Asia has been contour hedgerow systems with leguminous trees. This sloping agricultural land technology, commonly known as SALT, has become a common feature of extension programmes for sustainable agriculture on sloping uplands. These systems can control soil erosion effectively, even on steep slopes (Kiepe, 1995; Garrity, 1995). Extensive data from trials in six countries have confirmed that annual soil loss typically is reduced by 70 to 99 per cent when hedgerow systems are used correctly (Sajjapongse and Syers, 1995).

There are numerous reports of increased yield levels of annual crops when they are grown between hedgerows of leguminous trees. However, farmer adoption of these systems has been very low. Constraints include the tendency for the perennials to compete with crops for growth resources and hence to reduce crop yields, and the inadequate amounts of phosphorous cycled to the crop through the prunings. But the major problem is the extra labour needed to prune and maintain the hedgerows. We found that farmers' labour investment to prune their leguminous-tree hedgerows was about 31 days per hectare, or 124 days of annual labour for four prunings (ICRAF, 1996). This increased the total labour for upland rice by an average of 64 per cent. Labour for a maize crop was increased by 90 per cent with this system due to pruning operations. Such an increase in production costs was seldom rewarded by a commensurate increase in returns.

Tree legumes and fodder grasses were both tried and adopted by farmers in Claveria during the first years of the research and farmer-to-farmer training project (Fujisaka et al, 1994). Farmers who perceived soil erosion to be a problem were, however, much more interested in vegetative barrier techniques that minimized labour (Fujisaka et al, 1994). We observed that a few farmers independently tried the practice of laying out within their sloping fields contour strips that were left unplanted. These were re-vegetated naturally by native grasses and forbs.

Researchers found that these natural vegetative strips had many desirable qualities (Garrity, 1993). They needed much less maintenance compared with fodder grasses or tree hedgerows, and compared with the introduced species they offered little competition for nutrients and no competition for sunlight to the adjacent annual crops. They were very efficient in minimizing soil loss, and they had no tendency to cause greater weed problems for the associated annual crops. NVSs were also found to be an indigenous practice on a few farms in other localities, eg in Batangas and Leyte provinces.[3]

A key advantage of NVSs is their simplicity of establishment. Once contour lines have been laid out, no further investment in planting materials or labour is needed. The vegetative strips do not need to conform closely to the contour since they act as filter strips rather than as barriers. Their biomass production and their economic value as fodder is lower than with many other hedgerow options, but labour requirements are minimized. Vetiver grass (*Vetiver ziznoides*) can fill a similar niche as a low-value but effective hedgerow species. But for vetiver, or any other introduced hedgerow species, the planting materials must be obtained and planted out, which requires extra labour. As their designation indicates, NVSs are covered by the processes of nature.

There are a number of contour farming practices that work satisfactorily. The major advantage of NVSs is that they are less costly and less management-intensive to install and maintain than are other alternatives. In the Asian contour hedgerow vocabulary, the NVS system is equivalent to a 'simple

SALT'. This translates into wider and more rapid adoption, and less food crop loss due to competition.

One limitation of these low maintenance NVSs is that they do not greatly enhance the nutrient supply to the crops. In this respect they do not differ from many other hedgerow enterprises, including fodder grasses or perennial cash crops like coffee. Perennials with economic value are, nevertheless, an attractive option to many farmers. Those farmers who have established NVS are experimenting with producing a wide range of perennial crops in their hedgerows, including many types of fruits, coconuts, coffee, mulberry and even fast-growing timber species. With continuous cropping, NVS or other low-management hedgerow options can only be sustainable with organic or inorganic fertilization. But they have proven to be popular in Northern and Central Mindanao and have been adopted increasingly in recent years.

It is noteworthy that farmer organizations have evolved to share knowledge of these practices within the villages and municipalities of the region. The approach developed into a dynamic movement that now has some 300 self-governing groups, and over 5000 members in six municipalities (Mercado et al, 2000). The units of local government in the area were so impressed with the energy of this movement that they are now supporting the effort financially, with active involvement of village leaders. Early in 1999, the municipal council of Claveria passed legislation making it mandatory to establish contour buffer strips on all sloping fields within the municipality. A recent survey (Keil, 1999) found that more than 90 per cent of the sample households supported the implementation of this legislation.

RIDGE TILLAGE FOR SMALLHOLDER SYSTEMS

As we searched for practical ways to improve farming operations within contour buffer strip systems, so as to avoid the movement of soil downslope by tillage, we adapted the principles of permanent-ridge tillage systems to smallholder agriculture using animal draft power. Such systems maintain alternate strips of untilled and tilled land in a row-cropped field. The untilled strip (the ridge) is where the crop is planted in the same exact row position in each successive season; the inter-row area is where cultivation is practised for weed control, and hilling-up is done.

The ridges act as a partial barrier to the surface flow of water, but their major function is that they act as a zone of greater infiltration of water runoff. The no-tillage area tends to accumulate organic matter and macro-pores, due to soil biological activity and root channels. Since primary and secondary tillage operations are not practised for land preparation between crops, the land is less subject to erosion in the off-season. Labour and other expense in land preparation is eliminated. Pre-planting weed control can be accomplished by judicious use of a herbicide.

We recently completed a four-year study of permanent-ridge till systems that compared the conventional system with ridge tillage, with NVS and with a combination of the two (Thapa et al, 2000). The annual soil loss on bare, uncropped soil was 85.5t/ha. Ridge tillage reduced soil loss by 49 to 58 per cent. NVS reduced soil loss even more, by 90 to 97 per cent. When the two conservation tillage systems were combined, annual soil loss was reduced to an insignificant 0.3–1.1t/ha.

Clearly, both systems were effective measures to reduce erosion dramatically. When combined, they proved remarkably effective, reducing soil loss by 98.7 to 99.7 per cent. The permanent ridges in the ridge-till treatments had high infiltration rates, which reduced runoff from row to row in the ridge-till system, and they reduced runoff through the grass barriers in the NVS. Kiepe (1995) has demonstrated that a much higher infiltration rate in the vicinity of contour hedgerows is the major factor explaining the exceptional ability of contour hedgerows systems to reduce runoff and off-field soil losses.

Mean grain yields of six crops over the three-year period were the same for the conventional system and ridge-till. Thus, ridge tillage maintained maize yields while it drastically reduced the amount of labour invested in tillage and weed control, thereby reducing production costs and increasing profitability. We see ridge tillage as a practice that will complement the use of contour buffer strips in the future.

CONTRIBUTIONS TO FUTURE FOOD PRODUCTION AND FOOD SECURITY

Earlier it was stated that maize yields in Claveria had ranged between 1–2t/ha in 1984, and have increased to between 2 and 3t/ha currently. This increase resulted from a number of interacting changes in crop and land management. Particularly noteworthy are the shift to hybrid maize from local cultivars and the increasing use of lime and nitrogen and phosphorous fertilizers.

In order to isolate the production effects of contour buffer strip systems, Nelson et al (1998) modelled the long-term trends in maize yields for several alternative buffer strip systems compared with maize produced under the same set of conventional management practices but without the strips. This was done based on an extensive set of on-farm experiments conducted by Agus et al (1998). Yields began to diverge significantly after about five years, and by the 15th year, yields in the buffer-strip systems were about 0.5t/ha higher than in the open-field system. Yields of the three buffer-strip systems (NVS, planted grass strips, and leguminous tree hedgerows) were almost identical throughout the 15-year period.

The main driver of the differences between the buffer-strip systems and open-field farming was a much steeper decline in soil carbon and nitrogen in the open-field system. This was due to accelerated soil erosion. These effects were estimated assuming a constant application of 60kg/ha of nitrogen fertilizer per crop, which is now common practice.

Contour buffer strips were found to result in a gradual but increasing advance in yield due to reduced degradation in the soil resource base. But the situation is significantly more dynamic than this. The more favourable soil moisture environment typically observed in contour buffer strip systems was not factored into these calculations. Also, we observe that management practices and choice of enterprise tend to change dramatically upon installation of contour buffer strips.

It is also typical to see much higher rates of nutrient application (both inorganic and organic) to maize and other crops grown in fields where the buffer strips have been installed. Farmers attribute this to their greater confidence that an investment in fertilizer and manure (labour and cash) will not be eroded. Also, it is common for the system to switch to higher value crops or more productive crop cultivars. These changes should result in a much greater cumulative contribution to production and income from fields with buffer strips than the direct effects of 0.5t/ha, noted above, that are estimated as resulting from the installation of NVS alone.

What are the implications if these farmer-friendly contour buffer strips were adopted on a wider scale in Southeast Asia? Assuming that adoption were to occur on just 2 per cent of the 186 million hectares of strongly acid upland soils, with aggregate production effects estimated to be on the order of one ton per hectare, a production advantage of some 3.6 million tons of maize may be expected. Assuming per capita consumption of 100kg/person per year, this additional maize would provide for the basic food needs for 36 million people. Adoption on 10 per cent of the area would benefit 160 million people.

CONTRIBUTIONS TO ENVIRONMENTAL PROTECTION

Until recently, fallow rotation was the only feasible way for most upland farmers to produce annual crops. Now, with yield-conserving practices like contour buffer strips and reduced tillage, and with yield-enhancing practices such as fertilizers and new varieties, continuous intensified production is possible. These are gains obtained on sloping soils that could not be expected to remain productively farmed to annual crops without such conservation measures. These practices also induce the opportunity to release land for other more profitable and environmentally suitable enterprises, including vegetable production systems, perennial horticultural trees, timber production and livestock systems, all of which tend to have comparative advantage on sloping uplands. Since contour buffer systems can be adapted to a wide range of food and perennial crops, the NVS solution is broad-based and generalizable. We have no evidence to think that they will not be effective for an indefinite period of time.

It is conceivable that in the future, the upland areas of Asia will see a decline in rural populations. When this occurs, the intensity of landuse may decline, and some areas that are farmed with terraces can be transformed from food crops into tree crops. In limited areas, this is already happening. Perhaps from the perspective of food production, a shift to woody perennials may be considered unfortunate. But from the standpoint of rural income, and in terms of the evolution of environmentally favourable landuse systems, such a trend could be quite positive.

NVS systems do not fix and cycle nitrogen, as can be done by hedgerows of leguminous trees. However, in phosphorous-limited Asian upland environments, tree-based hedgerows are themselves not effective in cycling adequate amounts of phosphorous to meet crop demand, and they are therefore unable to sustain crop yields without phosphorous fertilization (Garrity, 1993). Some importation of nutrients through manure and/or fertilizers containing adequate amounts of crop-available phosphorous is therefore essential for maintaining, and especially for increasing yields.

The NVS practice is a technical innovation that opens up new possibilities for sustained farming on quite steep slopes, thus expanding the land base of food-crop agriculture. Adopters of NVS were recently surveyed around Claveria. One of the questions asked was whether they perceived that the installation of the buffer strips increased the value of their land. All respondents strongly believed that it had. When asked for their estimate of the amount that they expected land values had been elevated by NVS in their area, their answers ranged from 35 to 50 per cent (Stark, 1998; and personal communications).

In the future, contour hedgerow systems are fairly certain to shift towards low-labour alternatives like NVS, with soil fertility being maintained by nutrient importation. This is not different than in most other types of agricultural systems, except that NVS systems make sustainable annual cropping possible on steeply sloping lands otherwise prone to severe erosion.

SUMMING UP

This chapter has reviewed evidence that NVSs are a land-conserving technology that has the potential to make a substantial contribution to increased tropical food production in the coming decades. But where do NVSs fall on the spectrum of 'conventional' vs 'alternative' agriculture? They may be seen as an alternative agricultural practice in the sense that they are an innovation based wholly upon resources internal to the farm, and they were originated by farmers themselves. Also, they are an application of agroecological principles to the challenge of evolving simple, practical, low-external-input solutions for soil conservation. But it may also be claimed that NVSs reinforce conventional approaches, as they tend to stimulate indirectly the use of commercial fertilizers and modern cultivars.

Thus, NVSs are neither an alternative or conventional approach. Rather, they seem to be neutral in this classification. They can be employed by farmers practising low-input biological farming, or high-input conventional agriculture. They are conducive to the use of fertilizers, reduced tillage, organic matter recycling, green manures and cover crops, and other nutrient-providing practices. They add to the examples of management systems that cannot be defined unambiguously as alternative or conventional. Systems themselves are usually constructions of diverse elements, seeking to capitalize on as many synergies as possible.

It will be difficult to ascertain with any precision how much alternative forms of agriculture can increase the world's food supply. It is easier to agree that there are many fruitful pathways by which the application of agroecological knowledge to the development and refinement of farming practices will contribute to this goal. There is scope for combining ecological knowledge synergistically with much of the experience generated through advances in chemistry (fertilizers, pesticides), genetics (new varieties) and engineering (tools, equipment).

This underscores a point made in preceding chapters: that human capital is the fundamental resource in creating and adapting solutions to the myriad farming environments of the tropics. In particular, the potential of farmer-led organizations, such as Landcare in Australia and now in The Philippines, has not been given the attention it deserves for transforming both agricultural extension and research (Campbell and Woodhill, 1999; Campbell et al, 1999; Mercado et al, 2000; Sabio et al, 2001). Moving beyond a choice between 'alternative' vs 'conventional' agriculture will enable us to explore the common ground they share. The central issue is how to guide decision-makers to invest in research to realize a world agriculture that employs and benefits from all of these tools.

NOTES

1   Human pressure is by no means the only major driving force for deforestation and erosion since Southeast Asian landscapes tend to be geologically young, and exceptionally steep, which accelerates their degradation.
2   The uplands of Southeast Asia are dominated by strongly acidic soils and hill-slope topography. The geographic extent of acid upland soils is 188 million ha, or 39 per cent of the region's total land area (IRRI, 1986). More than half of this area has soils with pH less than 5. The acid uplands vary from one-third of total land area in Indonesia and The Philippines, to as much as two-thirds of the area in Laos.
3   For a brief history of the introduction, suppression and return to NVS in Swaziland, see Osunade and Reij (1996). The same considerations that made these strips popular in The Philippines applied in Swaziland.

Part 3

Advancing Agroecological Agriculture with Participatory Practices

Chapter 19

Exploiting Interactions Between Planned and Unplanned Diversity in Agroecosystems: What do We Need to Know?

Alison G Power and Peter Kenmore

The emergence of sustainability as an important goal for agriculture and development has stimulated increasing interest in understanding ecological processes within agriculture so that they can be managed more effectively for enhancing agricultural productivity and reducing negative environmental impacts of agricultural activities. Although the opportunities created by agroecological approaches to agricultural development have been demonstrated many times, as seen in Part 2 of this volume, there is still much to be learned about how such approaches may actually function. Understanding the functioning of practices such as described in this volume will allow for better design of agricultural practices and systems that enhance both productivity and environmental benefits. Increasingly, these benefits are not only sought within countries but are sometimes imposed upon them, as when the maximum residue levels of pesticides tolerated in fruit, vegetables and flowers imported from tropical countries were dramatically decreased by the European Union in mid-2001.

Over the last several decades, much agroecological research has focused on the role of diversity in agricultural systems (eg Vandermeer, 1989; 1995; Andow, 1991; Altieri, 1995). One should consider, at a minimum, the following complementary dimensions of genetic diversity, species diversity, structural diversity and functional diversity. Biodiversity within managed agroecosystems can be operationally classified as being either *planned* or *unplanned*. Planned diversity includes the spatial and temporal arrangement of domesticated plants and animals that farmers purposely include in their farming systems. This may also include beneficial organisms that are deliberately added to the agroecosystem, such as biological control agents or plant-associated nitrogen-fixing bacteria. Planned diversity typically involves all or several of the dimensions of biodiversity listed above.

Unplanned diversity includes all the other associated organisms that persist in the system after it has been converted to agriculture or that colonize it from the surrounding landscape. This aspect of diversity is likely to include a variety of herbivores, predators, parasites and micro-organisms that make up the majority of species in any ecosystem, even a simplified one like an agroecosystem. It has become increasingly clear that the two components of diversity are significantly linked; as planned diversity increases along any of the four dimensions listed above, so does the diversity of the associated biota. Although the evidence for above-ground diversity is strongest (eg Andow, 1991; Perfecto et al, 1996), below-ground diversity is also likely to correlate with plant diversity (Giller et al, 1997; Hooper et al, 2000).

Given this distinction – and relationship – between planned and unplanned diversity, how do both types of diversity influence agricultural productivity and other ecological processes that occur in agroecosystems? In mixed cropping systems, planned diversity often leads to higher productivity. The mechanisms for this are relatively well understood (Francis, 1989; Vandermeer, 1989). On the other hand, the links between planned and unplanned diversity are not well understood. Moreover, much remains to be elucidated about the role of unplanned diversity in agroecosystem functioning (Power, 2001). Can agroecosystems be designed to exploit the ecological services provided by unplanned diversity? This chapter focuses on gaps in our understanding of the linkages between planned and unplanned diversity and on the consequent effects of such linkages on ecosystem processes in agricultural systems, with particular attention to processes that appear to have the most potential to offer agricultural benefits.

PLANNED DIVERSITY AND ECOSYSTEM PROCESSES

Agroecosystems vary dramatically in their complexity and degree of planned diversity. The input-intensive, monoculture systems that dominate commercial agriculture worldwide have very low planned diversity in terms of crop species and thus have relatively constrained community assemblages. At the other extreme, traditional agricultural systems and home gardens of the tropics typically have high planned diversity and relatively unconstrained, complex assemblages of associated biota. Planned diversity has been shown to influence a variety of ecological processes that operate in agricultural systems, including primary production, pest regulation, decomposition and nutrient cycling (Power and Flecker, 1996).

Productivity and Stability

The tendency of diverse cropping systems to have higher productivity than would be expected on the basis of the productivity of their component crops grown separately in monoculture is referred to as 'overyielding'. This may result from a variety of mechanisms, such as more efficient use of resources (light, water, nutrients) or reduced pest damage. There have been a number of experimental studies that examine these mechanisms (Trenbath, 1974; 1976; Francis, 1989). When *inter*specific competition for a limiting factor is less than is *intra*specific competition for that factor, overyielding is easily understandable and predictable. Facilitation occurs when one crop modifies the environment in a way that benefits a second crop, for example, by lowering the population of a critical herbivore or by releasing nutrients that can be taken up by the second crop. Facilitation may result in overyielding even where direct competition between crops is substantial (Vandermeer, 1989). Ecologists have recently begun refocusing attention on facilitation and other beneficial influences among species as an essential foundation for the assembly of unmanaged, natural communities (Stachowicz, 2001), and its role in human-managed ecosystems is likely to be equally important.

There is some evidence that diverse cropping systems exhibit greater yield stability as well as higher productivity, suggesting that resistance to environmental perturbation may be higher in more diverse systems. Yield stability has been measured in at least three ways: by calculating coefficients of variation, by computing regressions of yield against an environmental index, and by estimating the probability of crop failure. Polycultures exhibit greater yield stability according to all three criteria (Rao and Willey, 1980; Francis and Sanders, 1978): polycultures tend to have lower coefficients of variation than crops grown as separate monocultures; the response of polycultures to environmental change tends to be as stable or more stable than the most stable component crop grown in monoculture; and polycultures tend to have a much lower probability of crop failure than the component crops grown in monoculture. The probability of crop failure is an estimate of farmers' risk, and lower probabilities of failure result from both the higher yields of polycultures and their putative yield stability. A number of studies – starting from the analytical work of Richards (1939) in Zambia, Howard (1940) in India, Kenya and South Africa, and de Schlippe (1956) in Congo and Southern Sudan – have indicated that, overall, diverse cropping systems are less risky and more stable than less diverse cropping systems such as monocultures, both agronomically and economically.

Several mechanisms may lead to greater yield stability in diverse systems. When one crop performs poorly, because of drought or pest epidemic, for example, the other crop(s) can compensate, using the space and resources made available by structural diversity. Such compensation is obviously not possible if the crops are grown separately. If the yield advantages of polyculture are greater under stress conditions, then yield stability is higher and economic risk is lower. This polyculture advantage has been demonstrated for crops under both nutrient stress and drought stress (eg Natarajan and Willey, 1986). Moreover, where polycultures lead to reduced pest attack, as they often do (Andow, 1991), greater yield stability may result from the dampening of pest outbreaks and disease epidemics.

Pest Regulation

The planned diversity of the agroecosystem has important effects on herbivorous insects and microbial communities that attack crops. Traditional agricultural systems often include substantial planned genetic and species diversity (Allan, 1965), and genetically diverse grain crops are used in many parts of the world to control pathogens (Finckh et al, 2000; Zhu et al, 2000, this latter example was cited in Chapter 1). In contrast, the low planned genetic and species diversity of most commercial monocultural systems often results in large crop losses from a relatively small number of pest species that are individually more abundant than in systems with higher planned genetic or species diversity. The trend for higher pest densities in monocultures compared with diverse cropping systems is especially strong for specialist insect herbivores with a narrow host range, ie those which feed on only certain plants (Andow, 1991).

As planned genetic and species diversity increases, population densities of these specialist herbivores decrease. Accumulating evidence is showing that host-finding behaviour and insect movement, both colonization and emigration, play important roles in herbivorous insects' response in different agroecosystems. Densities of specialists are likely to be lower in diverse systems for two reasons: they have difficulty locating hosts, due to interference with olfactory or visual cues; and they leave hosts more often due to lower plant quality and then have difficulty relocating them. These behaviours are significantly affected by the structural and functional (eg chemical and nutritional) diversity that accompanies planned plant species diversity.

Natural enemies make up an important component of the unplanned diversity that accompanies greater planned diversity in agroecosystems. Compared with monocultures, diverse systems are likely to have higher rates of predation and parasitism and higher ratios of natural-enemies-to-herbivores, in part because the herbivore densities are already constrained by plant-associated dimensions of diversity, all of which can contribute to lower pest densities.

Microbial pathogens also respond to planned diversity within an agroecosystem, but their response is more variable than that of herbivorous insects. Crop diversification can modify the microclimatic conditions that play an important role in the development and severity of plant disease. Pathogen growth and reproduction may be either encouraged or inhibited in more diverse cropping systems, depending on the particular requirements of the organism. The effects of diversity depend on a variety of dispersal processes, infection efficiency and the rate of disease progress. Much more research is needed before we could use planned species diversity to effectively manage plant pathogens.

However, we know that the genetic diversity of crops can dramatically reduce pathogen impacts on crop productivity. Mixtures of genotypes of a single species, such as multiline cultivars and varietal mixtures, have been used effectively to retard the spread and evolution of fungal pathogens in small grains and other crops (Finckh et al, 2000; Zhu et al, 2000). There is some evidence that they may also have lower densities of insect herbivores and lower incidence of plant viruses (eg Power, 1991). Typically these mixtures include both resistant and susceptible crop genotypes, though they may be mixtures of several different resistant genotypes. The reduction in pathogen spread is greater than would be expected simply on the basis of the proportion of resistant genotypes in the mixtures. The lower spread therefore appears to be due to the effects of diversity per se on the ability of pathogens to disperse (Garrett and Mundt, 1999).

High planned diversity that utilizes the dimensions of genetic diversity, species diversity and structural diversity thus has a strong influence on populations of herbivorous insects and plant pathogens. The intentional manipulation of planned diversity for the purpose of pest control is common in traditional agricultural systems, and it offers real potential for increasing the productivity of contemporary agricultural systems without the negative health and environmental impacts of pesticides (Altieri and Nichols, 1999). Even when farmers are pushed by macro-economic policies, market failures, political forces or local customs to practise monoculture, they can still encourage functional diversity by not applying pesticides, for example. This will facilitate unplanned diversity that can provide essential ecosystem services like pest population regulation.

Soil Processes

Planned diversity may also have significant impacts on the soil community, since the diversity of some groups of soil organisms is positively correlated with plant diversity (Giller et al, 1997; Wardle and Giller, 1997). Plant pathogens and their antagonists in the soil are well known to respond to crop diversity (Kennedy, 1999; Cook, 2000). Long-term, continuous monoculture can result in dramatic shifts in the competitive balance among microbial species and can increase the aggressiveness of plant pathogens. Conversely, populations of plant parasitic fungi, bacteria and nematodes often decline when the monocultures of their host plants are replaced by diverse cropping systems. Rhizosphere microbial communities, including a variety of bacteria and fungi, respond to plant species composition of the cropping system, plant phenology, plant nutrient status and even plant genotype.

Some ecological processes such as nitrogen cycling appear to be substantially controlled by the unplanned diversity of bacteria and fungi associated with crop species and genetic diversity. For example, the inclusion of legumes can increase rates of biological nitrogen fixation. Decomposition rates may be quite responsive to planned diversity, owing to the effects of litter diversity (Hooper et al, 2000). Recent work linking litter quality to rates of decomposition and nutrient release has increased our ability to predict the nutrient value of various organic inputs.[1] Planned diversity may usefully include plants that are selected for their litter quality, ie their nutrient value to other crops.

Despite a growing awareness of the likely importance of planned crop diversity for the soil community, our ability to predict the impacts of manipulating planned diversity is still limited. This is because our understanding of the complex interactions among soil organisms and between plants and soil microbes is still so limited. Greater attention to the ecological dynamics of these soil communities under different cropping systems will aid in determining the extent to which we can manage these communities to enhance agricultural productivity by varying planned diversity (Brussard, 1998).

UNPLANNED DIVERSITY AND ECOLOGICAL SERVICES

The unplanned diversity that accompanies planned diversity in agricultural systems can provide many ecological services to agriculture, so the conservation of biodiversity offers significant benefits to agriculture. Uncultivated species, including wild relatives of crops that occur in and around the agroecosystem, are an important source of germplasm for developing new crops and cultivars. Increasing planned crop diversity can augment the resources available to plant pollinators and to natural enemies such as parasitic wasps, resulting in higher populations of these beneficial organisms. Increasing planned diversity may also foster beneficial soil organisms and the conservation of functional processes such as decomposition and nutrient cycling.

The demand by crops for scarce soil nutrients changes over the course of crop development, and across seasons, so that the *synchronization* of nutrient availability with crop demand has been a major focus of the Tropical Soil Biology and Fertility Programme. The quality of leaf litter in a planned alley-cropping system (Henrot and Brussaard, 1997) and the quality of mulches from different species applied to conventionally grown maize (Tian et al, 1997) significantly affected earthworm densities, activity and nutrient supply to crops. This work suggests that it may become possible to make the levels of nutrients more optimal and efficient by manipulating the unplanned diversity of earthworms and their activity through the planned diversity of mulching and litter fall.

Pest Regulation

One of the ecological services provided by the unplanned diversity in agroecosystems is the regulation of herbivorous insects and microbial pathogens by competitors, predators and parasites. As discussed above, highly planned diversity often results in higher densities of predators and parasites. Even in agroecosystems with low planned diversity, the diversity of natural enemies that reduce predator populations may be quite high as long as pesticides are not used. For example, paddy rice monocultures managed without pesticides can have a surprisingly high diversity of herbivorous insects, predators and parasitoids, compared with similar monocultures in which pesticides are used. Even within planned monocultures, the unplanned species diversity of natural enemy populations can be encouraged.

By refraining from applying pesticides, tropical rice farmers can enable unplanned communities of natural enemies to build up to hundreds of species per hectare, which provide good economic pest population regulation (Way and Heong, 1994; Settle, 1996; Kenmore et al, 1984). Recent pest management programmes in Southeast Asian paddy rice systems have taken advantage of this diversity and drastically reduced pesticide inputs without sacrificing yields (Oka, 1997). In traditionally managed rice fields, the predators are likely to include fish and amphibians, which contribute to pest regulation and also provide additional nutritional resources for farm families.

Unplanned Diversity and Soil Processes

Despite some significant advances in our understanding of soil processes, we know relatively little about the soil biota and their impact on agricultural productivity. In natural ecosystems, decomposition and soil nutrient cycling are regulated by a diverse community of invertebrates and micro-organisms, such as termites, earthworms, nematodes, fungi and bacteria. The composition, abundance and activity levels of the soil biota in agricultural systems are markedly different from that found in surrounding natural ecosystems (Giller et al, 1997). For example, the diversity and abundance of soil insects and earthworms in tropical agroecosystems are typically significantly lower compared with a wide range of undisturbed tropical ecosystems. In those cases where abundance remains high in agricultural systems, the soil communities are usually dominated by a single or small number of species, highly adapted to the modified environment.

The changes in the soil community under agriculture result from a variety of perturbations to the soil environment. The initial conversion from undisturbed ecosystems typically involves the removal and/or burning of plant biomass followed by tillage, activities that have drastic impacts on soil structure and soil chemistry, which in turn affect soil biology. The physical changes at the soil surface amplify diurnal and seasonal fluctuations of temperature and moisture. In addition, organic inputs to the soil are significantly reduced as a result of plant biomass removal, and the chemical composition of organic inputs is altered. These extreme modifications of the soil environment can result in the elimination of some soil organisms and, at a minimum, are likely to change the competitive balance among species. Nutrient flows may switch from fungal-dominated channels in zero tillage to bacterial-dominated channels in conventional tillage systems (Hendrix et al, 1986; Swift and Anderson, 1994). To the extent that agricultural systems minimize these perturbations, by reducing tillage or burning, for example, then the impacts on the soil community will be less severe.

Despite these well known effects of agricultural conversion on the soil community, the link between loss of soil biodiversity and various ecological processes has not been extensively studied. Decomposition and nutrient mineralization, for example, are controlled by the activities of a diverse community of organisms. It is not at all clear that the loss of some species will result in significant changes to these functions, but the degree of functional redundancy among different species in natural as well as agricultural communities is still controversial (Swift et al, 1998; Hooper et al, 2000). On the other hand, some processes such as nitrogen fixation are carried out by very specific organisms, whose loss might substantially affect nitrogen cycling. Given the importance of soil biota for decomposition and nutrient cycling, it is essential that the link between this component of unplanned diversity and function be explored more fully.

Managers of agroecosystems have, for livelihood or market reasons, significantly simplified crop species diversity, crop genetic diversity and often structural diversity. The challenge of adaptively managing these agroecosystems, as sketched broadly by Anderson (1996) for oil palm plantations, is to reconstruct from this simplified base as much as possible of the functional diversity and ecological services found in more complex natural communities. The challenge is how to assemble the right combinations of functional groups of species by using planned diversity to induce unplanned diversity.

SYNERGY AND SYNCHRONY: SYNDROMES OF PRODUCTION

It is often suggested that planned diversity in agroecosystems can result in synergy and synchrony. Synergy is evident where the outcome of interactions among system components is more than the sum of its parts. In this context, synchrony implies that system components interact temporally in a way that maximizes the benefits of these interactions. Is there any evidence that synergy and synchrony occur in ecologically managed systems? And do they result in higher productivity?

Andow and Hidaka (1989) suggest that synergies can be crucial to the functioning of agricultural systems, and they describe a synergistic system as a 'syndrome of production'. A production syndrome is a set of management practices that are mutually adaptive and lead to high performance. Subsets of this collection of practices may be significantly less adaptive, that is, the interaction among practices can lead to improved system performance that cannot be explained by the additive effects of individual practices.

Alternative syndromes of production may be likened to peaks of yield (or profit) on an 'adaptive landscape' (in the sense of Wright, 1932) of management practices, such that moving to another, higher peak on the landscape requires travelling through non-adaptive valleys (Figure 19.1). Thus system performance may decline as farmers attempt the transition from conventional to sustainable systems, particularly if they adopt sustainable practices one by one.

This synergistic effect is often identified by practitioners as key to the benefits of sustainable systems (Uphoff, 1999). It may also help to explain why farmers rarely adopt new technologies or practices without modification. Synergies make it difficult to evaluate individual practices effectively, because experimental tests of individual practices or subsets of practices are unlikely to reveal the true potential of any production syndrome. Despite the success of many organic and low-input production systems in practice, low-input practices isolated from the context of other practices and environment often do not outperform conventional practices in standard experimental comparisons (Vandermeer, 1997).

The concept of production syndromes was illustrated in a comparative study of two types of Japanese rice production (Andow and Hidaka, 1989):

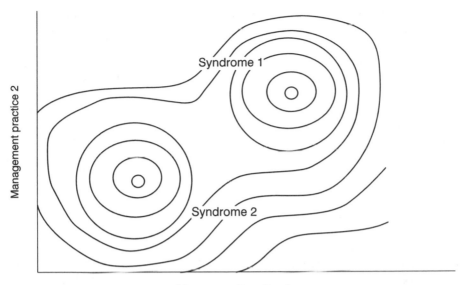

Management practice 1

Note: Contour lines represent a measure of success such as yield, which peaks where the two management practices together produce highest yields. Small deviations in either management practice lead away from the peak and to a decline in yield; so moving from one peak to a higher peak leads to a reduction before attaining greater output. A real production syndrome would involve many more management dimensions than are depicted here
Source: Adapted from Andow and Hidaka, 1989, and Power, 2000

Figure 19.1 *Production Syndromes Plotted in Two Dimensions of Management Practices*

conventional, high input production, and the *shizeñ* or 'natural' farming system that has become widely known through the efforts of Fukuoka (1978). Although rice yields were comparable in the two systems, management practices differed in almost every respect: irrigation practice, transplanting technique, plant density, fertility source and quantity, and management of insects, diseases and weeds. Along the dimension of genetic diversity, the same variety of rice was planted in both systems, although the *shizeñ* farmer used seed saved from previous seasons.

This set of cultural practices and pest management practices resulted in functional differences that could not be accounted for by any single practice. Andow and Hidaka (1989) argued that systems like *shizeñ* function in a qualitatively different way than conventional systems. Other rice production syndromes that lead to high performance, such as the rice intensification system (SRI) of Madagascar (Chapter 12), may form other peaks on the rice production landscape. At the national level, the experience of The Netherlands during the 1990s with its multi-year crop protection plan shows how a well supported policy commitment (and sufficient funds) to shifting from one production syndrome – heavy applications of soil pesticides – to another with less than 40 per cent of those pesticides can protect thousands of farmers as they make the trek from one adaptive peak to another.

The study of production syndromes is in its infancy, but it may provide an effective means of synthesizing the type of information discussed in this volume. The ecological processes that result from planned and unplanned diversity are complex, and exploiting them effectively for agricultural production is a significant challenge. However, investigating these processes in the context of a production syndrome should aid in contributing to practical improvements for smallholder agriculture, as well as enhancing our understanding of the functioning of complex ecological systems.

NOTE

1   Recently, the Tropical Soil Biology and Fertility Programme developed a database of information on organic inputs commonly found in tropical agroecosystems, the TSBFP Organic Resource Database (see Palm et al, 2001). In conjunction with decision tools, this database should help farmers or their advisors identify organic inputs that can be used by smallholders as a means of improving soil fertility.

Chapter 20

Human Dimensions of Agroecological Development

Jules Pretty and Norman Uphoff

Raising agricultural output sustainably and, in the process, improving people's livelihoods will depend on the establishment of approaches and capabilities that encourage personal and social learning. Technological change is always a complex process with both biophysical and socioeconomic aspects, and so it is not a matter of introducing new physical things that exist independently of human interaction. Rather, it results from changes in the thinking and activities of individuals, households and communities as well as in market and organizational relationships. In such transitions, learning applies to new systems of behaviour and valuation, not just techniques or methods.

In Chapter 21, we discuss a variety of policy reforms that can support the adoption and adaptation of sustainable agroecological practices. Along with the very evident processes of change that occur at individual, enterprise and community levels, there are larger, less visible forces that shape both the context of local decision-making and the opportunities available. Large-scale economic factors and institutional constraints will ultimately determine the success or failure of efforts to promote agroecological development beyond community levels. Local successes can be threatened by external changes – such as when cheaper imports compete with agroecologically-produced crops, or when new tax or land tenure policies favour larger landholders with capital-intensive systems. Further, the kinds of economic considerations analysed in Chapter 5 create the *context* as well as establish the *incentives* that influence choices and behaviour at both micro and macro levels.

However, the importance and complexity of these larger processes mean that there needs to be a great deal of learning to improve people's choices and actions, taking into account effects from one level to the other. In this chapter we focus on changes that affect the orientation and operation of public and other institutions which support the agricultural sector.

We should restate that participatory technology development is not necessarily coterminous with agroecological approaches. Agricultural development based on agrochemical or mechanical innovations can be undertaken in ways that are more participatory than in the past, and this may make these elements more acceptable. However, we have observed, and the case studies confirm, that participatory modes of development with an emphasis on learning are intrinsic to the success of agroecological approaches. They depend more on informed decision-making and on intensive management, adapting any innovation to make the best and most sustainable use of local resources. Agroecological development thus requires a thorough understanding of the human and social factors discussed in Chapter 4, including how they interact with elements of the biophysical domain. Material factors are analytically but not practically separable from the social and institutional elements of agricultural change.

LEARNING ENVIRONMENTS

Previously it has been assumed that the new and better knowledge needed to improve agriculture will originate within the domain of modern science, and that development follows from transmitting such knowledge to other actors within the agricultural sector for them to adopt. An agroecological vision, in contrast, involves many actors in a multipolar process of knowledge generation, with farmers and other practitioners acting as partners with researchers in this process – indeed, often taking initiative rather than being recipients only. This does not derogate scientists' role or make them redundant. It recasts their role so that many kinds of people are contributing to new knowledge by learning from and with each other. Multi-role collaborations draw together the experience, insights and intelligence of different categories of people in investigation and learning that is both cross-disciplinary and multi-institutional.

People make decisions and act not just as individuals but concurrently as members of organizations and communities where established roles, procedures, values and objectives shape behaviour. It is therefore important to consider how organizations (and we include communities in this category) can reorient themselves to become more hospitable to these new approaches to learning. Most organizations have traditionally functioned with explicit or unstated mechanisms that identify and correct any departures from standard operating procedures. These support what Argyris and Schön (1978) characterized as *single-loop learning*, which is concerned with achieving single and simple objectives most quickly and efficiently.

Organizations will benefit, however, from a more complex process that is described as *double-loop learning*. This involves questioning and possibly making changes in the wider values and procedures under which organizations operate, evaluating their ends as well as their means. Learning organizations, as described by Korten (1980), ensure that people become more aware of how they can learn, both individually and collectively, from their mistakes as well as their successes. This requires, in Korten's provocative phrase, 'embracing error'.

Institutions can improve their learning capabilities by encouraging people to develop greater awareness of how information can be acquired, evaluated, applied and exchanged (Leeuwis et al, 1999). To promote learning, people in organizations need to remain in close touch with their external environments, not becoming absorbed in *self-referential* cultures and processes. This means being committed to participative decision-making, backed up by ongoing analyses and evaluations of performance. (Engaging in participatory decision-making focused entirely within an organization will heighten self-referential tendencies.) Fortunately, being attuned to information from a variety of sources is becoming easier in an era of electronic communication, and participatory decision-making is increasingly acceptable with the growing worldwide commitment to democracy.

Learning organizations are more decentralized than centralized, with openness to new ideas and cross-disciplinary perspectives. Learning at organizational peripheries needs to be shared systematically so that other parts of the organization, and particularly decision-makers at the centre, benefit from new technologies, insights and principles. This will enable organizations to produce more varied outputs that respond to the demands and needs of members and clients. Multiple realities and complexities will be better understood if organizations have pluralistic linkages that sustain regular interaction among professional and public actors. With new professional norms and practices in place, agencies become better able to build up rather than draw down the natural and social forms of capital discussed in Chapter 4.

It is more than just research institutions and extension services that should become engaged in reorientation; so should ministries which give policy and technical guidance; universities, technical institutes and public schools that provide education at various levels and shape attitudes while they develop human capabilities; financial institutions providing saving and credit services; farm supply, marketing and other entities furnishing inputs and handling outputs; and the wholesale and retail components of the food supply system.

Such institutional reorientation is a complex process. It involves making changes in the situations in which people find themselves – presenting them with different opportunities, constraints, rewards, sanctions, etc – and inducing changes in the people involved: new attitudes, values and skills (Uphoff et al, 1991, pp170–197). Many professional institutions and interests currently reinforce the prevailing situation which emphasizes *teaching* rather than learning, by assuming that professionals are the custodians of knowledge that can be dispensed to (or withheld from) others who could learn it (Argyris et al, 1985; Pretty and Chambers, 1993). The latter are expected to *receive* knowledge rather than to *acquire* it. Improved knowledge is expected to be accepted and utilized intact rather than being adapted or, indeed, improved upon by users. Examples of this are documented for India, Kenya and the United States with regard to soil conservation by Pretty and Shah (1997).

If agricultural development institutions undertake to promote learning rather than continue teaching, the focus will be not just on *what* is to be learned, but also on *how* and *why* things are learned, and *with whom* they can best be learned. This implies new roles for development professionals

(Chambers, 1997). Unfortunately, the language for explaining this shift implies two polarized kinds of professionalism when what is really involved is an expansion of the *range* of theories and methodologies that professionals can draw upon when working with various groups. It means selecting methods that are most appropriate for particular tasks, guided by what the intended *learners* find most helpful and effective (Funtowicz and Ravetz, 1993).

NEW ROLE RELATIONSHIPS

Fortunately, experiences in a variety of countries have shown that *participatory* approaches to research and development, with innovative modes of interaction and mutual learning among farmers, scientists and technical personnel, can effectively develop sustainable agriculture (Chambers, 1997; Pretty, 1995; Thrupp, 1996; Veldhuizen et al, 1997). There is now more interest in the agricultural knowledge and information systems within which farmers and professionals interact in the design, adaptation and distribution of new technologies.

Merrill-Sands and colleagues (1990; 1995) have described the greater effectiveness of a *triangular* relationship among farmers, researchers and extension personnel which can be more effective than the conventional linear one. These relationships can also be portrayed as *circular*, with continuous interplay among various actors in an arena of information exchange (Thrupp, 1996). In both characterizations, at least some farmers interact directly with both researchers and extension agents, and the division of labour previously advocated (researchers generate knowledge, extension agents transmit it and farmers adopt it) is modified. All participants contribute something, although not necessarily equally, to all three functions.

Such connections enable farmers to learn (and trust) agricultural innovations more readily, while researchers and extensionists can produce more beneficial knowledge to improve agricultural practice by learning with and from farmers. In such a system, extension staff act more as *facilitators* of innovation than as *transmitters* of specific technologies. Also, researchers and extension personnel will interact more closely and collegially than they do now, once the latter are no longer regarded simply as conduits or transmitters for scientists' ideas.

Representative farmers, if not all farmers, should be explicitly and continuously engaged in processes of *problem identification*, in the formulation and assessment of *possible solutions* and in *testing and evaluation* of what works best under particular conditions, reflecting the diversity of farmers' needs, conditions and demands. Technology, rather than being something transferred and adopted as a self-contained package, is understood to result from frequent iteration between experimentation, evaluation and adaptation, with continuing cycles of testing and modification.

In this ongoing process of technology generation and dissemination, users become discoverers and adapters as well as adopters of new practices or material inputs. Institutions now need to be reoriented to support a process where farmers are *partners* in the enterprise of invention rather than recipients or beneficiaries – or worse yet, 'targets' to be served, enticed or manipulated to meet scientists' or policy-makers' goals. This role involves more than being 'clients' of the national agricultural research system, just one among many stakeholders (World Bank, 2000, p7).

Farmers' knowledge will not be sufficient by itself for solving all of their problems of production, processing and storage. The experience with rice intensification reported in Chapter 12 shows how common knowledge can mislead. However, it represents a good starting point in most cases. Knowledge created and organized in modes considered scientific certainly has much to contribute to processes of agricultural improvement. But it cannot contribute most to such improvement operating in isolation from its users; rather, it can benefit from being combined with local knowledge.[1] This is similar to the way in which external inputs such as inorganic fertilizers can make critical contributions to improving the productivity of farmers' own resources and thus should be regarded as complementary means to augment production rather than as stereotyped alternatives, as suggested in Chapter 2 (see also Ruben and Lee, 2000).

To develop effective relationships in participatory research and extension, it is important to observe certain principles of social interaction, such as reciprocity, mutual respect, transparency, equity and fairness in the sharing of resources and responsibility. This has been seen from a number of successful cases of agricultural innovation around the world (Thrupp, 1998; Pretty and Ward, 2001). Increasingly, agricultural scientists are appreciating that a synthesis of local and outside knowledge is likely to produce the most beneficial and sustainable results. The first type of knowledge is rich in detail, contributing site-specific information such as particular pest problems or variations in the water cycle, whereas the latter type has been formulated for its general validity.

Any innovation in agriculture should be tested and evaluated under a variety of local conditions, and for this, farmers are sources of ideas as well as evaluators. Sometimes local knowledge also contributes to expanded or qualified understanding of scientific principles. Possible solutions to problems can come from many different sources, some local and others from halfway around the world. But the identification of problems, determination of criteria for evaluating solutions, and weighing of advantages and disadvantages clearly need farmer involvement. When farmers themselves experiment, they increase their understanding of what does and does not work. Once they have satisfied themselves that a new crop or new practice is beneficial and feasible, they can and often will help to diffuse the innovation (Bunch, 1982).

Chambers et al (1989) highlighted new roles for scientists as partners with with farmers. Instead of bringing 'packages' of innovations to farmers, scientists and extension personnel should bring principles, methods and baskets of choices. Since few farmers are likely to engage in developing new methods and principles on an individual basis, group activities that support collective learning are vital. The transition to sustainable agriculture can be accelerated by supporting networks of farmers who engage in joint learning and experimentation and who provide mutual insurance against risks, as seen in Part 2. Such networks will benefit from stronger links with research institutions, supply organizations, mass communication media and adult education programmes, which support both farmers and other professionals working in their new roles.

THE CENTRALITY OF HUMAN RESOURCES IN TECHNOLOGICAL CHANGE

Although our concern with agroecological approaches began with a focus on biophysical factors and on ways to improve agricultural technology, the case studies presented in Part 2 indicate the extent to which development remains basically a social process with changes in people's ideas, attitudes and capabilities determining whether advances are feasible or not. Technology is itself a creation of the human mind, and whether it creates benefits or costs for people depends in practice on their cognitive assumptions and capacities.

Experience in Indonesia, one of the most successful countries in the Green Revolution, offers some instructive lessons in this regard. A former minister of agriculture in that country at a national workshop on integrated pest management (IPM) in 1994 suggested that 'the next Green Revolution' in Indonesia would be based more on human resource development than on technology (Baharsjah and Rasahan, 1994). The country had made great progress in its agricultural sector during the 1970s and 1980s, achieving self-sufficiency in rice after having been the world's largest importer of rice in the 1960s. The introduction and use of high-yielding rice varieties made a huge contribution to this turnaround. But even the first Green Revolution in Indonesia depended on having farmer organizations that were grounded in village realities (Dudung and Pakpahan, 1992). Such organizations complemented the efforts of researchers and administrators by making the most of ideas and material inputs.

The most impressive agricultural advances in Indonesia over the last ten years have been in the area of IPM to reduce rice crop losses as well as the costs of production. As one of the leaders of this programme has stressed, their strategy was conceived not as IPM *for* farmers, but as IPM *by* farmers (Oka, 1997, p190). Farmers have not been taught IPM practices in a didactic manner, but rather the principles of agroecology in a hands-on way, creating methodologies that have been extended and adapted to other Asian countries. Farmers thus learned to do more than apply pest control measures; they learned how to grow healthier crops, better able to resist pest and disease attacks. That the government could save US$150 million a year in pesticide subsidies while continuing to raise rice production through this participatory approach was a revelation to all. The programme's success has led to farmer-centred IPM programmes elsewhere, such as those reported in Chapters 16 and 17. In recent years, however, rice yield increases in Indonesia have been stagnating, even with modern input packages, so new approaches are needed, now more agroecologically-based.[2]

As efforts are directed towards finding the means to double food production over the next three to five decades, success will depend on identifying appropriate, productive and sustainable technologies, especially those that are low-cost, accessible, and synergistic with the environment. This will require much more broad-based human resource development than has been promoted in the past. Farmers do not need training or instruction so much as opportunities for engagement with scientists, policy-makers and administrators, and experimentation and evaluation that lead to plans of action agreed by all parties as likely produce greater benefits.

It is increasingly recognized that the most undervalued and underappreciated human resources in agriculture are those of women. Women must be involved in problem identification, experimentation and evaluation, since they perform in most countries the largest share of agricultural work, as well as being responsible for post-harvest and household tasks. Many demands are already being made on rural women's time, so it is important not to expect or impose yet more work. At the same time, this should not be taken as a reason for excluding women from involvement in technological improvement or bypassing them.

The IPM programme in Indonesia initially ignored women, assuming that they were too busy to participate in its training courses, and believing that social norms in a mostly Muslim country would constrain such participation. When the programme experimented with the inclusion of women in its courses, however, it found that women's voluntary attendance was higher than men's because women were anxious to reduce pest damage and became avid learners. Even more surprising was that the *men's* learning and performance was higher in mixed-gender classes than in men-only classes. Why was this? Women are more willing to ask questions and they promote a better group-learning atmosphere by being less individually competitive, according to an evaluation by Lestari (1993). As women have been so often neglected in technology improvement, working with them systematically and flexibly will be one of the keys to raising agricultural production in the future. Also, the direct involvement of – and control by – women provides greater assurance that improvements in technology will result in improved food security and household nutrition.

We want to emphasize further that young people – both young men and young women – need to be involved more directly in future efforts to accelerate agricultural improvements. They are often marginalized by the unacknowledged subordination of 'ageism'. When young people are excluded from decision-making and have lower status in rural areas, they have more reason to abandon agriculture and turn to occupations and lifestyles that may be less productive than those that could be pursued in rural communities if the next generation were to be given meaningful economic opportunities and social respect (Uphoff et al, 1998, pp84–85). Involving young farmers in processes of agricultural development based on experimentation and evaluation can attract and engage intellectual talent that would be otherwise squandered. It can make farming a more modern and valued enterprise bringing greater status and self-respect, as well as greater remuneration.

An agroecological vision of natural processes stresses the inter-relatedness of all factors. For sustainability, we need to satisfy the needs both of people and the natural resource systems on which they depend for their livelihoods.

This cannot be achieved by the decisions and actions of individuals, or even just by local groups of people. Decisions and actions emanating from levels above and beyond the community – and particularly from the policies of governments – will create or constrain the possibilities for optimization and synergy. So we need to consider which reforms and reconfigurations can make such outcomes more attainable.

NOTES

1   Some good examples of how indigenous technical knowledge can contribute valuable insights to the research process have been given by Ramakrishnan (1994). In the *jhum* system, a form of shifting cultivation in Northeastern India, farmers conserve bamboo and alder on their farmlands because they think these improve the soil. Research has shown that bamboo indeed accumulates potassium, a nutrient easily lost via leaching, and alders fix atmospheric nitrogen, thereby restoring soil fertility. Fernandes and Nair (1986) have documented how smallholding farmers' strategies for diversifying species in their home gardens can contribute to the design of more efficient home gardens and agroforestry systems once their species-site matching, IPM and nutrient cycling are understood scientifically.

2   The Indonesian Agency for Agricultural Research and Development at its Sukamandi rice research station has begun evaluating the system of rice intensification reported in Chapter 12. A yield of 9.5t/ha was achieved in the wet season in 2000, through changes in management alone. These yields were higher than those achieved using high-yielding varieties and modern inputs at the same site (Sunendar Kartaatmadja, personal communication, April 2000). It is expected that when the new methods are used more skilfully, still greater increases should be attainable. The human factor is a large part of the productivity of any rice management system.

Chapter 21

Institutional Changes and Policy Reforms

Jules Pretty, Ruerd Ruben and Lori Ann Thrupp

This chapter considers how institutional and policy environments can better support agroecological approaches. We do not expect that such approaches will displace all present farming methods for increasing productivity. However, if market and institutional 'playing fields' are made more level, and if environmental and equity concerns are weighed along with those of food security and human welfare, we expect that agroecological farming systems will spread. Increasing consumer concerns about the quality and safety of food will further enhance public interest in having more healthy production methods.

For agricultural systems to be economically, environmentally and socially sustainable, national governments need to encourage more participatory systems of research and extension as well as favourable institutional settings. Without appropriate policy support, innovations will remain at best localized in extent, and at worst they will wither away. Some progress has been made on getting more favourable policy content, for example the Indonesian government's withdrawal of subsidies for pesticides to support integrated pest management. In sub-Saharan Africa, exchange rate adjustments within the framework of structural adjustment programmes have improved some opportunities for nutrient management based on local resources, as external inputs have become more expensive without subsidies. However, many positive effects were also lost when agricultural extension systems were adversely affected by heavy budget cuts, also part of structural adjustment (Kuyvenhoven et al, 1999).

Fortunately, there have been some notable examples of effective policy support in recent years, similar to the Indonesia case discussed in the preceding chapter, so the prospects for achieving policy change are not entirely daunting (Pretty and Ward, 2001).

- In India and Nepal, the granting to community groups in the early 1990s by national governments of access rights to local forest resources and concessions for forest products contributed to the emergence of 20,000 new forest resource user groups in those countries (Malla, 1997; Raju, 1998; Shrestha, 1998).

- In Sri Lanka, participatory irrigation management with water users' groups became national policy in 1988, leading to a national system of irrigation associations with about 250,000 members today (Brewer, 1994; Uphoff, 1996).
- In Kenya, the government's soil conservation programme became more successful following a policy introduced in the late 1980s that permitted community groups to plan and set priorities for themselves (Pretty et al, 1995).
- In Bangladesh, the emergence of microfinance institutions primarily targeted to benefit the poorest and most excluded groups of society is in part due to policy reforms that permitted groups to receive credit on the basis of social collateral (Gibbons, 1996).
- In Australia, the government's decade of Landcare, launched in 1989, helped propel a movement which now encompasses over 4500 local Landcare groups that involve about one-third of rural households nationwide (Campbell and Woodhill, 1999).

Sustainable agriculture, as seen in the case studies, does not derive from a set of practices that are fixed in time and space. It depends on both individual and collective capacities to adapt and alter technologies and, indeed, whole farming systems as external and local conditions change. At a minimum, governments need to rescind and eliminate policies that *prescribe* certain practices which farmers should use; to the extent possible, they should create *enabling conditions* for the beneficial use of a variety of technologies, including ones that are locally devised and adapted.

In the past, thousands of initiatives for soil and water conservation, rangeland management, protected area management, irrigation development and modern crop dissemination have followed a common pattern. Technical prescriptions are developed and evaluated under controlled, uniform conditions, supported by limited cases of success; they are then applied widely with little regard for diverse local needs and conditions (Ostrom, 1990; Benhke and Scoones, 1992; Pretty and Pimbert, 1995; Ghimire and Pimbert, 1997; Pretty and Shah, 1997). When these practices are rejected locally, governments then promote them by manipulating social, economic and ecological conditions so that the technologies would be accepted intact. Then if this did not work, there remained the alternative of outright enforcement (Pretty and Shah, 1999). This has not, however, led to sustainable agricultural innovation. Accordingly, we need to think about policy processes using different verbs. Rather than deciding, promulgating, implementing and enforcing, governments should formulate (after consultation), test and evaluate, expand and adapt, and periodically assess and modify. Policy in this mode becomes the result of social learning rather than an exercise of political authority.

INCENTIVE REGIMES

Most farm households involved in the development of agroecological farming systems are engaged in local factor and commodity markets that are characterized by a large degree of what economists call 'imperfection'. Among the imperfect market conditions confronting rural households are: a lack of credit at competitive rates of interest; limited access to good quality farm inputs at reasonable prices; lack of market information, transport and other means of communication; limited numbers of marketing outlets and commercial agents who can make markets competitive; and precarious conditions for renting in land. Consequently, local networks of exchange relations tend to be highly complex, and prices reflect the power relations between the agents involved more than supply and demand interactions (de Alcantara, 1993).

Rural producers make their decisions on input purchase and marketing influenced by relative prices, opportunity costs and risks. Low-external-input technologies will be rejected by farmers if their additional labour requirements cannot be met, for instance, when a household derives an important share of its income from off-farm employment (Ruben and Lee, 2000). Similarly, fixed investments in land improvements, such as tree planting or anti-erosion practices, are unattractive when long-term land ownership is insecure.

The market environment creates a variety of incentives for or against investments in sustainable agriculture. Efforts to affect these incentives will be more successful if they relate to the 'pull' of farmer needs rather than to the 'push' of technology (Conway and Barbier, 1990; Pretty, 1995). Market prices should accordingly reflect real scarcity relations and not be altered by interventions by the state or non-governmental organizations (NGOs). There is ample evidence that offering credit or short-term subsidies on inputs seldom leads to lasting adoption of agroecological practices. Providing free inputs has proved to be a costly and unsustainable policy that benefits only a limited number of farmers, and usually not the poorest ones. Subsidized demonstration plots are generally less convincing to farmers than are experiments conducted on their own fields with investments of their own resources. Similarly, financial support systems based on farmers' own savings are far more sustainable than entirely subsidized credit systems (Adams et al, 1984; Kiriwandeniya, 1997).

Suitable interventions for improving the market environment in favour of agroecological production practices could include the following, according to de Graaff (1993):

- Increasing *prices* of imported inputs like fertilizers and other agrochemicals through adjustment of the foreign exchange rate (devaluation) and elimination of input subsidies.
- Improving the efficiency of input-delivery and output-marketing systems, reducing the *transaction costs* involved in market exchange, through public and private investments in services and through provision of infrastructure.
- Introducing *user charges and fees* for water, roads and technical assistance that facilitate the channelling of scarce resources towards the farmers able

to make the most productive use of them, at the same time seeking to ensure the institutional sustainability and maintenance of these services – experience shows that subsidized scarce services tend to be captured by the more powerful but not necessarily most efficient farmers.

- Creating more competitive markets by reducing *market entry costs*, for example by establishing farmers' commercial cooperatives or market information services.
- Improving opportunities for the creation of *value added* in agricultural production and marketing through investments in agroprocessing, trade centres, certification of organically grown produce, farmers' markets, etc.
- Enhancing backward and forward linkages within the agricultural sector, including integrated *agrocommodity chains* based on long-term contractual relations between agents.
- Supporting *diversification* of factor and commodity markets to give farmers more access to non-farm and off-farm income opportunities that enable them to intensify their farming systems.

A market environment that creates a favourable context for sustainable resource management will be based on clearly defined and complementary institutional roles among public, private and civil organizations. These roles should operate according to principles of accountability in public service provision, transparency in the management of voluntary agencies and broad participation of farmer organizations (Picciotto, 1995).

POLICIES SUPPORTING COLLECTIVE ACTION

For many years, extension advice and material inputs have been focused on 'progressive farmers', producers who were expected to create good examples and set the pace for their neighbours. This approach has many drawbacks, however, including a propensity to further concentrate opportunities and wealth. But the main problem is that people do not make decisions simply or strictly as individuals. They make decisions mindful of their circumstances, which are in large part created by the activities, ideas and evaluations of other farmers (Grunig, 1972).

Group extension methods have become popular in many countries as a cost-saving approach, disseminating information to farmers in groups rather than as individuals. This approach can contribute more to changing agricultural practices than by giving economies of scale in communication and lowering costs. When groups assess opportunities and constraints, looking at new technologies in the light of shared knowledge and evaluations, new ideas that can solve recognized problems are more likely to gain acceptance.[1]

A major advantage of making local organizations a central part of agricultural development strategy is that the collective capabilities developed for dealing with agricultural problems are often expanded upon by leaders and members who proceed to tackle other problems in the social and economic domains of rural development (Thrupp, 1996; Uphoff et al, 1998, pp205–213). Thus, the payoffs from developing local capacities for agricultural innovation can be diverse and unexpected.

There is certainly need and scope for individual initiative and evaluation in agricultural development. Innovation is usually a personal creation, and some people are more inclined, whether by personality or circumstances, to conceive and try new things than are others. Progressive farmers, in the sense of being willing and able to innovate, are essential for the kind of agroecological advances envisioned here. Most agree that their learning should be 'socialized', that is, become common knowledge and common practice. This is more feasible where there are structures (roles, procedures, networks, etc) and ways of thinking that facilitate sharing of information and cooperation, things discussed as social capital in Chapter 4, predisposing people towards mutually beneficial collective action and/or facilitating this.

INSTITUTIONAL ALTERNATIVES

The generation and diffusion of new knowledge and information systems to support sustainable agriculture has long been considered the exclusive domain of governments. Considerable amounts of public funds have been invested in experimental stations and extension services. In practice, these have then been oriented more towards larger commercial producers in the most fertile areas, with smaller farmers in more remote regions remaining virtually untouched. Local agencies, NGOs and farmers' associations have sometimes filled in the gap, identifying appropriate technologies for the development of farming systems that would be less input-intensive, but their means and access to technology are limited.

In response to the general budgetary crisis in many developing countries, one of the policy reforms in recent years has been to downsize public extension, input delivery and rural financial services. This has in turn reduced the capacity to support agroecological development for small farmers. A concomitant process of *privatization* has reassigned technical assistance to private firms, based on certain bidding and contracting procedures, to serve specific types of producers or particular regions (World Bank, 2000a). In principle, such systems should make technical support more *demand oriented*, but since the purchasing power of poor farmers is invariably low, their demand is also low. Occasionally, local organizations have been able to participate in the bidding process to access public funding to operate supportive activities.

Another aspect of institutional reform is a process of *decentralization* involving regional and local governments in decisions on public services, infrastructure provision and spatial planning. This can favour the advancement of sustainable agriculture if the local expression of needs is better heard and more respected (World Bank, 2000). In practice, however, decision power is often delegated without provision for sufficient resources to meet local needs. Moreover, power and dependency relations at local levels tend to be complex and unequal so that small farmers often still face severe difficulties in getting their needs addressed.[2]

One positive result of the process of privatization and decentralization can be an increased awareness among farmers that the state and its agents are not on their own sufficient or reliable sources of support. Accordingly, there is growing sophistication about advancing or defending farmers' interests and less expectation that government provision of (subsidized) services is the key to agricultural improvement. Farmer cooperatives can evolve into businesses that connect their members with market opportunities on efficient and equitable terms. Once earlier attitudes and relationships of dependency are severed, new possibilities of self-reliant and self-directed development are opened up, consistent with a strategy of agroecological development that derives from farmers' evaluation of new and alternative opportunities.

POLICY CONDITIONS AND REFORMS

Some sustainable production technologies and practices are already being promoted by farmer groups and NGOs seeking to reduce dependence on external inputs and reinforce agroecological sustainability, eg the Landcare movement in The Philippines reported in Chapter 18. Donor-assisted projects can make advances by developing alternative nutrient sources and by covering the sunk costs of soil conservation measures. In the long run, however, genuine sustainability will require that these practices be economically feasible and independent of external support, with policies reinforcing appropriate incentives and conditions.

Most current and past policies, in both more and less developed countries, have *not* supported agricultural practices and technologies that give weight to environmental and social considerations as well as purely economic ones. Common examples of such policies include the following:

- Economic incentives and subsidies for pesticides and fertilizers that tend to perpetuate agrochemical dependency even when economic returns and environmental impacts are negative.
- Subsidies or policy prescriptions for planting certain uniform varieties, along with requirements to use associated inputs, that reduce biodiversity and create vulnerability to pests and disease, also reinforcing chemical dependence.
- Extension programmes and agricultural policies that encourage monoculture of crops that may be poorly adapted to local conditions.[3]
- Land tenure policies that undermine small farmers' security of ownership and consequently discourage investments in sustainable practices.
- Trade and marketing policies that promote investments in crops not suitable for production by poorer farmers, or that create inequitable market opportunities.
- Price support policies that affect use of inputs and/or choice of crops in unsustainable directions.[4]
- Incentive policies, such as granting land rights for deforestation and agricultural clearing of land, that are unsuited for sustained production.

- Economic and trade policies that favour export-oriented agricultural development to the neglect of food security considerations.

Many of these types of policies have inherent biases that tend to aggravate economic inequities as well as induce adverse ecological impacts, particularly in the long run. Such results have been seen in both northern and southern countries (eg Faeth, 1993, 1995; Repetto, 1988; Thrupp, 1995, 1999; Pretty et al, 2000). Changing such policies is generally difficult since they have entrenched economic support. Such policies have been reversed at least partially, however, in countries as diverse as Indonesia, Cuba, Brazil and several European countries (Pretty, 1998).

Beyond changing existing policies that have adverse implications for agroecological development, there are additional needs for new policies that put in place economic incentives and effective institutions to facilitate ongoing adoption of sustainable practices and technologies. Some policy instruments that are available to accomplish this are listed in Table 21.1, following the analysis of criteria for sustainable agricultural intensification presented in Chapter 5.

Profitability can flow from farmers' involvement in market exchange and institutional networks that enhance access to inputs and information. Input price subsidies frequently induce perverse or skewed effects, leading to excessive use (from a social or environmental perspective) of purchased inputs on marginal fields. Stable and remunerative market prices for agricultural products and access to market outlets are far more effective as incentives to mobilize resources into sustainable production systems.

Contrary to prevailing opinion, there is evidence that public investment in rural infrastructure can be helpful in reducing transaction costs and enhancing market development in remote regions. Fan and Hazell (1999) have showed that marginal returns to investments in developing areas can be favourable because, with current production levels being low, yield increases can be attained at a relatively low cost. Policies that promote sustainable agricultural intensification and poverty alleviation in marginal areas need to enhance input-use efficiency by establishing selective incentive regimes attuned to local conditions. They countervail the tendency for general incentives to be capitalized on more quickly and thoroughly by those in more advantaged areas. Training, education and extension are important to enhance farmers' access to knowledge and information regarding appropriate technologies and feasible marketing strategies.

Rural financial systems should facilitate farmers' borrowing for investment, input purchase and insurance purposes. Formal banks are usually reluctant to lend to smallholders, however, local credit and savings schemes can reduce transaction costs and the risks of rural investment. But when access to rural financial institutions is constrained, farmers are likely to increase their engagement in off-farm employment as a means to secure investment funds. Given the prospective profitability of numerous agroecological technologies and practices, the creation and provision of investment funds seems well warranted.

Table 21.1 *Policy Instruments Available for Supporting Sustainable Agricultural Practices*

Criteria to be met	Policy instruments	Implications
Profitability	Changing input costs and market prices	Higher average returns
	Infrastructure investment	Lower transaction costs that expand market opportunities
Input efficiency	Provision of complementary inputs and relative prices that reflect marginal productivity	Higher marginal returns on input investments
		Greater access to knowledge and information
	Training and extension	
Factor substitution	Improved rural financial systems	Greater access to financial resources
	Support of social capital formation	Greater resource-sharing
	New sustainable agriculture technologies	
Risk management	Price stabilization policies	More stable returns
	Support for more off-farm employment opportunities	Greater willingness to invest
		Income diversification and consumption smoothing
	Greater number and quality of rural schools	Access to non-agricultural employment
Sustainability	Enhanced security of property rights	More collateral for lending
		Greater income certainty

Technological change in agriculture is strongly influenced by the dynamics of rural labour markets, as shown by Hayami and Ruttan (1985). Agrochemically-based intensification is most likely to occur when labour is relatively scarce and its opportunity costs are high. The amount of labour put into land conservation practices is most likely to increase when there are good commercial cropping activities in response to new market opportunities (Tiffen et al, 1994). Diversification of household labour allocations into non-farm activities is a common procedure for portfolio and risk management, enabling farmers to finance agricultural input purchases from non-agricultural sources. In particular, land use technologies should economize on wage labour and raise the marginal productivity of family labour if they are to have a good chance of adoption.

Agricultural intensification depends on the effective mobilization and combination of land, labour and capital resources. Well defined land owner-ship and use and transfer rights permit farmers to invest in land improvements and input purchase, and provide suitable collateral for loans (Besley, 1995; Ruben et al, 2001). Although private ownership provides the most demonstrable incentives for farmers' welfare and sustainability, secure land rights can also be found under common property regimes; so secure access and usufruct rights are more critical than ownership.

Because many small and marginal farm households are constrained when it comes to adopting labour-intensive technologies, sustainable agricultural intensification will contribute most to poverty alleviation when the returns to land and labour are simultaneously increased, as possible with the system of rice intensification developed in Madagascar (Chapter 12). Even in such situations, however, poorer households can encounter labour constraints that limit adoption because they have urgent needs for income that can be met more assuredly by off-farm employment than by investing labour in agriculture, however remunerative, because its returns are deferred until the end of the season.

Promotion of sustainable agricultural intensification requires also *institutional support* by various types of agents (Picciotto, 1995). NGOs can play a key role in the development of communication networks for sharing knowledge and information regarding sustainable landuse practices. Informal cooperative arrangements and other types of social capital provide local frameworks for risk sharing that encourage private investment. The role of the state is essential in the creation and enforcement of legal property rights. Moreover, public investment in physical infrastructure is required to develop local factor and commodity markets, especially in the more underdeveloped areas. Private agents and financial institutions are likely to follow once sufficient market surpluses become available. This is an example of how incentive regimes can change over time. Appropriate policy is not a matter of getting incentives and institutional arrangements right once and for all time, but of continually adapting incentive regimes to encourage the productive and sustainable use of resources, considering all forms of capital as proposed in Chapter 4.

Concern with issues of technology adoption directs attention to 'the food security puzzle' suggested in Chapter 9. This showed how alterations in farming practices are impeded by the interdependence and embeddedness of farming system components. All participants in agricultural development decision-making, from farmers to policy-makers, need to have a good sense of this complexity since few changes will succeed on their own, and most require rethinking the systems themselves.

NOTES

1   The training and visit (T&V) method of extension promoted for many years by the World Bank ostensibly follows a group approach, providing information on new and improved practices to 'contact farmers' who in turn are supposed to convey this information to groups of 'follower farmers' (Benor and Harrison, 1977). This system has, however, failed to achieve its potential for a variety of reasons, including the lack of a genuine group at the base. 'Follower farmers' were usually a hastily constructed list with no sociological reality, while contact farmers were more often than not selected by extension agents rather than by fellow farmers, so they were not representative of the majority of farmers. See Chambers (1997) and Salmen (2000).

2   The World Bank 'good practice note' on decentralizing agricultural extension (2000) in its reporting on experience in Ghana, Colombia and the Philippines notes more shortcomings than successes in these decentralization examples to date.

3   An example was given in Chapter 14 of how farmer experimentation and evaluation headed off such a recommendation in Bolivia.

4   As was the case in Indonesia with subsidized pesticides before integrated pest management was introduced; also policies required farmers to grow rice and off-season (*palawija*) crops with prescribed agrochemical use.

Chapter 22

A More Productive Synthesis for Agriculture

Norman Uphoff

Agriculture is one of the most intractably material vocations there is. By their labour, farmers – women and men, young and old – have for 10,000 years sought to produce food and fibre directly or indirectly from the soil, using seeds, animals, fodder, compost, fertilizer, water and other material inputs. Knowledge and skill are, of course, extremely important, informing and improving these practices. But the material nature of the biophysical world that defines agriculture is always evident. Climate, hydrology, geology and topography, as well as proximity to markets, transportation infrastructure and storage facilities, underscore the physical nature of the agricultural enterprise.

Yet we should never forget that 'agriculture' includes 'culture'. The enterprise is inextricably bound up in concepts, values, traditions, language and social relationships. How we think about agriculture greatly affects what can and will be produced, as discussed in Chapter 2. Over the previous century, continuous efforts have been made, many of them very fruitful, to improve agriculture by conceiving it and practising it in industrial terms. Such a view of agriculture emphasizes the role of capital investment and new technologies, promoting large-scale production and uniformity of operations to gain economies of scale.

This way of thinking downplays the biological nature of agriculture, which keeps inputs and outputs from being linked in fixed, mechanical relationships. Rather than offer predictable production results, agriculture can produce huge returns – or none all – depending on how successful and bountiful are the life processes being nurtured. (Returns are further made variable by the contingencies of supply and demand forces operating through product and factor markets.) Anthropologists and sociologists use the term 'culture' to refer to shared symbols and meanings, established roles, institutions and values. In the biological realm, however, culture refers to cultivation and rearing. The two meanings meet in the practice of agriculture, where the ways in which cultivation and rearing are carried out is profoundly shaped by what social scientists have understood by this term.

As further evidence accumulates on the benefits of an agriculture that is ecologically attuned and reinforced – and on the costs of ignoring these dynamics and relationships – we expect there will be more willingness to modify the mechanistic paradigm that has guided research, investment and policy in this sector for the past century. The development process is often seen as one where capital is substituted for labour, and labour inputs are reduced. This view, however, is too simplistic, since development is driven is by increases in total factor productivity, rather than by the invariant substitution of one factor for another. If labour can be made more productive by capitalizing on agroecological processes, it would be unwise to let the paradigm of substituting capital for labour direct our attention to scarce resources rather than to abundant ones.[1]

As stated in the Introduction, agroecological approaches are not proposed as antithetical to the present practice of agriculture, though they do challenge some of its assumptions and seek to remedy some of its problems, such as environmental pollution, dependence on fossil-fuel products, and relevance to small farmers. We present below a characterization of the agroecological approach that contrasts it with current thinking and modes of operation. However, the real-world practice of agriculture clearly requires attempts at synthesis and optimization to make the best use of available resources, given current knowledge, to meet human needs. The more that environmental and equity problems are found to be associated with the first column, the more attractive will be the alternatives characterized in the second column. Table 22.1 summarizes many of the ideas and themes presented in the second chapter and elaborated in succeeding chapters.

The agriculture of the 21st century will be some kind of hybrid, including approaches and technologies that are not yet understood or even dreamed of today.[2] The agroecological vision of a more productive and sustainable agriculture has been evolving over recent decades, building on some classic works such as by King (1911), Howard (1943) and Balfour (1943). It will continue to evolve as experience grows and its scientific foundations become better established.

In this process of knowledge generation, in contrast to the strategy of modern agriculture, substantial contributions will come from farmers, whose powers of observation and analysis are greater than has been conceded in a culture as preoccupied with formal education as ours.[3] Such education has been immensely powerful and valuable, and the contributors to this volume have themselves benefited greatly from the knowledge and opportunities that formal education provides. We wish it to become more widely available throughout the world, especially for women, who in most countries have limited access to educational opportunities.

The integrating theme throughout this assessment of how the foreseeable needs for food in this world can be met with new approaches has been taking a *systems perspective*. This examines connections among apparently diverse actors and conceives of the benefits to be created in multiple terms.

Table 22.1 *Alternative Conceptions of Agricultural Development*

	'Modern' Agriculture	*'Sustainable' Agriculture*
Regarded as:	Basically *mechanical*	Basically *biological*
Efforts Focus on:	Agriculture as *field culture*, emphasizing crops, especially staple crops – rice, wheat maize	Agriculture based on *many flora and fauna*: crops, trees, animals, fish, fruits, vegetables, tubers, grasses, other crops and livestock
	Component/crops, to be improved through disciplines	*Farming systems*, to be improved through interdisciplinary work
	Fields regarded as units of analysis and action	*Agroecosystems*, containing fields, as units of analysis and action
	Farms regarded as *management systems*	*Households* regarded as *livelihood systems*
	Larger farms, seeking economies of scale	*Smaller farms*, seeking economies of detail and stability of diversification
	Most-favored areas, to boost production most quickly and increase GNP	*All areas*, to raise production in ways that preserve the environment and meet human needs, thereby reducing pressure on most-favored areas
Strategies:	*Annualism*, with annual cycle of plowing, planting, and harvesting of crops	*Permaculture*, especially agroforestry including trees, shrubs and grasses, along with animals and fish, complementing annual crops
	Two-dimensional cultivation, breaking the plane of the soil with a plough; soil is considered as *surface* or as a thin layer	*Three dimensional* cultivation, manipulating and forming soil through terraces, raised beds, drains etc; soil is seen as *volume*
	Monocropping, with 'clean' fields considered ideal	*Polyculture*, with high value attached to ground cover
	Genetic improvements plus *modern inputs*	Better *management of all inputs*, including genetic diversity
	Market production and *global exchange*	*Niche production* and *contractual exchange*
Views of Components:	*Deep tillage*, is best done by mechanical ploughing	*Minimum or no tillage*, now called 'conservation tillage'
	Fallowing is a *passive gap* in the agricultural cycle	Fallowing is a period for *active management*
	Mulch is a *nuisance*, unnecessary or unsightly	Mulch is *essential* for soil protection and enrichment
	Chemical fertilizers are needed to improve and maintain soil fertility	*Integrated nutrient management* uses organic and mineral inputs to maintain soil fertility; *nutrient cycling* replaces nutrients removed by harvest

	'Modern' Agriculture	'Sustainable' Agriculture
	Chemical control of pests and diseases	Integrated pest management using biological controls and other means to minimize use of chemicals
	Irrigation to provide water for crops; find and deliver supply to meet demand	Water conservation and harvesting are important; adjust cropping system and soil management practices
	Packages of purchased inputs are recommended, usually requiring external borrowing	Use of own resources as much as possible; use own savings supplemented by rural financial institutions
Research Focus on:	Soil chemistry, and also soil physics	Soil biology in particular; also soil structure
	Crop growth above ground	Root growth and function below ground
Objectives:	Highest possible yields per unit of land area	Sustainable livelihoods for rural farm households
	Production of food with malnutrition seen as a medical problem; focus on food supply	Production of food should contribute to people's health; malnutrition is to be reduced through redesigned agricultural systems; concern with people's purchasing power and effective demand
Approaches:	Research and extension are in sequential, linear relationship; scientific knowledge is to be transmitted through extension; rely on modern inputs to 'transform' agriculture	Farmer-centred research and extension in triangular relation-ship; draw on local knowledge along with scientific knowledge; utilize local resources as much as possible for incremental improvement
	Project approach to development, which is achieved by implementing project designs ('blueprints')	Partnership and process approaches to development through multi-actor information-sharing and pervasive learning process (Korten 1980)
Emphasis on:	Technology and material resources; use incentives to get desired behaviour	Organization, knowledge and human resources; create and use ideas that can produce new technologies, refashion existing ones, and change expectations and incentives

Such a perspective admittedly makes evaluation more difficult, but it contributes to the positive-sum dynamics that are essential for sustainability:

- For example, the greater and more empowered participation of *women* will benefit also *men* if one looks beyond narrow interests, as well as to the next generation.
- Involving *rural people* more centrally in governmental and non-governmental initiatives to promote rural development will benefit *scientists and policy-makers* and *urban residents*, as well as rural women, men and children.
- Integrating *socioeconomic and biophysical factors* will be essential for making further progress in agriculture. Actually, each of these sets of factors encompasses a complex bundle of complementary disciplinary endeavours that should communicate and cooperate more than they do now. Each discipline can benefit from cross-fertilization with others, gaining new concepts, analytical methods and perspectives.
- The agricultural sector should be concerned with making available the kinds and quality of food that enhance *human health and wellbeing*, not just with *producing more food*. All food should be produced in ways that contribute to environmental and social objectives. This is a lofty aim, but attainable as both science and practice advance.

Many scientific and technological accomplishments were made in the 20th century, some of the most important being in the agricultural area. The insights of systems thinking make it relevant to the connections between the agricultural sector and all other sectors, seeking better integration and complementarity across sectors. It argues further that the more prosperous parts of the world should not take their present advantages for granted, assuming that their comfort and security can be sustained even if the food and other basic needs of the poorer parts are not attended to.[4]

Meeting these needs will require multiple partnerships: between richer and poorer countries; among decision-makers at local, regional, national and international levels; among researchers, farmers, extension personnel and others who are generating and applying new knowledge; between public and private sector actors and between them and NGOs; between men and women; between those with more and those with less education; and across generations, to include youth in more active roles.

Our subject is by its nature future-oriented, so this last partnership is particularly crucial. We work always in the present, building on and learning from the past but concerned about the future. Things done now should make life for humans and other living creatures more secure and satisfactory in the generations to come. The food system, with its husbanding, recycling and distribution of nutrients, holds the key to the future. We have tried through this collaborative, interdisciplinary venture to map out constructive possibilities for making this system serve better the needs of people in ways that are environmentally, economically and socially sustainable.

NOTES

1   Where biological nitrogen fixation can be promoted by different soil management practices, this taps a huge 'free' resource of atmospheric nitrogen; soil aeration practices that furnish more oxygen to the rhizosphere can similarly increase production by stimulating microbiological processes that aid plants' access to soil nutrients. As was suggested in Chapter 2, agricultural research has grossly under-invested in microbiology, so our knowledge of its contributions to both more productive and more sustainable agriculture is very limited. But what has become known indicates considerable potential to utilize abundantly renewable natural resources that repay labour and management very well, especially in developing countries where labour costs are currently lower.

2   For an example of how a synthesis of methods for apple production can surpass 'modern' methods relying on chemical inputs in terms of output as well as profitability, see Reganold et al (2001).

3   The *Mukibat* system for increasing yields of cassava, noted in Chapter 2, was devised by an uneducated farmer in Indonesia, after whom the system is now named. It can raise yields by five to ten times through a simple grafting operation (Foresta et al, 1994). Why this technology continues to be ignored (though it has recently been publicized by ICRAF) is hard to understand, except that it was not invented through 'modern' institutional channels.

4   We have not addressed, because it is beyond our expertise, the impending impacts of HIV/AIDS and other diseases such as malaria, tuberculosis and other life-threatening illnesses that are becoming more serious because of antibiotic-resistance in pathogens no longer controllable by pharmacologic means. These threats need to be factored into everyone's thinking about the future, especially for Africa and South Asia, areas where unmet food needs are currently greatest. Several countries in Africa are already expected to experience absolute population declines as well as falling life expectancy due to mortality from AIDS and other diseases. While this would reduce food requirements, it will more than proportionally reduce the labour force for producing food supplies. Already, agricultural production in South Africa and Zimbabwe is predicted to decline by 20–25 per cent over the next decade because of mortality and morbidity in the agricultural labour force. Labour constraints may make more difficult the adoption of agroecological methods that are more labour-intensive. However, given the constraints of poverty and adverse climate, such methods are likely to be more accessible and resilient than modern input-intensive agriculture. The 'modern' strategy does not necessarily become more feasible as a result of public health disasters, especially if global warming, shifting patterns of rainfall and weather variability (extreme events) also increase in the decades ahead.

References

Acharya, M S (1997) 'Integrated vermiculture for rural development', *International Journal of Rural Studies*, vol 4, no 1, pp8–10

Adams, Dale W, D H Graham and J D von Pischke (eds) (1984) *Undermining Rural Development with Cheap Credit*, Boulder, CO, Westview Press

Agus, Fachmuddin (2000) 'Intensification of indigenous fallow rotations using *Leucaena leucocephala*', in M F Cairns (ed) *Voices from the Forest: Farmer Solutions Towards Improved Fallow Husbandry in Southeast Asia*, Bogor, International Centre for Research in Agroforestry, Southeast Asian Regional Research Programme Proceedings of a regional conference on Indigenous Strategies for Intensification of Shifting Cultivation in Southeast Asia, Bogor, Indonesia, 23–27 June (1997, pp291–297

Agus, F, D P Garrity, D K Cassel and A Mercado (1998) 'Grain crop response to contour hedgerow systems on sloping oxisols', *Agroforestry Systems*, vol 42, pp107–120

Allmaras, R R and R H Dowdy (1985) 'Conservation tillage systems and their adoption in the United States', *Soil and Tillage Research*, vol 7, pp197–222

Almeida, F S and B N Rodrigues (1985) *Guia de Herbicidas: Contribuicao para o Uso Adequado em Plantio Direto e Convencional*, Londrina, IAPAR

Altieri, M A (1987) *Agroecology: The Scientific Basis of Alternative Agriculture*, Boulder, Westview Press

Altieri, M A (1990) 'Agroecology and rural development in Latin America', in M Altieri and S Hecht (eds) *Agroecology and Small Farm Development*, Boca Raton, FL, CRC Press

Altieri, M A (1994) *Biodiversity and Pest Management in Agroecosystems*, New York, Hayworth Press

Altieri, M A (1995) *Agroecology: The Science of Sustainable Agriculture* (second edition), Boulder, CO, Westview Press

Altieri, M A (1999) 'Enhancing the productivity of Latin American traditional peasant farming systems through an agroecological approach', paper delivered to the Conference on Sustainable Agriculture, Bellagio, Italy, 26–30 April

Altieri, M A and U D Doll (1978) 'The potential of allelopathy as a tool for weed management in crop fields', *Panscience*, vol 24, no 4, pp495–502

Altieri, M A, D K Letourneaour and J R Davis (1983) 'Developing sustainable agro-ecosystems', *BioScience*, vol 33, pp45–49

Altieri, M A and C I Nicholls (1999) 'Biodiversity, ecosystem function and insect pest management in agricultural systems' in W W Collins and C O Qualset (eds), *Biodiversity in Agroecosystems*, Boca Raton, FL, CRC Press

Altieri, M A and P Rosset (1995) 'Agroecology and the conversion of large-scale conventional systems to sustainable management' *International Journal of Environmental Studies*, vol 50, pp165–185

Amarasinghe, Upali A, R Saktivadivel and Hammond Murray-Rust (1998) *Impact Assessment of Rehabilitation Interventions in the Gal Oya, Left Bank*, Research Report 18, Colombo, International Irrigation Management Institute

Andow, David A (1991) 'Vegetational diversity and arthropod population response', *Annual Review of Entomology*, vol 36, pp561–586

Andow, David A and K Hidaka (1989) 'Experimental natural history of sustainable agriculture: syndromes of production', *Agriculture, Ecosystems and Environment*, vol 27, pp447–462

Andriakaja, Andry Heritiana (2001) *Mise en evidence des opportunités de développment de la viziculture par adoption du SRI, et evaluation de la fixation du l'azote*, Mémoire de fin d'etudes, Ecole Supérieure des Sciences Agronomique, University of Antananarivo

Appelhof, M, K Webster and J Buckerfield (1996) 'Vermicomposting in Australia and New Zealand', *BioCycle*, vol 37, no 6, pp63–66

Araújo, A G, R Casão Jr, F Skora Neto, G Merten, F F Fernandes and M Siqueira (1991) 'Pesquisa do IAPAR: o plantio direto na pequena propriedade', *Jornal do Plantio Direto*, vol 3, no 1, p3

Argyris, Chris, R Putnam and D M Smith (1985) *Action Science*, San Francisco, Jossey-Bass

Audirac, Y (1997) *Rural Sustainable Development in America*, New York, John Wiley

Avery, Dennis T (1995) *Saving the Planet with Pesticides and Plastic: The Environmental Triumph of High-Yield Farming*, Indianapolis, IN, Hudson Institute

Avery, Dennis T and Alex Avery (1996) 'Farming to sustain the environment', Hudson Briefing Paper no 190, May, Indianapolis, IN, Hudson Institute

Aziz, Murad Bin and Rakibul Hasan (2000) *Evaluation of System of Rice Intensification (SRI) in Bangladesh*, Dhaka, CARE-Bangladesh, Locally Intensified Farming Enterprises Project

Baharsjah, Sjarifudin and Chairil A Rasahan (1994) 'IPM training in Indonesia: moving towards the second generation of Green Revolution', paper delivered to the National Workshop on the Socioeconomic Impact of Integrated Pest Management, Bogor, 7–9 March

Bairrão, J F M, L F D Goelzer and A Bego (1988) 'Comportamento de alternativas de inverno com vista a integracao em rotacao de culturas' in *Resultados de Pesquisa na Safra de Inverno*, Cascavel, OCEPAR, pp112–113

Baldani, J I and J Döbereiner (1995) 'Biological nitrogen fixation associated with sugar cane and rice: contributions and prospects for improvement', *Plant and Soil*, vol 174, pp195–209

Baldani, J I, L Caruso, V L D Baldani, S R Goi and J Döbereiner (1997) 'Recent advances in BNF with non-legume plants', *Soil and Biological Biochemistry*, vol 29, pp911–922

Balfour, E B (1943) *The Living Soil*, London, Faber and Faber

Bannister, M E and P K R Nair (1990) 'Alley cropping as a sustainable agricultural technology for the hillsides of Haiti: experience of an agroforestry outreach project', *American Journal of Alternative Agriculture*, vol 5, no 2, pp51–59

Barbier, B and G Bergeron (1998) 'The impact of policy interventions on land management in Honduras: results of a bio-economic model', *Agricultural Systems*, vol 60, no 1, pp1–16

Barzman, Marco S and Laila Banu (2000) *Project Implementation Report: New Options for Pest Management, January to June 2000*, Dhaka, CARE-Bangladesh, Agriculture and Natural Resources Sector

Barzman, Marco S and Luther Das (2000) 'Ecologising rice-based systems in Bangladesh', *ILEIA Newsletter*, vol 16, no 4, pp16–17

Bawden, R J, R Macadam, R Packam and I Valentine (1984) 'Systems thinking and practices in the education of agriculturalists', *Agricultural Systems*, vol 13, pp205–225

Bebbington, A (1991) 'Sharecropping agricultural development: the potential for GSO–government cooperation', *Grassroots Development*, vol 15, p2

Beingolea, J (1993) 'Utilización del tarwi como a bono verde en el programa de Chiroqasa de Norte De Potosí, Bolivia' in D Buckles (ed) *Gorras y Sombreros: Caminos hacia la colaboración entre técnicos y campesinos*, México, International Centre for Maize and Wheat Improvement (CIMMYT), pp101–104

Bene, J G, H W Beall and A Cote (1977) *Trees, Food and People*, Ottawa, International Development Research Centre

Benhke, R and I Scoones (1992) 'Rethinking range ecology: implications for rangeland management in Africa', Drylands Programme Issues Paper No 33, London, International Institute for Environment and Development

Benor, Daniel and James Harrison (1977) *Agricultural Extension: The Training and Visit System*, Washington, World Bank

Berg, H, P Michelsen, C Folke, N Kautsky and M Troell (1996) 'Managing aquaculture for sustainability in tropical Lake Kariba, Zimbabwe', *Ecological Economics*, vol 18, pp141–159

Berkes, Fikret (ed) (1989) *Common Property Resources: Ecology and Community-Based SustainableDevelopment*, London, Belhaven Press

Berry, R Albert and William Cline (1979) *Agrarian Structure and Productivity in Developing Countries*, Baltimore, Johns Hopkins University Press

Besley, T (1995) 'Property rights and investment incentives: theory and evidence from Ghana', *Journal of Political Economy*, vol 103, no 5, pp903–937

Binswanger, Hans and Klaus Deininger (1997 'Explaining agricultural and agrarian policies in developing countries', Policy Research Working Paper No 1764, Washington, World Bank

Boddy, R M, O C de Oliveira, S Urquiaga, V M Reis, F L de Olivares, V L D Baldani and J Döbereiner (1995) 'Biological nitrogen fixation associated with sugar cane and rice: contributions and prospects for improvement', *Plant and Soil*, vol 174, pp195–209

Borlaug, N and C R Dowswell (1994) 'Feeding a human population that increasingly crowds a fragile planet', in International Society of Soil Science, *Supplement to Transactions of the 15th World Congress of Soil Science, Acapulco, Mexico*, Chapingo, Mexico

Boserup, E (1965) *The Conditions of Agricultural Growth: The Economics of Agrarian Change under Population Pressure*, London, Allen and Unwin

Boswell, J (1993) *Journal of a Tour to the Hebrides with Samuel Johnson LLD* (first published 1786), London, Penguin Books

Bourdieu, Pierre (1986) 'The forms of capital' in J Richardson (ed) *Handbook of Theory and Research for the Sociology of Education*, Westport, CT, Greenwood Press

Boyte, H (1995) 'Beyond deliberation: citizenship as public work', paper delivered at PEGS conference, 11–12 February. Civic Practices Network at http://wwwcpnorg

Brady, N C (1982) 'Chemistry and world food supplies', *Science*, vol 218, pp847–853

Breman, H and K Sissoko (1998) *L'intensification agricole au Sahel: Mythe ou realite?*, Paris, Karthala

Brewer, Jeffrey (1994) 'The participatory irrigation management system policy', *Economic Review*, Colombo, vol 20, no 6

Bromley, Daniel G (1993) 'Common property as metaphor: systems of knowledge, resources and the decline of individualism', *The Common Property Digest*, vol 27, pp1–8

Bruijn, G H de and T S Dharmaputra (1974) 'The Mukibat system: a high-yielding method of cassava production in Indonesia', *Netherlands Journal of Agricultural Science*, vol 22, no 1, pp89–100

Brummett, R E (1995) 'The context of smallholding integrated aquaculture in Malawi: a case study for Sub-Saharan Africa', *Naga*, vol 18, no 4, pp8–10

Brummett, R E (1997) 'Why Malawian smallholders don't feed their fish', paper presented to the First SADC Regional Conference on Aquaculture, Bunda College of Agriculture, Lilongwe, Malawi, 17–19 November

Brummett, R E (2000) 'Factors influencing fish prices in Southern Malawi aquaculture', *Aquaculture*, vol 186, pp3–4 and 243–251

Brummett, R E and F J K Chikafumbwa (1995) 'Management of rainfed aquaculture on Malawian smallholdings', paper presented at the PACON Conference on Sustainable Aquaculture, Honolulu, Hawaii, 11–14 June

Brummett, R E and B A Haight (1997) 'Research–development linkages' in M Martinez-Espinosa (ed) *Report of the Expert Consultation on Small-Scale Rural Aquaculture*, FAO Fisheries Report 548, Rome, Food and Agriculture Organization of the United Nations

Brummett, R E and R P Noble (1995) 'Aquaculture for African smallholders', ICLARM Technical Report no 46, Manila, International Centre for Living Aquatic Resources Management

Buck, L E, J P Lassoie and E C M Fernandes (eds) (1999) *Agroforestry in Sustainable Agricultural Systems*, Boca Raton, FL, CRC Press

Buckles, D, B Triomphe and G Sain (1997) *Exploring the Limits of Hillside Agriculture*, Mexico, CIMMYT

Budd , J W (1993) 'Changing food prices and rural welfare: a nonparametric examination of the Côte d'Ivoire', *Economic Development and Cultural Change*, vol 41, no 3, pp587–603

Bulte, E (1997) 'Essays in economics of renewable resources', unpublished PhD thesis, Wageningen, Netherlands, Wageningen Agricultural University

Bunch, Roland (1977) 'Better use of land in the highlands of Guatemala', in Elizabeth Stamp (ed) *Growing Out of Poverty*, Oxford, UK, Oxford University Press

Bunch, Roland (1982) *Two Ears of Corn: A Guide for People-Centered Agricultural Development*, Oklahoma City, OK, World Neighbors

Bunch, Roland (1988) 'Guinope integrated development program, Honduras' in Czech Conroy and Miles Litvinoff (eds) *The Greening of Aid: Sustainable Livelihoods in Practice*, London, Earthscan Publications

Bunch, Roland (1999) 'Soil recuperation in Central America: how innovation was sustained after project intervention' in F Hinchcliffe et al (eds) *Fertile Ground: The Impacts of Participatory Watershed Development*, London, Intermediate Technology Publications, pp32–41

Bunch, Roland (2001) 'Nutrient quantity or nutrient access? A new understanding of how to maintain soil fertility in the tropics', unpublished paper, COSECHA, Tegucigalpa, Honduras

Bunch, Roland and Gabiño Lòpez (1995) 'Soil recuperation in Central America: sustaining innovation after intervention', Gatekeeper Series no 55, London, International Institute for Environment and Development

Buresh, R J, P A Sanchez, and F Calhoun (eds) (1997) *Replenishing Soil Fertility in Africa*, SSSA Special Publication no 51, Madison, WI, Soil Science Society of America

Cairns, Malcolm (1999) 'Improving shifting cultivation in Southeast Asia by building on indigenous fallow management strategies' in R J Buresh and P J M Cooper (eds)

The Science and Practice of Short-Term Improved Fallows, Dordrecht, Netherlands, Kluwer Academic Publishers

Cairns, M F (ed) (2000)*Voices from the Forest: Farmer Solutions Towards Improved Fallow Husbandry in Southeast Asia*, Bogor, International Centre for Research in Agroforestry, Southeast Asian Regional Research Programme. Proceedings of a Regional Conference on Indigenous Strategies for Intensification of Shifting Cultivation in Southeast Asia, held in Bogor, Indonesia, 23–27 June 1997

Calegari, A (1995) 'Leguminosas para Adubação Verde de Verão no Paraná', Circular 80, Londrina, IAPAR

Calegari, A and M A Pavan (1995) 'Efeitos da rotação de milho com adubos verdes de inverno na agregação do solo', *Arq Biol Tecnol*, vol 38, no 1, pp45–53

Calegari, A, A Mondardo, E A Bulisani, L do P Wildner, M B B Costa, P B Alcântara, S Miyasaka and T J C Amado (1993) *Adubação Verde no Sul do Brasil* (second edition), Rio de Janeiro, AS-PTA

Campbell, Andrew, Phil Grice and Justin Hardy (1999) 'Local conservation action in Western Australia' in F Hinchcliffe et al (eds) *Fertile Ground: The Impacts of Participatory Watershed Management*, London, Intermediate Technology Publications, pp340–353

Campbell, Andrew, and Jim Woodhill (1999) 'The policy landscape and prospects of landcare' in F Hinchcliffe et al (eds) *Fertile Ground: The Impacts of Participatory Watershed Management*, London, Intermediate Technology Publications, pp194–208

CARE/Sri Lanka (1998) *Household Livelihood Security Assessment in the Plantation Sector*, Colombo, CARE International

Carney, D (1998) *Sustainable Rural Livelihoods*, London, Department for International Development

Carrol, C Ronald, John H Vandermeer and Peter M Rosset (1990) *Agroecology*, New York, McGraw Hill

Carter, J (1995) 'Alley cropping: have resource-poor farmers benefited?', *ODI Natural Resource Perspectives*, no 3, London, Overseas Development Institute

Central Bank (1998) *Economic Progress of Independent Sri Lanka 1948–1998*, Colombo, Central Bank of Sri Lanka

Cernea, Michael M (1987) 'Farmer organisations and institution building for sustainable development', *Regional Development Dialogue*, vol 8, pp1–24

Cernea, Michael M (1991) *Putting People First: Sociological Variables in Rural Development* (second edition), Oxford, Oxford University Press

Cernea, Michael M (1993) 'The sociologist's approach to sustainable development', *Finance and Development*, vol 35, no 12, pp11–13

Chambers, Robert (1997) *Whose Reality Counts? Putting the First Last*, London, Intermediate Technology Publications

Chambers, Robert and B P Ghildyal (1985) 'Agricultural research for resource-poor farmers: the farmer-first-and-last model', *Agricultural Administration*, vol 20, no 1, pp15–30

Chambers, Robert, Arnold Pacey and Lori Ann Thrupp (eds) (1989) *Farmer First: Farmer Innovation and Agricultural Research*, London, Intermediate Technology Publications

Charreau, C (1974) 'Soils of tropical dry and dry–wet climatic areas of West Africa, and their use and management', Agronomy Department Mimeo 74–26, Ithaca, NY, Cornell University

Chatterji, Jaya et al (1999) 'Scaling up soil and water conservation efforts in Chotanagpur Plateau, Eastern India' in F Hinchcliffe et al (eds) *Fertile Ground:*

The Impacts of Participatory Watershed Management, London, Intermediate Technology Publications, pp259–272

Cheatle, R J (1999) 'Current work of the ABLH in Kenya' in T F Shaxson (ed) *Assessing the Effects of Better Land Husbandry*, summary record for the workshop held at University of Bradford, 6–7 January

Chikafumbwa, F J K (1994) 'Farmer participation in technology development and transfer in Malawi' in R E Brummett (ed) *Aquaculture Policy Options for Integrated Resource Management in Sub-Saharan Africa*, ICLARM Conference Proceedings 46, Manila, International Centre for Living Aquatic Resources Management

Chimatiro, S K and U F Scholz (1995) 'Integrated aquaculture–agriculture farming systems: a sustainable response towards food security for small-scale poor farmers in Malawi', paper presented to the Bunda College Aquaculture Symposium, Lilongwe, Malawi, 11 February

CIIFAD (1996) *CIIFAD Annual Report, 1995–96*, Ithaca, NY, Cornell International Institute for Food, Agriculture and Development

CIIFAD (1997) *CIIFAD Annual Report, 1996–97*, Ithaca, NY, Cornell International Institute for Food, Agriculture and Development

Club du Sahel (1996) *Finding Problems to Suit the Solutions: Introduction to a Critical Analysis of Methods and Instruments of Aid to the Sahel*, Paris, Club du Sahel

Cogle, A L, W M Strong, P G Saffigna, J N Ladd and M Amato (1987) 'Wheat straw decomposition in subtropical Australia, II: effect of straw placement on decomposition and recovery of added 15N-urea', *Australian Journal of Soil Research*, vol 25, p481

Coleman, James S (1988) 'Social capital and the creation of human capital', *American Journal of Sociology*, vol 94, Supplement S95–S120

Coleman, James S (1990) *Foundations of Social Theory*, Cambridge, MA, Harvard University Press

Combs, Gerald F et al (eds) (1996) *Food-Based Approaches to Preventing Micronutrient Malnutrition: An International Research Agenda*, Ithaca, NY, Cornell International Institute for Food, Agriculture and Development

Conway, Gordon R (1987) 'The properties of agroecosystems', *Agricultural Systems*, vol 24, no 1, pp95–117

Conway, Gordon R (1997) *The Doubly Green Revolution: Food for All in the 21st Century*, London, Penguin Books

Conway, Gordon R and E B Barbier (1990) *After the Green Revolution: Sustainable Agriculture for Development*, London, Earthscan Publications

Cook, R J (2000) 'Advances in plant health management in the twentieth century', *Annual Review of Phytopathology*, vol 38, pp95–116

Cook, R L and B G Ellis (1987) *Soil Management: A World View of Conservation and Production*, New York, John Wiley

Crop Protection Compendium (1997) *Module 1*, Wallingford, UK, CAB International

Crosson, Pierre and Jock R Anderson (1999) 'Technologies for meeting future global demands for food', paper for the Conference on Sustainable Agriculture, Bellagio, Italy, 26–30 April

Current, D, E Lutz and S Scherr (eds) (1995) 'Costs, benefits and farmer adoption of agroforestry: project experiences in Central America and the Caribbean', Environment Paper no 14, Washington, World Bank

Dancette, C and P Sarr (1985) *Degradation et Regeneration des Sols dans les Regions Centre-Nord du Senegal*, Dakar, Institut Senegalaise du Recherche Agriculture

DANIDA (1994) *Agricultural Sector Evaluation: Lessons Learned*, Copenhagen, Ministry of Foreign Affairs

Darolt, M R (1998) *Plantio Direto: Pequena Propriedade Sustentável*, Circular 101, Londrina, IAPAR

Dasgupta, Partha and Ismail Serageldin (eds) (2000) *Social Capital: A Multifaceted Perspective*, Washington, World Bank

de Alcantara, C H (ed) (1993) *Real Markets: Social and Political Issues of Food Policy Reform*, London and Geneva, Frank Cass and United Nations Research Institute for Social Development

de Graaff, J (1996) *The Price of Soil Erosion: An Economic Evaluation of Soil Conservation and Watershed Development*, Mansholt Studies no 3, Wageningen, Netherlands, Wageningen Agricultural University

de los Reyes, R and S G Jopillo (1986) *An Evaluation of the Philippines Participatory Communal Irrigation Program*, Quezon City, Philippines, Institute of Philippine Culture, Ateneo de Manila

Deaton, A (1992) *Understanding Consumption*, Oxford, Clarendon Press

del Castillo, Dennis and Dai Peters (1994) 'Paddy rice research' in *Final Report for the Agricultural Development Component of the Ranomafana National Park Project in Madagascar* Raleigh, NC, Soil Science Department, North Carolina State University, pp14–27

Delgado, C L, J Hopkins and V A Kelly (1998) 'Agricultural growth linkages in Sub-Saharan Africa', Research Report no 107, Washington, International Food Policy Research Institute

Derpsch, R (1986) 'Erosion problems in Paraná, Brazil: research results and strategies for the implementation of efficient soil conservation measures' unpublished dissertation, Agricultural Extension and Rural Development Centre, Reading University, UK

Derpsch, R and A Calegari (1992) *Plantas para Adubação Verde de Inverno*, Circular 73, Londrina, IAPAR

Derpsch, R, C H Roth, N Sidiras and U Köpke (1991) *Controle da erosão no Paraná, Brasil: Sistemas de cobertura do solo, plantio direto e preparo conservacionista do solo*, Eschborn, GTZ, and Londrina, IAPAR

Derpsch, R, M A Florentin and K Moriya (2000) *Importancia de la Siembra Directa para Alcanzar la Sustentabilidad Agricola*, Proyecto Conservación de Suelos San Lorenzo, Paraguay, MAG-GTZ

Deybe, D (1994) 'Vers un agriculture durable', unpublished PhD thesis, Montpellier, CIRAD

Diaz Cisneros, Heliodoro et al (1997) 'Plan Puebla: an agricultural development program for low-income farmers in Mexico' in Krishna et al, *Reasons for Hope: Instructive Experiences in Rural Development*, West Hartford, CT, Kumarian Press, pp120–136

Döbereiner, Johanna (1987) *Nitrogen-Fixing Bacteria in Non-Leguminous Crop Plants*, Berlin, Springer Verlag

Doolette, J B, and J W Smyle (1990) 'Soil and moisture conservation strategies: review of the literature' in J B Doolette and J Magrath (eds) *Watershed Development in Asia: Strategies and Technologies*, World Bank Technical Paper no 127, Washington, World Bank, pp37–70

Drew, Malcolm C (1997) 'Oxygen deficiency and root metabolism: injury and acclimation under hypoxia and anoxia', *Annual Review of Plant Physiology and Plant Molecular Biology*, vol 48, pp223–250

Dudung, Abdul Adjid and Agus Pakpahan (1992) 'Social engineering to enhance agricultural and rural development in Indonesia' in E Pasandaran et al (eds) *Poverty Alleviation with Sustainable Agricultural and Rural Development in Indonesia,* Bogor, Centre for Agro-Socio-Economic Research, and Ithaca, NY, CIIFAD, pp198–209

Dyer, N and A Bartholomew (1995) 'Project completion reports: evaluation synthesis study', Evaluation Report no 583, London, Overseas Development Administration

EC (1994) *Evaluation des Projets de Developpement Rural Finances durant les Conventions de Lomé I, II, et III,* Brussels, European Commission

Ekanayake, I J, D P Garrity and J C O'Toole (1986) 'Influence of deep root density on root-pulling resistance in rice', *IRRI Reports 1986,* pp1181–1186

Elster, Jon (1989) *The Cement of Society: A Study of Social Order,* Cambridge, UK, Cambridge University Press

Elton, C S (1958) *The Ecology of Invasions by Animals and Plants,* London, Chapman & Hall

Erenstein, O C A (1999) 'The economics of soil conservation in developing countries: the case of crop residue mulching', unpublished PhD thesis, Wageningen, Netherlands, Wageningen Agricultural University

Esman, Milton J and Norman Uphoff (1984) *Local Organizations: Intermediaries in Rural Development,* Ithaca, NY, Cornell University Press

Fafchamps, M (1993) 'Sequential labor decisions under uncertainty: an estimable household model of West African farmers', *Econometrica,* vol 61, no 5, pp1173–1197

Fall, Abdou (2000) 'Makueni district profile: livestock development 1990–1998', Working Paper no 8, Crewkerne, UK, Drylands Research

Fan, S and P Hazell (1999) 'Are returns to public investment lower in less-favored rural areas? An empirical analysis of India', IFPRI-EPTD Discussion Paper no 43, Washington, International Food Policy Research Institute

FAO (1994) *Water Harvesting for Improved Agricultural Production,* Rome, Food and Agriculture Organization

FAO (1995) *FAOSTAT PC,* Rome, Food and Agriculture Organization

Feder, E, R E Just and D Zilberman (1985) 'Adoption of agricultural innovations in developing countries: a survey', *Economic Development and Cultural Change,* vol 33, pp254–297

Fernandes, E C M and J C Matos (1995) 'Agroforestry strategies for alleviating soil chemical constraints to food and fiber production in the Brazilian Amazon' in P R Seidl et al (eds) *Chemistry of the Amazon: Biodiversity, Natural Products and Environmental Issues,* Washington, American Chemical Society, pp34–50

Fernandes, E C M, P Motavalli, C Castilla and L Mukurumbira (1997) 'Management control of soil organic matter dynamics in tropical land-use systems' *Geoderma,* vol 79, no 1, pp49–67

Fernandes, E C M and P K R Nair (1986) 'An evaluation of the structure and function of tropical homegardens', *Agricultural Systems,* vol 21, no 4, pp279–310

Finch, C V and C W Sharp (1976) *Cover Crops in California Orchards and Vineyards,* Washington, US Department of Agriculture, Soil Conservation Service

Finkh, M R, E S Gacek, H Goyeau, C Lannou, U Merz, C C Mundt, L Munk, J Nadziak, A C Newton, C de Vallavieille-Pope and M S Wolfe (2000) 'Cereal variety and species mixtures in practice, with emphasis on disease resistance', *Agronomie,* vol 20, pp813–837

Flores, M (1995) 'Why do farmers go for the fertilizer bean?' *ILEIA Newsletter For Low External Input and Sustainable Agriculture,* no 7 (March)

Foresta, Hubert de, A Basri and Wiyono (1994) 'A very intimate agroforestry association: cassava and improved homegardens – the Mukibat technique', *Agroforestry Today,* vol 6, no 1, pp12–14

Francis, Charles A (1986) *Multiple Cropping Systems,* New York, Macmillan

Francis, Charles A (1989) 'Biological efficiencies in multiple cropping systems', *Advances in Agronomy,* vol 42, pp1–36

Francis, Charles A and Sanders, J H (1978) 'Economic analysis of bean and maize systems: monoculture versus associated cropping', *Field Crops Research,* vol 1, pp319–335

Franzel, S, R Coe, P J Cooper, F Place and S J Scherr (1998) 'Assessing the adoption potential of agroforestry practices: ICRAF's experiences in sub-Saharan Africa', paper presented at international symposium of the Association for Farming Systems Research-Extension, Pretoria, South Africa, 29 November– 4 December

Freeman, P (1982) *Land Regeneration and Agricultural Intensification in Senegal's Groundnut Basin,* Dakar, US Agency for International Development

Fujisaka, S (1994) 'Learning from six reasons why farmers do not adopt innovations intended to improve sustainability of upland agriculture', *Agricultural Systems,* vol 46, pp409–425

Fujisaka, S and D P Garrity (1989) 'Developing sustainable food crop farming systems for the sloping acid uplands: a farmer-participatory approach' in *Proceedings of the SUAN IV Regional Symposium on Agroecosystems Research,* Khon Kaen, Thailand, Khon Kaen University, pp182–193

Fujisaka, S, E Jayson and A Dapusala (1994) 'Trees, grasses, and weeds: species choice in farmer-developed contour hedgerows' *Agroforestry Systems,* vol 25, pp13–22

Fukuoka, M (1978) *The One-Straw Revolution,* Emmaus, PA, Rodale Press

Fukuyama, Francis (1995) *Trust: The Social Values and the Creation of Prosperity,* New York, Free Press

Funtowicz, S O and J R Ravetz (1993) 'Science for the post-normal age', *Futures,* vol 25, no 7, pp739–755

Gachengo, C N, C A Palm, B Jama and C Othieno (1998) 'Tithonia and senna green manures and inorganic fertilizers as phosphorous sources for maize in western Kenya', *Agroforestry Systems,* vol 44, pp21–36

Garrett, K A and C C Mundt (1999) 'Epidemiology in mixed host populations', *Phytopathology,* vol 89, pp884–900

Garrity, D P (1993) 'Sustainable land-use systems for sloping uplands in Southeast Asia' in J Ragland and R Lal (eds) *Technologies for Sustainable Agriculture in the Tropics,* Madison, Wisconsin, American Society of Agronomy, pp41–66

Garrity, D P (1995) 'Improved agroforestry technologies for conservation farming: pathways toward sustainability' in A Malinao and R Sajjapongse (eds) *Proceedings of International Workshop on Conservation Farming for Sloping Uplands in Southeast Asia: Challenges, Opportunities and Prospects,* Proceedings no 14, Bangkok, International Board for Soil Research and Management, pp145–168

Garrity, D P and P C Agustin (1995) 'Historical land use evolution in a tropical acid upland agroecosystem', *Agriculture, Ecosystems, and Environment* vol 53, pp83–95

Garrity, D P, D M Kummer and E S Guiang (1993) 'The Philippines' in *Sustainable Agriculture in the Humid Tropics,* Washington, National Research Council, pp549–624

Garrity, D P and P E Sajise (1992) 'Sustainable land use systems research in Southeast Asia: a regional assessment' in R D Hart and M W Sands (eds) *Sustainable Land Use Systems Research and Development,* Emmaus, PA, Rodale Press, pp59–76

Geertz, Clifford (1963) *Agricultural Involution: The Process of Ecological Change in Indonesia*, Berkeley, University of California Press

Ghimire, Krishna and Michel Pimbert (1997) *Social Change and Conservation*, London, Earthscan Publications

Gibbons, D S (1996) 'Resource mobilisation for maximising MFI outreach and financial self-sufficiency', Issues Paper no 3, Bank-Poor '96 conference, Kuala Lumpur, 10–12 December

Giller, K E, M H Beare, P Lavelle, A-M Izac and M J Swift (1997) 'Agricultural intensification, soil biodiversity, and agroecosystem function', *Applied Soil Ecology*, vol 6, pp3–16

Gliessman, Steven R (1990) 'The ecology and management of traditional farming systems' in M Altieri and S Hecht (eds) *Agroecology and Small Farm Development*, Boca Raton, FL, CRC Press

Gliessman, Steven R (1998) *Agroecology: Ecological Processes in Sustainable Agriculture*, Ann Arbor, MI, University of Michigan Press

Goetz, S J (1992) 'A selectivity model of household food marketing behaviour in Sub-Saharan Africa', *American Journal of Agricultural Economics*, vol 64, pp444–452

Goldman, Abe and Joyotee Smith (1995) 'Agricultural transformations in India and Northern Nigeria: exploring the nature of Green Revolutions', *World Development*, vol 23, no 2, pp243–63

Goodell, G E, P E Kenmore, J A Litsinger, J P Bandong, C G de la Cruz and M D Lumaban (1982) 'Rice insect pest management technology and its transfer to small-scale farmers in the Philippines' in R E Rhoades et al (eds) *Report of an Exploratory Workshop on the Role of Anthropologists and Other Social Scientists in Interdisciplinary Teams Developing Improved Food Production Technology*, Los Baños, Philippines, International Rice Research Institute, pp25–42

Graaff, J de (1993) *Soil Conservation and Sustainable Land Use: An Economic Approach*, Amsterdam, Royal Tropical Institute

Greenland, D J (1981) 'Soil management and soil degradation', *Journal of Soil Science*, vol 32, pp301–322

Grootaert, Christian (1998) 'Social capital: the missing link', Social Capital Initiative Working Paper no 5, Washington, World Bank

Grosenick, G, A Djegal, J King, E Karsh and P Warshall (1990) 'Senegal natural resources management assessment: report for USAID/Senegal' PDC-5517-1-13-7136-00, Dakar, US Agency for International Development

Grunig, James (1972) 'Communication and the economic decision-making processes of Colombian peasants', *Economic Development and Cultural Change*, vol 19, no 4, pp580–597

Guerra, L C, S I Bhuiyan, T P Thuong and Randolph Barker (1998) 'Producing more rice with less water in irrigated systems', SWIM Paper no 5, Colombo, International Water Management Institute

Hagmann, J, E Chuma, K Murwira and M Connolly (1999) 'Putting process into practice: operationalising participatory extension', Network Paper no 94, London, Overseas Development Institute

Hamilton, N A (1995) 'Learning to learn with farmers', unpublished PhD thesis, Wageningen, Netherlands, Wageningen Agricultural University

Hargrove, W L, P B Ford and Z C Somda (1991) 'Crop residue decomposition under controlled and field conditions' in *Proceedings of 12th Conference of International Soil Tillage Research Organizations, Ibadan, Nigeria*, Columbus, Ohio State University, p99

Harrison, E, J A Stewart, R L Stirrat and J Muir (1994) 'Fish farming in Africa: what's the catch?', summary report of the ODA-sponsored research project on Aquaculture Development in Sub-Saharan Africa Stirling, Scotland, Institute of Aquaculture, and Brighton, UK, University of Sussex

Harrison, Paul (1987) *The Greening of Africa*, New York, Penguin Books

Hart, Gillian (1986) *Power, Land and Livelihood: Processes of Change in Rural Java*, Berkeley, University of California Press

Hassan, Mian Sayeed (2000) *Yield Maximization through System of Rice Intensification (SRI)*, Comilla, Bangladesh Rice Research Institute Regional Station

Hatta, S (1967) 'Water consumption in paddy field and water saving rice culture in the tropical zone', *Japanese Journal of Tropical Agriculture*, vol 11, no 3, pp106–112

Hayami, Y and V W Ruttan (1985) *Agricultural Development: An International Perspective*, Baltimore, Johns Hopkins University Press

Hazell, Peter and E Lutz (1998) 'Integrating environmental and sustainability concerns into rural development policies' in E Lutz (ed) *Agriculture and the Environment: Perspectives on Sustainable Rural Development*, Washington, World Bank

Hazell, Peter and Shenggen Fan (2001) 'Balancing regional development priorities to achieve sustainable and equitable agricultural growth' in David R Lee and Christopher B Barrett (eds) *Critical Tradeoffs: Agricultural Intensification, Economic Development and the Environment in Developing Countries*, Wallingford, UK, CAB International

Heerink, N B M and R Ruben (1996) 'Economic approaches for the evaluation of low external input agriculture', *Tijdschrift voor Sociaalwetenschappelijk Onderzoek in de Landbouw*, vol 11, no 4, pp255–261

Hendrix, P H, D A Crossley Jr and D C Coleman (1990) 'Soil biota as components of sustainable agroecosystems' in C A Edwards, R Lal, P Madden, R Miller and G House (eds) *Sustainable Agricultural Systems*, Ankeny, IA, Soil and Water Conservation Society

Herter, L I T (1998) 'Avaliação da produtividade de um latossolo, sob plantio direto', *Revista Plantio Direto*, Passo Fundo, RS, Edição 44, pp20–25

Herzog, D C and J E Funderburk (1986) 'Ecological bases for habitat management and pest cultural control' in M Kogan (ed) *Ecological Theory and Integrated Pest Management Practice*, New York, Wiley, pp217–250

Hobbes, Peter and Michael Morris (1996) 'Meeting South Asia's future food requirements from rice-wheat cropping systems: priority issues facing researchers in the post-green revolution era', NRG Paper no 96-01, Mexico, DF, CIMMYT

Hodge, Angela et al (1999) 'Plant, soil fauna and microbial responses to N-rich organic patches of contrasting temporal availability', *Soil Biology and Biochemistry*, vol 31, p1517

Hooper, D U, D E Bignell, V K Brown, L Brussaard, J M Dangerfield, D H Wall, D A Wardle, D C Coleman, K E Giller, P Lavelle, W H Van der Putten, P C De Ruiter, J Rusek, W L Silver, J M Tiedje and V Wolters (2000) 'Interactions between aboveground and belowground biodiversity in terrestrial ecosystems: patterns, mechanisms and feedbacks', *Bioscience*, vol 50, pp1049–1061

Horn, D J (1988) *Ecological Approach to Pest Management*, London, Elsevier

Howard, Albert (1943) *An Agricultural Testament*, London, Oxford University Press

ICLARM and GTZ (1991) *The Context of Small-Scale Integrated Agriculture–Aquaculture Systems in Africa: A Case Study of Malawi*, Manila, International Center for Living Aquatic Resources Management, and Eschborn, Germany, Deutsche Gesellschaft für Technische Zusammenarbeit

ICRAF (1995) *Annual Report for 1994*, Nairobi, International Centre for Research in Agroforestry
ICRAF (1996) *Annual Report for 1995*, Nairobi, International Centre for Research in Agroforestry
ICRAF (1998) *Annual Report for 1997*, Nairobi, International Centre for Research in Agroforestry
IRRI (1986) 'Area and distribution of acid upland soils in Southeast Asia' in *IRRI Annual Report 1985*, Los Baños, Philippines, International Rice Research Institute, pp214–215,
Izac, A-M N and P A Sanchez (2001) 'Towards a natural resource management research paradigm: an example of agroforestry research', *Agricultural Systems*, vol 69, nos 1–2, pp5–25
Jama, B, C A Palm, R J Buresh, A I Niang, C Gachengo, G Nziguheba and B Amadalo (2000) '*Tithonia diversifolia* as a green manure for soil fertility improvement in western Kenya: a review', *Agroforestry Systems*, vol 49, pp201–221
Janvry, A de, M Fafchamps and E Sadoulet (1991) 'Peasant household behaviour with missing markets: some paradoxes explained', *The Economic Journal*, vol 101, pp1400–1417
Jiggins, Janice and H de Zeeuw (1991) 'Participatory technology development in practice: process and methods', in *Farming for the Future*, Wageningen, Netherlands, ILEIA
Jodha, N S (1992) *Common Property Resources: A Missing Dimension of Development Strategies*, Washington, World Bank
Joelibarison (1998) *Perspective de Développement de la Region de Ranomafana: Les mechanismes physiologiques du riz sur de bas-fonds – Case du SRI*, Memoire de fin d'etudes, Ecole Superieure des Sciences Agronomiques, Departement Agriculture, University of Antananarivo, Antananarivo
Johnson, Bruce K (1994) 'Soil survey' in *Final Report for the Agricultural Development Component of the Ranomafana National Park Project in Madagascar*, Raleigh, NC, Soil Science Department, North Carolina State University, pp5–12
Johnson, Nancy L and Vernon R Ruttan (1994) 'Why are farms so small?', *World Development*, vol 24, pp691–706
Jones, K A (1996) 'IPM in developing countries: the Sri Lankan experience', *Pesticide News*, vol 31, pp4–5
Jones, K A (1999a) 'Integrated pest management in rice, vegetables and other food crops in Sri Lanka' in *Integrated Pest Management in Developing Countries: The Experience of CARE International*, Atlanta, CARE International
Jones, K A (1999b) *Integrated Training and Safe Use of Pesticides in Sri Lanka (INTEGRATED): Final Report to the European Community and Department for International Development*, Colombo, CARE International
Kahn, L P and A Diaz-Hernandez (2000) 'Tannins with antihelmintic properties' in J D Brooker (ed) *Tannins in Livestock and Human Nutrition*, ACIAR Proceedings no 92, Canberra, Australian Centre for International Agricultural Research, pp130–139
Kamp, K and V Scarborough (1996) 'Teaching the teacher to fish: a case study', Network Paper no 59b, London, Overseas Development Institute, pp14–17
Kang, B T, G F Wilson and T L Lawson (1984) *Alley Cropping: A Stable Alternative to Shifting Agriculture*, Ibadan, International Institute for Tropical Agriculture
Kangmin, L and L Peizhen (1995) 'Integration of agriculture, livestock and fish farming in the Wuxi region of China' in J-J Symoens and J-C Micha (eds) *The*

Management of Integrated Freshwater Agro-Piscicultural Ecosystems in Tropical Areas, Wageningen, Netherlands, Technical Centre for Agricultural and Rural Cooperation, and Brussels, Royal Academy of Overseas Sciences

Kapetsky, J M (1994) 'A strategic assessment of warm-water fish farming potential in Africa', CIFA Technical Paper no 27, Rome, UN Food and Agriculture Organization

Kapetsky, J M (1995) 'A first look at the potential contribution of warm-water fish farming to food security in Africa' in J-J Symoens and J-C Micha (eds) *The Management of Integrated Freshwater Agro-Piscicultural Ecosystems in Tropical Areas,* Wageningen, Netherlands, Technical Centre for Agricultural and Rural Cooperation, and Brussels, Royal Academy of Overseas Sciences

Kar, S, S B Varade, T K Subramanyam and B P Ghildyal (1974) 'Nature and growth pattern of rice root system under submerged and unsaturated conditions', *Riso* (Italy), vol 23, no 2, pp173–179

Katayama, T (1951) *Ine mugi no bungetsu kenkyu* (Studies on Tillering in Rice, Wheat and Barley), Tokyo, Yokendo Publishing

Kenmore, Peter E (1986) 'Management of planthopper pests in tropical rice', proceedings of Second International Conference on Plant Protection in the Tropics, Los Baños, pp34–35

Kenmore, Peter E (1997) A perspective on IPM, *ILEIA Newsletter*, vol 3, no 4, pp8–9

Kenmore, Peter E, F O Cariño, C A Perez, V A Dyck and A P Gutierrez (1984) 'Population regulation of the rice brown planthopper (*Nilapavarta lugens* Stal) within rice fields in the Philippines', *Journal of Plant Protection in the Tropics*, vol 1, pp19–37

Kennedy, A C (1999) 'Microbial diversity in agroecosystem quality' in W W Collins and C O Qualset (eds) *Biodiversity in Agroecosystems,* Boca Raton, FL, CRC Press

Kerkhof, P (1990) *Agroforestry in Africa: A Survey of Project Experience,* London, Panos Institute

Khan, Z R, J A Pickett, J van den Berg, L J Wadhams and C M Woodcock (2000) 'Exploiting chemical ecology and species diversity: stem borer and striga control for maize and sorghum in Africa', *Pest Management Science*, vol 56, pp957–962

Khush, G S (1996) 'Prospects of and approaches to increasing the genetic yield potential of rice' in R E Evenson, R W Herdt and M Hossain (eds) *Rice Research in Asia: Progress and Priorities*, Wallingford, UK, CAB International, pp59–71

Khush, G S, S Peng and S S Virmani (1998) 'Improving yield potential by modifying plant type and exploiting heterosis' in J C Waterlow, D G Armstrong, L Fowden and R Riley (eds) *Feeding a World Population of More than Eight Billion People: A Challenge to Science*, Oxford, UK, Oxford University Press, pp150–170

Kiepe, P (1995) *No Runoff, No Soil Loss: Soil and Water Conservation in Hedgerow Barrier Systems,* Wageningen, Netherlands, Wageningen Agricultural University

King, F H (1911) *Farmers over Forty Centuries, or Permanent Agriculture in China, Korea and Japan,* Emmaus, PA, Rodale Press (republished 1973)

King, K F S (1968) 'Agri-silviculture', Bulletin no 1, Ibadan, Nigeria, Department of Forestry, University of Ibadan

Kiriwandeniya, P A (1997) 'SANASA: The Savings and Credit Cooperative Movement in Sri Lanka' in Krishna et al *Reasons for Hope: Instructive Experiences in Rural Development*, West Hartford, CT, Kumarian Press, pp57–74

Kirk, G J D and D R Bouldin (1991) 'Speculations on the operation of the rice root system in relation to nutrient uptake' in F W T Penning de Vries, H H van Laar and M J Kroff (eds) *Simulation and Systems Analysis for Rice Production*, Wageningen, Pudoc, 195–203

Kirk, G J D and J L Solivas (1997) 'On the extent to which root properties and transport through the soil limit nitrogen uptake by lowland rice', *European Journal of Soil Science*, vol 48, pp613–621

Kladivko, E J, D R Griffith and J V Mannering (1986) 'Conservation tillage effects on soil properties and yield of corn and soya beans in Indiana', *Soil and Tillage Research*, vol 8, pp277–287

Koenig, R (1990) 'Cowpea and millet inter-cropping in Senegal: 1989 on-farm research results', unpublished thesis report, Rodale Institute/USAID Contract #685-0294, Dakar

Korten, David C (1980) 'Community organization and rural development: a learning process approach', *Public Administration Review*, vol 40, no 5, pp480–511

Krishna, Anirudh (1997) 'Participatory watershed development and soil conservation in Rajasthan, India' in Krishna et al *Reasons for Hope: Instructive Experiences in Rural Development*, West Hartford, CT, Kumarian Press, pp255–272

Krishna, Anirudh and Norman Uphoff (2000) 'Mapping and measuring social capital through assessment of collective action to conserve and develop watersheds in Rajasthan, India', in C Grootaert and T van Bastelaar (eds) *The Role of Social Capital in Development: An Empirical Assessment*, Cambridge, Cambridge University Press

Krishna, Anirudh, Norman Uphoff and Milton J Esman (eds) (1997) *Reasons for Hope: Instructive Experiences in Rural Development*, West Hartford, CT, Kumarian Press

Kronzucker, H J, M Y Siddiqui, D M Glass and G J D Kirk (1999) 'Nitrate-ammonia synergism in rice', *Plant Physiology*, vol 119, pp1041–1045

Kruseman, G and J Bade (1998) 'Agrarian policies for sustainable land use: bioeconomic modelling to assess the effectiveness of policy instruments', *Agricultural Systems*, vol 58, no 3, pp465–481

Kruseman, G, H Hengsdijk and R Ruben (1993) 'Disentangling the concept of sustainability: conceptual definitions, analytical framework and operational techniques in sustainable land use', DLV Report no 2, Wageningen, Netherlands, Wageningen Agricultural University, AB-DLO

Kruseman, G, R Ruben, H Hengsdijk and M K van Ittersum (1996) 'Farm household modelling for estimating the effectiveness of price instruments in land use policy', *Netherlands Journal of Agricultural Sciences*, vol 43, pp111–123

Kudagamage, C, H B Senarath and H J P Fernando (1992) 'National integrated pest management program in Sri Lanka' in PAC Ooi, G S Lim, T H Ho, P L Manalo and J Waage (eds) *Integrated Pest Management in the Asia-Pacific Region*, Wallingford, UK, CAB International, pp163–180

Kuyvenhoven, A, J A Becht and R Ruben (1998a) 'Financial and economic evaluation of phosphate rock use to enhance soil fertility in West Africa: is there a role for government?' in G A A Wossink, G C van Kooten and G A Peters (eds) *Economics of Agrochemicals*, Aldershot, Ashgate, pp249–261

Kuyvenhoven, A, N B M Heerink and R Ruben (1999) 'Economic policies in support of soil fertility: which interventions after structural adjustment?' in E M A Smaling, O Oenema and L O Fresco (eds) *Nutrient Disequilibria in Agroecosystems: Concepts and Case Studies*, Wallingford, UK, CAB International, pp119–140

Kuyvenhoven, A, R Ruben and G Kruseman (1998b) 'Technology, market policies and institutional reform for sustainable land use in Southern Mali', *Agricultural Economics*, vol 19, pp53–62

Kwesiga, F R and R Coe (1994) 'The effect of short rotation *Sesbania sesban* planted fallows on maize yields', *Forest Ecology and Management*, vol 64, pp199–208

Kwesiga, F R, D Phiri and A-L Raunio (1997) 'Improved fallows with sesbania in eastern Zambia' in *Summary Proceedings of a Consultative Workshop*, Chipata, Zambia/Nairobi, ICRAF

Kwesiga, F R, S Franzel, F Place, D Phiri and C P Simwanza (1999) '*Sesbania sesban* improved fallows in Eastern Zambia: their inception, development and farmer enthusiasm', *Agroforestry Systems*, vol 47, pp49–66

Ladha, J K, G J D Kirk, J Bennett, S Peng, C K Reddy, P M Reddy and U Singh (1998) 'Opportunities for increased nitrogen-use efficiency from improved lowland rice germplasm', *Field Crops Research*, vol 56, pp41–71

Lal, Rattan (1989) 'Agroforestry systems and soil surface management of a tropical altisol I: soil moisture and crop yields', *Agroforestry Systems*, vol 8, pp7–29

Lal, Rattan (1990) *Soil Erosion in the Tropics: Principles and Management,* New York, McGraw-Hill

Larson, K J, K G Cassman and D A Phillips (1989) 'Yield, nitrogen balance in irrigated white lupine in a Mediterranean climate', *Agronomy Journal*, vol 81, pp538–543

Laulanié, Henri de (1993) 'Le systèm de riziculture intensíve malgache', *Tropicultura*, Brussels, vol 1, no 3, pp110–114

Laulanié, Henri de (1993a) 'Technical presentation on the system of rice intensifica-tion, based on Katayama's tillering model', unpublished paper translated from French, available from Cornell International Institute for Food, Agriculture and Development, Ithaca, NY

Leakey, R R B and P A Sanchez (1997) 'How many people use agroforestry products?' *Agroforestry Today*, vol 9, no 3, pp4–5

Leakey, R R B, A B Temu, M Melnyk and P Vantomme (eds) (1996) *Domestication and Commercialization of Non-timber Forest Products in Agroforestry Systems: Non-Wood Forest Products,* Rome, Food and Agriculture Organization

Leakey, R R B and T P Tomich (1999) 'Domestication of tropical trees: from biology to economics and policy' in L E Buck, J P Lassoie and E C M Fernandes (eds) *Agroforestry in Sustainable Agricultural Systems*, Boca Raton, FL, CRC Press, 319–335

Leeuwis, Cees (ed) (1999) 'Integral design: innovation in agriculture and resource management', Mansholt Studies no 15, Wageningen, Netherlands, Wageningen Agricultural University

Lestari, Alifah (1993) 'Women and IPM field schools' in *IPM Farmer Training: The Indonesia Case*, Jakarta, National IPM Program, BAPPENAS

Lewontin, Richard (2000) *The Triple Helix: Gene, Organism and Environment,* Cambridge, MA, Harvard University Press

Liebman, M and A S Davis (2000) 'Integration of soil, crop and weed management in low-external-input farming systems', *Weed Research*, vol 40, pp27–47

Liebman, M and T Ohno (1998) 'Crop rotation and legume residue effects on weed emergence and growth: implications for weed management' in J L Hotfield and B A Stewert (eds) *Integrated Weed and Soil Management*, Ann Arbor, MI, University of Michigan Press

Lightfoot, C, M A P Bimbao, J P T Dalsgaard and R S V Pullin (1993) 'Aquaculture and sustainability through integrated resources management', *Outlook on Agriculture*, vol 22, no 3, pp143–150

Lightfoot, C and D R Minnick (1991) 'Farmer-first qualitative methods: farmer diagrams for improving methods of experimental design in integrated farming systems', *Journal for Farming Systems Research-Extension*, vol 2, no 1, pp57–69

Lightfoot, C and R Noble (1993) 'A participatory experiment in sustainable agricul-ture' *Journal for Farming Systems Research-Extension*, vol 4, no 1, pp11–34

Lightfoot, C and R S V Pullin (1995) 'Why an integrated resource management approach?' in R E Brummett (ed) *Aquaculture Policy Options for Integrated Resource Management in Sub-Saharan Africa,* ICLARM Conference Proceedings 46, Manila, International Center for Living Aquatic Resources Management

Lloyd, R W and D R Krieg (1987) 'Cotton development and yield as affected by insecticides', *Journal of Economic Entomology,* vol 80, pp854–858

Lockeretz, W (1989) 'Problems in evaluating the economics of ecological agriculture', *Agriculture, Ecosystems and Environment,* vol 27, pp67–75

Lòpez, Gabiño et al (nd) 'Adopción de Tecnologías de Conservación de Suelos y Agua en el Distrito de Guinope, El Paraíso, Honduras', study for the Silsoe Research Institute, UK

Low, A R C (1993) 'The low-input, sustainable agriculture (LISA) prescription: a bitter pill for farm households in southern Africa', *Project Appraisal,* vol 8, no 2, pp97–101

Lund, Susan and Marcel Fafchamps (1997) 'Risk-sharing networks in rural Philippines', unpublished paper, Department of Economics, Stanford, University

Lundgren, Bjorn O and J B Raintree (1982) 'Sustained agroforestry' in B Nestel (ed) *Agricultural Research for Development: Potentials and Challenges in Asia,* The Hague, International Service for National Agricultural Research, pp37–49

Lutz, E, S Pagiola and C Reiche (eds) (1994) 'Economic and institutional analysis of soil conservation projects in Central America and the Caribbean', Environment Paper no 8, Washington, World Bank

Magbanua, R D and D P Garrity (1990) 'Agroecosystems analysis of a key upland farming systems research site' in *Proceedings of the 1988 Acid Upland Design Workshop,* Los Baños, Philippines, International Rice Research Institute

Magdoff, Fred R (1992) *Building Soils for Better Crops: Organic Matter Management,* Lincoln, NE, University of Nebraska Press

Magdoff, F R, and D R Bouldin (1970) 'Nitrogen fixation in submerged soil-sand-energy material media and the aerobic-anaerobic interface', *Plant and Soil,* vol 33, no 1, pp49–61

Maher, C (1937) 'Soil erosion and land utilisation in the Ukamba Reserve (Machakos)', report to the Department of Agriculture, Rhodes House, Oxford, UK, Mss Afr S755

Malla, Y B (1997) 'Sustainable use of communal forests in Nepal', *Journal of World Forest Resource Management,* vol 8, pp51–74

Mandac, A M, J C Flinn and M P Genesilda (1986) 'Developing technology for upland farms in northern Mindanao, Philippines', *Philippine Journal of Crop Science,* vol 11, pp69–79

Mannering, J V and LD Meyer (1963) 'The effects of various rates of surface mulch on infiltration and erosion', *Soil Science Society of America Proceedings 1963,* vol 27, pp84–86

Matson, P A, R Naylor and I Ortiz-Monasterio (1998) 'Integration of environmental, agronomic, and economic aspects of fertilizer management', *Science,* vol 280, pp112–115

Matson, P A, W J Parton, A G Power and M J Swift (1997) 'Agricultural intensification and ecosystem properties', *Science,* vol 277, pp504–509

Mausolff, C and S Farber (1995) 'An economic analysis of ecological agricultural technologies among peasant farmers in Honduras', *Ecological Economics,* vol 12, pp237–248

Mbogoh, Stephen (2000) 'Makueni district profile: crop production 1989–1998', Working Paper no 7, Crewkerne, UK, Drylands Research

Medeiros, G B, A Calegari and C Gaudêncio (1989) 'Rotacao de culturas' in *Manual Tecnico do Sub-programa de Manejo e Conservacao do Solo*, Curitiba, Brazil, Secretaria da Agricultura e do Abastecimento, pp189–195

Mercado, A R, M Patindol and D P Garrity (2000) 'The landcare experience in the Philippines: technical and institutional innovations for conservation farming', paper presented at the International Landcare Conference, Melbourne, Australia, March; Bogor, International Centre for Research in Agroforestry, Southeast Asia Regional Programme

Merrill-Sands, Deborah and David Kaimowitz, with others (1990) *The Technology Triangle: Linking Farmers, Technology Transfer Agents, and Agricultural Researchers*, The Hague, International Service for National Agricultural Research

Merrill-Sands, Deborah, and Marie Helene Collion (1995) 'Farmers and researchers: the road to partnership', *Agriculture and Human Values*, vol 11, nos 2–3, pp26–37

Meyer, L D, W H Wischmeyer, and G R Foster (1970 Mulch rates required for erosion control on steep slopes *Soil Science Society of America Proceedings, 1970*, vol 34, pp928–931

Michon, G (1997) 'Indigenous gardens: re-inventing the forest' in T Whitten and J Whitten (eds) *The Indonesian Heritage, Vol I: Plants*, Singapore, Grolier, pp88–89

Michon, Genevieve and Hubert de Foresta (1996) 'Agroforests as an alternative to pure plantations for the domestication and commercialization of NTFPs' in R R B Leaky et al (eds) *Domestication and Commercialization of Non-Timber Forest Products for Agroforestry*, Rome, Food and Agriculture Organization, pp160–175

Milliman, J D and J Meade (1983) 'World-wide delivery of river sediment to the oceans', *Journal of Geology*, vol 91, pp1–21

Miyazawa, M, M A Pavan and A Calegari (1994) 'Efeitos de materiais vegetais na acidez do solo', *R Bras Ci Solo Campinas*, vol 17, pp411–416

Mollison, Bill (1990) *Permaculture: A Practical Guide for a Sustainable Future*, Washington, Island Press

Morgan, R P C (1992) 'Soil conservation options in the UK', *Soil Use and Management*, vol 8, no 4, pp176–180

Muller-Samann, K M and J Kotschi (1994) *Sustaining Growth: Soil Fertility Management in Tropical Smallholdings*, Eschborn, GTZ/Margraf Verlag

Murray, Gerald (1997) 'A Haitian peasant tree chronicle: adaptive evolution and institutional intrusion' in A Krishna et al (eds) *Reasons for Hope: Instructive Experiences in Rural Development*, West Hartford, CT, Kumarian Press, pp241–254

Murton, J (1997) 'Coping with more people: population growth, non-farm income and economic differentiation in Machakos District, Kenya', unpublished PhD thesis, Cambridge, UK, University of Cambridge

Muzilli, O, M J Vieira and M S Parra (1980) 'Adubacao verde' in *Manual Agropecuario para o Paraná*, Londrina, IAPAR, pp77–93

Nair, P K R (1982) *Soil Productivity Aspects of Agroforestry*, Nairobi, International Centre for Research in Agroforestry

Nair, P K R (ed) (1989) *Agroforestry Systems in the Tropics*, Dordrecht, Netherlands, Kluwer Academic Publishers

Narayan, Deepa (1999) 'Bonds and bridges: social capital and poverty', Policy Research Working Paper no 2167, Washington, World Bank

Natarajan, M and R W Willey (1986) 'The effects of water stress on yield advantages of intercropping systems', *Field Crops Research* no 13, pp117–131

Neher, D and M E Barbercheck (1999) 'Diversity and function in soil mesofauna' in W W Collins and C O Qualset (eds) *Biodiversity in Agroecosystems*, Boca Ratan, FL, CRC Press

Neill, S and D R Lee (2000) 'Explaining the adoption and disadoption of sustainable agriculture: the case of cover crops in Northern Honduras', *Economic Development and Cultural Change*, vol 49, no 4, pp793–820

Nelson, R, R A Cramb, K M Manz and M A Mamicpic (1998) 'Bioeconomic modeling of alternative forms of hedgerow intercropping in the Philippine uplands using SCUAF', *Agroforestry Systems*, vo 39, no 3, pp241–262

Nemoto, K, S Morita and T Baba (1995) 'Shoot and root development in rice related to the phyllochron', *Crop Science*, vol 35, no 1, pp24–29

Netting, R M (1965) 'Household organisation and intensive agriculture: the Kofyar case', *Africa*, vol 35, pp422–429

Netting, R M and M P Stone (1996) 'Agro-diversity on a farming frontier: Kofyar smallholders on the Benue Plains of Central Nigeria', *Africa,* vol 66, no 1, pp52–70

Netting, R M, M P Stone and G D Stone (1989) 'Kofyar cash-cropping: choice and change in indigenous agricultural development', *Human Ecology*, vol 17, no 3

Niang, A, J de Wolf, M Nyasimi, T Hansen, R Romelsee and K Mdewa (1998) 'Soil fertility replenishment and recapitalization project in western Kenya', *Progress Report February 1997– July 1998*, Pilot Project Report no 9, Maseno, Kenya, Regional Agroforestry Research Centre

Nijland, G O and J Schouls (1997) 'The relation between crop yield, nutrient uptake, nutrient surplus, and nutrient application', *Wageningen Agricultural University Papers*, vol 97, pp1–151

Nziguheba, G, C A Palm, R J Buresh and P J Smithson (1998) 'Soil phosphorous fractions and adsorption as affected by organic and inorganic sources', *Plant and Soil*, vol 198, pp59 –168

Nzioka, Charles (2000) 'Makueni district profile: human resource management 1989–1998', Working Paper no 9, Crewkerne, UK, Drylands Research

Oka, Ido Nyoman (1997) 'Integrated crop pest management with farmer participation in Indonesia' in Krishna et al *Reasons for Hope: Instructive Experiences in Rural Development*, West Hartford, CT, Kumarian Press, 184–199

Ong, C and P A Huxley (eds) (1996) *Tree–Crop Interactions: A Physiological Approach,* Wallingford, UK, CAB International

Ooi, P A C (1996) 'Experiences in educating farmers to understand biological control' in *Technology Transfer in Biological Control,* proceedings of conference at Montpellier, France, 9–11 September, *1996 IOBC Bulletin*, vol 19, no 8, pp5–6

Ostrom, Elinor (1990) *Governing the Commons: The Evolution of Institutions for Collective Action,* Cambridge, UK, Cambridge University Press

Ostrom, Elinor (1998) 'Social capital: a fad or fundamental concept?', unpublished paper, Bloomington, IN, Center for the Study of Institutions, Population and Environmental Change, Indiana University

Osunade, M and C Reij (1996) 'Back to the grass strips: a history of soil conservation policies in Swaziland' in C Reij et al *Sustaining the Soil: Indigenous Soil and Water Conservation in Africa*, London, Earthscan Publications, pp151–155

Ouedraogo, Mattieu and Vincent Kabore (1996) 'The *zai*: a traditional technique for the rehabilitation of degraded land in the Yatenga, Burkina Faso' in C Reij et al *Sustaining the Soil: Indigenous Soil and Water Conservation in Africa*, London, Earthscan Publications, pp80–84

Palm, Cheryl A, R J K Myers and S M Nandwa (1997) 'Combined use of organic and inorganic nutrient sources for soil fertility maintenance and replenishment' in R J Buresh, P A Sanchez and F Calhoun (eds) *Replenishing Soil Fertility in Africa*, Soil

Science Society of America Special Publication 51 Madison, WI, Soil Science Society of America, pp193–217

Palm, Cheryl A, C N Gachengo, R J Delve, G Cadisch and K E Giller (2001) 'Organic inputs for soil fertility management in tropical agroecosystem: application of an organic resource database', *Agricultural Ecosystems and Environment*, vol 83, pp27–42

Pearce, D and R K Turner (1990) *Economics of Natural Resources and the Environment*, New York, Harvester, Wheatsheaf

Pearson, C J and R L Ison (1987) *Agronomy of Grassland Systems,* Cambridge, UK, Cambridge University Press

Penning de Vries, F W T and H H van Laar (eds) (1982) *Simulation of Plant Growth and Crop Production,* Wageningen, Pudoc

Perfecto, I, R A Rice, R Greenberg and M E van der Voort (1996) 'Shade coffee: a disappearing refuge for biodiversity', *BioScience*, vol 46, no 8, pp598–608

Petersen, P, J M Tardin and F Marochi (2000) 'Participatory development of no-tillage systems without herbicides for family farming: the experience of the center-south region of Paraná', *Environment, Development and Sustainability*, vol 1, nos 3–4, pp235–252

Phillips, S II and H M Young Jr (1973) *No-Tillage Farming,* Milwaukee, WI, Reiman Associates

Picciotto, Robert (1995) 'Putting institutional economics to work: from participation to governance', Discussion Paper no 304, Washington, World Bank

Pieri, C M G (1998) 'Soil fertility improvement: key connection between sustainable land management and rural well being', *16th World Congress of Soil Science* (CD-ROM), Montpellier, France

Piggin, Colin (2000) 'The role of *Leucaena* in swidden cropping and livestock production in Nusa Tenggara Timur, Indonesia' in M F Cairns (ed) *Voices from the Forest: Farmer Solutions Towards Improved Fallow Husbandry in Southeast Asia,* proceedings of a regional conference on Indigenous Strategies for Intensification of Shifting Cultivation in Southeast Asia, held in Bogor, Indonesia, 23–27 June 1997; Bogor, International Centre for Research in Agroforestry, Southeast Asian Regional Research Programme, pp278–290

Pingali, Prabhu, M Hossein and R V Gerpacio (1995) *Asian Rice Bowls: The Returning Crisis,* Wallingford, UK, CAB International with IRRI

Pingali, Prabhu and R V Gerpacio (1997) 'Living with reduced insecticide use for tropical rice in Asia', *Food Policy*, vol 22, no 2, pp107–118

Pinstrup-Anderson and Marc Cohen (1999) 'World food needs and the challenge to sustainable agriculture', paper for Conference on Sustainable Agriculture, Bellagio, Italy, 26–30 April

Piper, J K (1994) 'Neighborhood effects on growth, seed yield, and weed biomass for three perennial grains in polyculture ', *Journal of Sustainable Agriculture,* vol 4, no 2, pp11–31

Piper, J K and P A Kulakow (1994) 'Seed yield and biomass allocation in sorghum bicolor and F1 and backcross generations of S bicolor x S halepense hybrids', *Canadian Journal of Botany,* vol 72, no 4, pp468–474

Pohl, G and D Mihaljek (1992) 'Project evaluation and uncertainty in practice: a statistical analysis of rate-of-return divergences of 1015 World Bank projects', *World Bank Economic Review*, vol 6, no 2, pp255–277

Postel, Sandra (1996) 'Dividing the water: food security, ecosystem health, and the new politics of scarcity', Worldwatch Paper no 132, Washington, DC, Worldwatch Institute

Power, A G (1999) 'Linking ecological sustainability and world food needs', *Environment, Development and Sustainability*, vol 1, nos 3–4, pp598–608

Power, A G (2001) 'The ecology of agriculture' in S A Levin (ed) *Encyclopedia of Biodiversity*, New York, Academic Press

Power, A G and A S Flecker (1996) 'The role of biodiversity in tropical managed ecosystems' in G H Orians, R Dirzo and J H Cushman (eds) *Biodiversity and Ecosystem Processes in Tropical Forests*, New York, Springer-Verlag, pp173–194

Pretty, Jules N (1995) *Regenerating Agriculture: Policies and Practice for Sustainability and Self-Reliance,* London, Earthscan Publications, and Washington, DC, National Academy Press

Pretty, Jules N (1995a) 'The ecological basis of alternative agriculture', *Annual Review of Ecological Systems*, vol 26, pp201–224

Pretty, Jules N (1995b) 'Participatory learning for sustainable agriculture', *World Development*, vol 23, no 8, pp1247–1263

Pretty, Jules N (1997) 'The sustainable intensification of agriculture', *Natural Resources Forum*, vol 21, no 4, pp247–256

Pretty, Jules N (1998) *The Living Land: Agriculture, Food and Community Regeneration in Rural Europe,* London, Earthscan Publications

Pretty, Jules N (1999) 'Soil and water conservation: a brief history of coercion and control' in F Hinchcliffe et al *Fertile Ground: The Impacts of Participatory Watershed Management*, London, Intermediate Technology Publications, pp1–12

Pretty, Jules and Robert Chambers (1993) 'Towards a learning paradigm: new professionalism and institutions for sustainable agriculture', Discussion Paper no 334, Brighton, UK, Institute of Development Studies

Pretty, Jules N and R Hine (2000) 'The promising spread of sustainable agriculture in Asia', *Natural Resources Forum*, vol 24, pp107–121

Pretty, Jules N and R Hine (2001) *Reducing Food Poverty with Sustainable Agriculture: A Summary of New Evidence*, final report from SAFE-World Research Project, University of Essex, Colchester, UK

Pretty, Jules N and Michel Pimbert (1995) 'Beyond conservation ideology and the wilderness myth', *Natural Resources Forum*, vol 19, no 1, pp5–14

Pretty, Jules N and Parmesh Shah (1997) 'Making soil and water conservation sustainable: from coercion and control to partnerships and participation', *Land Degradation and Development*, vol 8, pp39–58

Pretty, Jules N and John Thompson (1996) 'Sustainable agriculture and the Overseas Development Administration', report for Natural Resources Policy Advisory Department, DFID London, Department for International Development

Pretty Jules N, John Thompson and J K Kiara (1995) 'Agricultural regeneration in Kenya: the catchment approach to soil and water conservation', *Ambio*, vol 24, no 1, pp7–15

Pretty, Jules N and Hugh Ward (2001) 'Social capital and the environment', *World Development*, vol 29, no 2, pp209–227

Primavesi, Ana (1984) *Manejo Ecologico del Suelo: Le Agriculture en Regiones Tropicales,* Buenos Aires: El Ateneo, Pedro Garcia SA Translation of O *Manejo Ecologico do Solo*, published by Livraria Nobel SA, Sao Paulo, Brazil, (1980

Puard, M, P Couchat and G Lasceve (1986) 'Importance de l'oxygenation des racines du riz (*Oryza sativa*) en culture inondée', *L'Agronomie Tropicale*, vol 41, no 2, pp119–123

Puard, M, P Couchat and G Lasceve (1989) Etude des mecanismes d'adaptation du riz aux contraintes du milieu I: Modification de l'anatomie cellulaire', *L'Agronomie Tropicale*, vol 44, no 2, pp156–173

Putnam, Robert D, with R Leonardi and R Y Nanetti (1993) *Making Democracy Work: Civic Traditions in Modern Italy,* Princeton, NJ, Princeton University Press

Putnam, Robert D (1995) 'Bowling alone: America's declining social capital', *Journal of Democracy,* vol 6, no 1, pp65–78

Putterman, L (1995) 'Economic reform and smallholder agriculture in Tanzania: a discussion of recent market liberalization, road rehabilitation and technology dissemination efforts', *World Development,* vol 23, no 2, pp311–326

Raju, G (1998) 'Institutional structures for community-based conservation' in A Kothari, N Pathak, R V Anuradha and B Taneja (eds) *Communities and Conservation: Natural Resource Management in |South and Central Asia,* New Delhi, Sage Publications

Rakotomalala, Holiarison William (1998) *Comparaison entre la Riziculture Traditionelle et le Systeme de Riziculture Intensive dan la Region de Ranomafana,* Memoire de fin d'etudes, Ecole Supérieure des Sciences Agronomiques, Departement Agriculture, University of Antananarivo, Antananarivo

Rajaonarison, Jean de Dieu (2000) *Contributions a l'amelioration des rendements de 2eme saison de la double riziculture par SRI sous experimentations multifacto-rielles – Cas des sols sableaux de Morondara,* Memoire de fin d'etudes, Ecole Supérieure des Sciences Agronomiques, University of Antananarivo, Antananarivo

Ramakrishnan, P S (1994) 'The *jhum* agroecosystem in northeastern India: a case study of the biological management of soils in a shifting agricultural system' in P L Woomer and M J Swift (eds) *The Biological Management of Soil Fertility,* New York, John Wiley, 189–207

Ramasamy, S, H F M ten Berge and S Purushothaman (1997) 'Yield formation in rice in response to drainage and nitrogen application', *Field Crops Research,* vol 51, pp65–82

Rao, M R, A Niang, F Kwesiga, B Duguma, S Franzel, B Jama and R J Buresh (1998) 'Soil fertility replenishment in sub-Saharan Africa: new techniques and the spread of their use on farms', *Agroforestry Today,* vol 10, no 2, pp3–8

Rao, M R and R W Willey (1980) 'Evaluation of yield stability in intercropping: studies on sorghum/pigeonpea', *Experimental Agriculture,* vol 16, pp105–116

Reardon, T (1995) 'Sustainability issues for agricultural research strategies in the semi-arid tropics: focus on the Sahel', *Agricultural Systems,* vol 48, no 3, pp345–360

Reardon, T (1997) 'Using evidence of household income diversification to inform study of the rural nonfarm labor market in Africa', *World Development,* vol 25, no 5, pp735–748

Reardon, T, C Barrett, V Kelly and K Savadogo (1999) 'Policy reforms and sustainable agricultural intensification in Africa', *Development Policy Review,* vol 17, no 4, pp293–313

Reardon, T, J Berdegué and G Escobar (2001) 'Rural nonfarm employment and incomes in Latin America: overview and policy implications', *World Development,* vol 29, no 3, pp395–410

Reardon, T, E Crawford and V Kelly (1994) 'Links between nonfarm income and farm investments in African households: adding the capital market perspective', *Agricultural Economics,* vol 76, no 5, pp1172–1176

Reardon, T, P Matlon and C Delgado (1988) 'Coping with household-level food insecurity in drought-affected areas of Burkina Faso', *World Development,* vol 16, no 10, pp1065–74

Reganold, J P, J D Glover, P K Andrews and H R Hinman (2001) 'Sustainability of three apple production systems', *Nature,* vol 410, pp926–930

Reij, Chris, Ian Scoones and Camilla Toulmin et al (eds) (1996) *Sustaining the Soil: Indigenous Soil and Water Conservation in Africa*, London, Earthscan Publications

Reijntjes, C B, B Haverkort and A Waters-Bayer (1992) *Farming for the Future: An Introduction to Low-External-Input and Sustainable Agriculture*, London, ILEIA/Macmillan

Rodale Institute (1989) 'Soil degradation and prospects for sustainable agriculture in the Peanut Basin of Senegal', report submitted to USAID/Dakar, 15 August, Emmaus, PA, Rodale Institute

Roger, P A, K L Heong and P S Teng (1991) 'Biodiversity and sustainability of wetland rice production: role and potential of microorganisms and invertebrates' in D L Hawksworth (ed) *The Biodiversity of Microorganisms and Invertebrates: Its Role in Sustainable Agriculture*, Wallingford, UK, CAB International, pp117–136

Röling, Niels (1995) 'Towards interactive agricultural science', inaugural address at occasion of appointment to Extra-Ordinary Chair in Agricultural Knowledge Systems, 21 September, Wageningen Agricultural University, Netherlands

Röling, N and F de Jong (1998) 'Learning: shifting paradigms in education and extension studies', *Journal of Agricultural Extension and Education*, vol 5, no 3, pp143–161

Röling, N and E Van de Fliert (1994) 'Transforming extension for sustainable agriculture: the case of integrated pest management in rice in Indonesia', *Agriculture and Human Values*, vol 11, nos 2–3, pp96–108

Roose, E (1977) 'Érosion et ruissellement en Afrique de L'Oest: Vingt années de mesures en petit parcelles expérimentales', work and documents from ORSTOM, no 78, Paris

Roth, C H (1985) 'Infiltrability of latosol-roxo soils in North Paraná, Brazil, in field trials for erosion control under different tillage systems', unpublished dissertation, Institut für Bodenwissenschaften, University of Göttingen

Rowley, J (1999) *Working with Social Capital*, London, Department for International Development

Ruben, R, and M v d Berg (1999) 'Farmers' selective participation in rural markets: off-farm employment in Honduras' in R Ruben and J Bastiaensen (eds) *Rural Development in Central America: Markets, Livelihoods and Local Governance*, Houndsmills, Macmillan, pp189–209

Ruben, R, A Kuyvenhoven and G Kruseman (2001) 'Bio-economic models for eco-regional development: policy instruments for sustainable intensification' in D R Lee and C B Barrett (eds) *Critical Tradeoffs: Agricultural Intensification, Economic Development and the Environment in Developing Countries*, Wallingford, UK, CAB International

Ruben, R and N Heerink (1995) 'Economic evaluation of low external input farming', *ILEIA Newsletter*, vol 11, p2

Ruben, R, G Kruseman and H Hengsdijk (1994) 'Farm household modelling for estimating the effectiveness of price instruments on sustainable land use in the Atlantic Zone of Costa Rica', DLV Report no 4, Wageningen, Netherlands, Wageningen Agricultural University, AB-DLO

Ruben, R, P v d Berg, M S van Wijk and N Heerink (1997) 'Evaluacion economica de sistemas de produccion con alto y bajo uso de recursos externos: el caso del frijol abono en la agricultura de ladera', Seminar IFPRI/CIAT/CIMMYT Hillside Programme, EAP Zamorano, Honduras, February

Ruben, Ruerd and David R Lee (2000) 'Combining internal and external inputs for sustainable agricultural intensification', IFPRI 2020 Brief no 65, Washington, International Food Policy Research Institute

Ruchijat, E and T Sukmaraganda (1992) 'National integrated pest management in Indonesia: its successes and challenges' in P A C Ooi, G S Lim, T H Ho, P L Manalo and J Waage (eds) *Integrated Pest Management in the Asia-Pacific Region*, Wallingford, UK, CAB International, pp329–347

Ruddell, E (1995) 'Growing food for thought: a new model of site-specific research for Bolivia', *Grassroots Development*, vol 19, no 2, pp18–26

Ruddell, E, J Beingolea and H Beingolea (1996) 'People-centered development empowers small indigenous farmers to double food production', *Journal of International Agricultural and Extension Education*, vol 3, no 2, pp33–41

Ruddell, E, J Beingolea and H Beingolea (1997) 'Empowering farmers to conduct experiments' in L Veldhuizen, A Waters-Bayer, R Ramirez, D Johnson and J Thompson (eds) *Farmers' Research in Practice*, London, Intermediate Technology Publications, pp199–208

Russell, E J (1961) *Soil Conditions and Plant Growth*, ninth edition, London, Longmans

Ruthenberg, Hans (1980) *Farming Systems in the Tropics*, Oxford, Clarendon Press

Ruttan, Vernon R (1999) 'Biotechnology and agriculture: a skeptical perspective', World Wide Web AgBioForum (http:wwwagbioforummissouriedu/BioForum/General/archiveshtml)

Sá, J C M (1993) *Manejo da Fertilidade do Solo no Plantio Direto*, Castro, PR, Fundacao ABC, Aldeia Norte Editora

Sabio, Eduardo A, Dennis P Garrity and Agustin R Mercado Jr (2001) 'The landcare approach to conservation farming in Mindanao', *CIIFAD Annual Report, 1999–2000*, Ithaca, NY, Cornell International Institute for Food, Agriculture and Development, pp106–107

Sajjapongse, A and K Syers (1995) 'Tangible outcomes and impacts from the ASIALAND management of sloping lands network' in *Proceedings of International Workshop on Conservation Farming for Sloping Uplands in Southeast Asia: Challenges, Opportunities and Prospects*, Proceedings no 14, Bangkok, International Board for Soil Research and Management, pp3–14

Salmen, Lawrence F (2000) 'The voice of the farmer in agricultural extension', AKIS Discussion Paper, Washington, World Bank

Sanchez, P A (1995) 'Science in agroforestry' *Agroforestry Systems*, vol 30, pp5–55

Sanchez, P A and R R B Leakey (1997) 'Land-use transformation in Africa: three determinants for balancing food security with natural resource utilization', *European Journal of Agronomy*, vol 7, pp15–23

Sanchez, P A, B Jama and A I Niang (2001) 'Soil fertility, small-farm intensification and the environment in Africa' in C Barrett and D Lee (eds) *Tradeoffs or Synergies? Agricultural Intensification, Economic Development and the Environment in Developing Countries* Wallingford, UK, CAB International

Sanchez, P A, R J Buresh and R R B Leakey (1997b) 'Trees, soils and food security', *Philosophical Transactions of the Royal Society of London*, Series B, no 353, pp949–961

Sanchez, P A, K D Shepherd, M I Soule, F M Place, R J Buresh, A-M N Izac, A U Mokwunye, F R Kwesiga, C G Ndiritu and P L Woomer (1997a) 'Soil fertility replenishment in Africa: an investment in natural resource capital' in R J Buresh, P A Sanchez and F Calhoun (eds) *Replenishing Soil Fertility in Africa*, SSSA Special Publication no 51, Madison, WI, Soil Science Society of America, p46

Sanchez, Pedro A and J R Benites (1987) 'Low-input cropping for acid soils in the humid tropics', *Science*, vol 238, pp1521–1527

Sanders, John H (1997) 'Developing technology for agriculture in Sub-Saharan Africa: evolution of ideas, some critical questions, and future research', Discussion Paper, Washington, International Food Policy Research Institute

Savadogo, K, T Reardon and K Pietola (1998) 'Adoption of improved land-use technologies to increase food security in Burkina Faso: relating animal traction, productivity and non-farm income', *Agricultural Systems*, vol 58, no 3, pp441–464

Sayre, K D and O H Moreño Ramos (1997) 'Applications of raised-bed planting systems to wheat', Wheat Program Special Report no 31, Mexico, DF, CIMMYT

Schlather, Ken (1998) 'The dynamics and cycling of phosphorous in mulched and unmulched bean production systems indigenous to the humid tropics of Central America', unpublished PhD thesis, Cornell University, Ithaca, NY

Scholz, U F, S K Chimatiro and M Hummel (1997) 'Status and prospects of aquacul ture development in Malawi: a case study of MAGFAD – is there sustainability and a future?', paper presented to the First SADC Regional Conference on Aquaculture, Bunda College of Agriculture, Lilongwe, Malawi, 17–19 November

Schorlemer, Dietmar (1999) 'Continuous adaptation for soil and water conservation: the case of PATECOR in Burkina Faso' in F Hinchcliffe et al (eds) *Fertile Ground: The Impacts of Participatory Watershed Management*, London, Intermediate Technology Publications, pp42–47

Scoones, I (1996) *Hazards and Opportunities: Farming Livelihoods in Dryland Africa*, London, Zed Books

Settle, W H, H Ariawan, E T Astuti, W Cahyana, A L Hakim, D Hindayana, A S Lestari, Pajarningsih and Sartanto (1996) 'Managing tropical rice pests through conservation of generalist natural enemies and alternative prey', *Ecology*, vol 77, pp1975–1988

Shrestha, B (1998) 'Involving local communities in conservation: the case of Nepal' in A Kothari, N Pathak, R V Anuradha and B Taneja (eds) *Communities and Conservation: Natural Resource Management in South and Central Asia*, New Delhi, Sage Publications

Sidiras, N and M A Pavan (1985) 'Influência do sistema de manejo de solo no seu nível de fertilidade', *Rev Bras Ci Solo Campinas*, vol 9, no 3, pp249–254

Sidiras, N and C H Roth (1984) 'Medicoes de infiltracao com infiltrometros e um simulador de chuvas em Latossolo roxo distrofico, Paraná, sobre varios tipos de cobertura de solo e sistema de preparo', *Congresso Brasileiro de Conservacao do Solo*, 5 Porto Alegre, RS, Anais

Simon, Herbert A (1957) *Administrative Behavior*, New York, Macmillan

Singh, I, L Squire and J Strauss (1986) *Agricultural Household Models: Extensions, Applications and Policy*, Baltimore, Johns Hopkins University Press

Sissoko, K (1998) 'El demain l'Agriculture? Options techniques et mesures politiques pour un developpement agricole durable en Afrique subsaharienne', Documents sur la Gestion des Resources Tropicales no 23, Wageningen, Wageningen Agricultural University

Smil, Vaclav (2000) *Feeding the World: A Challenge for the Twenty-First Century*, Cambridge, MA, MIT Press

Smillie, Joe and Grace Gershuny (1999) *The Soul of Soil: A Soil-Building Guide for Master Gardeners and Farmers*, fourth edition, White River Junction, VT, Chelsea Green Publishing

Smith, Adam (1991) *The Wealth of Nations*, London, Random House (first published 1775)

Snrech, S (1995) *West Africa Long-Term Perspective Study: Synthesis Report*, Paris, OECD, Club du Sahel

Somda, Z C, J M Powell, S Fernández-Rivera and J D Reed (1970) 'Feed factors affecting nutrient excretion by ruminants and the fate of nutrients when applied to soil' in J M Powell, S Fernández-Rivera, T O Williams and C Renard (eds) *Livestock and Sustainable Nutrient Cycling in Mixed Farming Systems of Sub-Saharan Africa*, Addis Ababa, International Livestock Centre for Africa, pp227–243

Sorrenson, W J and L J Montoya (1984) 'Economic implications of soil erosion and soil conservation practices in Paraná, Brazil', report on a consultancy Londrina, IAPAR, and Eschborn, GTZ

Steiner, K G (1982) 'Intercropping in tropical smallholder agriculture with special reference to West Africa', GTZ Publication no 137, Eschborn, Gesellschaft für Technische Zusammenarbeit

Steppler, H A and P K R Nair (eds) (1987) *Agroforestry: A Decade of Development*, Nairobi, ICRAF

Stocking, M and N Abel (1989) 'Labour costs: a critical element in soil conservation', paper presented to the VI International Soil Conservation Conference, Addis Ababa, Ethiopia, November

Stott, D E and J P Martin (1989) 'Organic matter decomposition and retention in arid soils', *Arid Soil Research and Rehabilitation*, vol 3, p115

Sturdy, D (1939) 'Leguminous crops in native agricultural practice', *East African Agricultural Journal*, vol 5, pp31–33

Subedi, K D (1998) 'El conocimiento local de los agricultores concuerda con los resultados de las experimentaciones formales', *Boletin de ILEIA*, March

Sumner, D R (1982) 'Crop rotation and plant productivity' in M Recheigl (ed) *CRC Handbook of Agricultural Productivity, Vol I*, Boca Raton, FL, CRC Press

TAC (1999) *Second External Review of ICRAF* Rome, FAO, Technical Advisory Committee of the Consultative Group on International Agricultural Research

Tanner, J, Holden, S J, Winugroho, M, Owen, E, and Gill M (1995) 'Feeding livestock for compost production: a strategy for sustainable upland agriculture on Java' in J M Powell, S Fernández-Rivera, T O Williams and C Renard (eds) *Livestock and Sustainable Nutrient Cycling in Mixed Farming Systems of Sub-Saharan Africa*, Addis Ababa, International Livestock Centre for Africa, pp115–128

Taylor, J E and I Adelman (1996) *Village Economies: The Design, Estimation and Use of Village-wide Economic Models*, Cambridge, UK, Cambridge University Press

Taylor, M (1982) *Community, Anarchy and Liberty*, Cambridge, UK, Cambridge University Press

TG-HDP (1995) *Thai–German Highland Development Project Annual Report*, Chiang Mai, Thailand

Thies, C and T Tscharntke (1999) 'Landscape structure and biological control in agroecosystems', *Science*, vol 285, pp893–895

Thrupp, Lori Ann (1996) *New Partnerships in Sustainable Agriculture*, Washington, World Resources Institute

Thrupp, Lori Ann (1998) *Cultivating Diversity: Agrobiodiversity and Food Security*, Washington, World Resources Institute

Thurston, H David et al (eds) (1994) *Slash/Mulch: How Farmers Use It and What Researchers Know About It*, Ithaca, NY, Cornell International Institute for Food, Agriculture and Development, and Turrialba, Costa Rica, CATIE

Tiffen, Mary (1976) *The Enterprising Peasant: Economic Development in Gombe Emirate, North Eastern State, Nigeria, 1900–1968*, London, Her Majesty's Stationery Office

Tiffen, Mary (1998) 'Demographic growth and sustainable land use' in H-P Blume et al (eds) *Towards Sustainable Land Use, Vol II: Advances in Geoecology*, Reiskirchen, Germany, Catena Verlag

Tiffen, M, M Mortimore and F Gichuki (1994) *More People, Less Erosion: Environmental Recovery in Kenya,* Chichester, UK, John Wiley

Toscano, N C, F C Sances, M W Johnson and L F LaPre (1982) 'Effects of various pesticides on lettuce physiology and yield', *Journal of Economic Entomology,* vol 75, pp738–741

Trenbath, B R (1974) 'Biomass productivity of mixtures', *Advances in Agronomy,* vol 26, pp177–210

Trenbath, B R (1976) 'Plant interactions in mixed crop communities' in R I Papendick, P A Sanchez and G B Triplett (eds) *Multiple Cropping,* Madison, American Society of Agronomy pp129–170

Triomphe, B L (1996) 'Seasonal nitrogen dynamics and long-term changes in soil properties under the mucuna/maize cropping system on the hillsides of Northern Honduras', unpublished PhD thesis, Ithaca, Cornell University

Turner, B L, and P M Haygarth (2001) 'Phosphorous solubilization in rewetted soils', *Nature,* vol 411, p258

U N Commission on Sustainable Development (1997) 'Report of the CSD on its 5th Session Addendum: Promoting sustainable agricultural and rural development', Chapter 14 of *Agenda 21,* New York, United Nations

Udry, C (1990) 'Credit markets in Northern Nigeria: credit as insurance in a rural economy', *World Bank Economic Review,* vol 4, pp251–270

Uhm, K B, J S Hyun and J M Choi (1985) 'Effects of different levels of nitrogen fertilizer and plant spacing on the population growth of the brown planthopper (*Nilapavarta lugens* Stal)', *Research Report RDA (P M & U),* vol 27, pp79–85

Unger, P W and J J Parker Jr (1968) 'Residue placement effects on decomposition, evaporation, and soil moisture distribution', *Agronomy Journal,* vol 60, p469

Uphoff, Norman (1986) *Improving International Irrigation Management with Farmer Participation: Getting the Process Right,* Boulder, Westview Press

Uphoff, Norman (1993) 'Grassroots organizations and NGO in rural development: opportunities with diminishing states and expanding markets', *World Development,* vol 21, no 4, pp607–622

Uphoff, Norman (1996) *Learning from Gal Oya: Possibilities for Participatory Development and Post-Newtonian Social Science,* London, Intermediate Technology Publications

Uphoff, Norman (1996a) 'Collaborations as an alternative to projects: Cornell experience with university-NGO-government networking', *Agriculture and Human Values,* vol 13, no 2, pp42–51

Uphoff, Norman (1999) 'Agroecological implications of the System of Rice Intensification (SRI) in Madagascar', *Environment, Development and Sustainability,* vol 1, nos 3–4, pp297–313

Uphoff, Norman (2000) 'Understanding social capital: learning from the analysis and experience of participation' in Partha Dasgupta and Ismail Serageldin (eds) *Social Capital: A Multifaceted Perspective,* Washington, World Bank, pp215–249

Uphoff, Norman and Milton J Esman (1974) *Local Organization for Rural Development: Analysis of Asian Experience,* Ithaca, NY, Rural Development Committee, Cornell University

Uphoff, Norman with Prithi Ramamurthy and Roy Steiner (1991) *Managing Irrigation: Analyzing and Improving the Performance of Bureaucracies,* New Delhi, Sage Publications

Uphoff, Norman and C M Wijayaratna (2000) 'Demonstrated benefits from social capital: the productivity of farmer organizations in Gal Oya, Sri Lanka', *World Development,* vol 28, no 11, pp1875–1890

USAID (1991) *Senegal Agricultural Sector Analysis*, Dakar, USAID

Vallois, Patrick (1997) *Malagasy Early Rice Planting System: For a Ton Increase per Acre in Developing Countries*, Antananarivo, Institut de Promotion de la Nouvelle Riziculture

Van Berkum, Peter and Charles Sloger (1983) 'Interaction of combined nitrogen with the expression of root-associated nitrogenase activity in grasses and with the development of N_2 fixation in soy bean (*Glycine max* L Mon)', *Plant Physiology*, vol 72, 7pp41–745

Van den Bosch, R, P S Messenger and A P Guttierez (1982) *An Introduction to Biological Control*, New York, Plenum Press

van Keulen, H (1982) 'Graphical illustration of annual crop response to fertilizer application', *Agricultural Systems*, vol 19, pp113–126

van Keulen, H and H D J van Heemst (1982) 'Crop response to the supply of macronutrients', Wageningen/Amsterdam, CABO/SOW

van Pelt, M F and A Kuyvenhoven (1994) 'Sustainability-oriented appraisal for agricultural projects' in N Maddock and F A Wilson (eds) *Project Design for Agricultural Development*, Aldershot, Avebury, pp143–172

Vandermeer, John (1989) *The Ecology of Intercropping*, Cambridge, UK, Cambridge University Press

Vandermeer, John (1995) 'The ecological basis of alternative agriculture', *Annual Review of Ecology and Systematics*, vol 26, pp201–224

Vandermeer, John (1997) 'Syndromes of production: an emergent property of simple agroecosystem dynamics', *Journal of Environmental Management*, vol 51, pp59–72

Vandermeer, J and B Schulz (1990) 'Variability, stability and risk in intercropping: some theoretical explorations', in S R Gliessman (ed) *Agroecology: Researching the Sustainable Basis for Sustainable Agriculture*, New York, Springer-Verlag, pp201–229

Veldhuizen, L, A Waters-Bayer, R Ramirez, D Johnson and J Thompson (1997) *Farmers' Research in Practice*, London, Intermediate Technology Publications

Vergara, O, S Druck and D S Assis (1991) 'Estudo "ex-ante" da redução de perdas de solo em decorrência do emprego de tecnologias conservacionistas para o período de 1990 a 2007' in *Programa e Resumos, Congresso Brasiliero de Ciência do Solo*, Porto Alegre, Sociedade Brasileira de Ciência do Solo / Universidade Federal do Rio Grande do Sul

Vines, Gail (2000) 'Follow that food: all around you, plants are actively seeking out the nutrients they need', *New Scientist*, 27 May, pp28–31

Wagger, M G, D E Kissel and S J Smith (1985) 'Mineralization of nitrogen from nitrogen-15 labeled crop residues under field conditions', *Soil Science Society of America Journal*, vol 49, p1220

Wardle, D A and K E Giller (1996) 'The quest for a contemporary ecological dimension to soil biology: discussion', *Soil Biology and Biochemistry*, vol 18, pp1549–1554

Way, M J and K L Heong (1994) 'The role of biodiversity in the dynamics and management of insect pests of tropical irrigated rice: a review', *Bulletin of Entomological Research*, vol 84, pp567–587

Wedum, Joanne, Yaya Doumbia, Boubacar Sanogo, Gouro Dicko and Ousssoumana Cisse (1996) 'Rehabilitating degraded land: *zai* in the Djenne Circle of Mali' in C Reij et al *Sustaining the Soil: Indigenous Soil and Water Conservation in Africa*, London, Earthscan Publications, pp62–68

Westley, K (1997) 'Women, men, and manure: assessment of gender and wealth inter-actions in a soil restoration project in Northern Senegal', *Journal of the Tropical Resources Institute*, vol 16, no 1, pp9–11

Wightman, John A (1998) 'International developments in IPM for vegetable crops', paper presented at National Symposium on Emerging Scenarios in Vegetable Research and Development, Varanasi, India, December

Wijayaratna, C M and Norman Uphoff (1997) 'Farmer organization in Gal Oya: improving irrigation management in Sri Lanka' in Krishna et al *Reasons for Hope: Instructive Experiences in Rural Development*, West Hartford, CT, Kumarian Press, pp166–183

Wijesinghe, Dushyantha and Michael Hutchings (1999) The effects of environmental heterogeneity on the performance of *Glechoma hederacia*: the interactions between patch contrast and patch scale', *Journal of Ecology*, vol 87, p860

Wilson, E O (1998) *Consilience: The Unity of Knowledge*, New York, Alfred A Knopf

Winkleman, D L (1998) *CGIAR Activities and Goals: Tracing the Connections – Issues in Agriculture*, Washington, World Bank, Consultative Group on International Agricultural Research

Winrock International (1992) *Assessment of Animal Agriculture in Sub-Saharan Africa*, Morrilton, Arkansas, Winrock International Institute for Agricultural Development

Wit, C T de (1992) 'Resource use efficiency in agriculture', *Agricultural Systems*, vol 40, pp125–151

Woolcock, Michael (1998) 'Social capital and economic development: towards a theoretical synthesis and policy framework', *Theory and Society*, vol 27, pp151–208

World Bank (1993) *Agricultural Sector Review*, Washington, World Bank, Agriculture and Natural Resources Department

World Bank (1995a) *Sustainable Agriculture: Combining Human and Ecological Needs with Economic Development*, Washington, World Bank

World Bank (1995b) *Sri Lankan Poverty Assessment*, Washington, World Bank

World Bank (1998) *Reforming Agricultural Research Organizations: Creating Autonomous Bodies and Managing Change*, Agricultural Knowledge and Information Systems (AKIS), Washington, World Bank

World Bank (2000) *Decentralizing Agricultural Extension: Lessons and Good Practice*, Agricultural Knowledge and Information Systems (AKIS), Washington, World Bank

World Bank (2000a) *Contracting for Extension: Review of Emerging Practices*, Agricultural Knowledge and Information Systems (AKIS), Washington, World Bank

World Neighbors (1993) *World Neighbors Strategic Planning Conference Report*, Oklahoma City, OK, World Neighbors

Wright, S (1932) 'The roles of mutation, inbreeding, crossbreeding, and selection in evolution', *Proceedings of the XIth International Congress of Genetics*, vol 1, pp356–366

Ying, J, S Peng, Q He, H Yang, C Yang, R M Visperas and K G Cassman (1998) 'Comparison of high-yield rice in tropical and subtropical environments, I: deter-minants of grain and dry matter yields', *Field Crops Research*, vol 57, pp71–84

Young, A (1997) *Agroforestry for Soil Management*, second edition, Wallingford, UK, CAB International

Zhu, Y, H Chen, J Fan, Y Wang, Y Li, J Chen, J Fan, S Yang, L Hu, H Leung, T W Mew, P S Teng, Z Wang and C C Mundt (2000) 'Genetic diversity and disease control in rice', *Nature*, vol 406, pp718–722

Index

Related titles available from Earthscan

FARMER INNOVATION IN AFRICA
A Source of Inspiration for Agricultural Development
Edited by *Chris Reij* and *Ann Waters-Bayer*

One of Africa's major untapped resources is the creativity of its farmers. This book presents a series of clear and detailed studies that demonstrate how small-scale farmers, both men and women, experiment and innovate in order to improve their livelihoods, despite the adverse conditions and lack of appropriate external support with which they have to contend.

The studies are based on fieldwork in a wide variety of farming systems throughout Africa, and have been written primarily by African researchers and extension specialists. Numerous lively examples show how a participatory approach to agricultural research and development that builds on local knowledge and innovation can stimulate the creativity of all involved – not only the farmers. This approach, which recognizes the farmers' capacity to innovate as the crucial component of success, provides a much-needed alternative to the conventional 'transfer of technology' paradigm.

This book is a rich source of case studies and analyses of how agricultural research and development policy can be changed. It presents evidence of the resilience and resolution of rural communities in Africa and will be an inspiration for development workers, researchers and policy-makers, as well as for students and teachers of agriculture, environment and sustainable development.

Chris Reij is a Fellow of the International Cooperation Centre of the Vrije Universiteit, Amsterdam and co-editor (with Ian Scoones and Camilla Toulmin) of *Sustaining the Soil* (Earthscan, 1996).

Ann Waters-Bayer is an agricultural sociologist with the development agency ETC Ecoculture Netherlands.

Paperback   £18.95   ISBN 1 85383 816 0

Orders to: Earthscan, Freepost 1, 120 Pentonville Road, London N1 9BR
Fax: +44 (0)20 7278 0433

www.earthscan.co.uk

DYNAMICS AND DIVERSITY
Soil Fertility and Farming Livelihoods in Africa

Edited by *Ian Scoones*

'*Highly recommended reading*' ERIC SMALING, Wageningen Agricultural University, The Netherlands

'Dynamics and Diversity *is to be recommended not just for the information and insights it provides with respect to the specific issue of soil fertility management, but also because of the major questions it provokes about the application of scientific research to the challenges of sustainable agriculture in Africa*' MIKE SWIFT, Director, Tropical Soil Biology and Fertility Programme, Nairobi

The management of Africa's soils is one of the major challenges facing agriculture and livelihoods in the 21st century. There is a common assumption that soils are being degraded on a large scale, and that farmers' practices often contribute to a 'downward spiral' of degradation and poverty – a familiar narrative of negative environmental change.

However, based on a series of detailed case studies from Ethiopia, Mali and Zimbabwe, this book explores the complex dynamics of soil fertility change from an interdisciplinary perspective, looking at the way farmers actually manage their soils and the social and environmental processes that determine their transformation. Through this analysis, *Dynamics and Diversity* suggests new, more positive ways of thinking about agricultural development policy and practice.

Ian Scoones is a Fellow at the Institute of Development Studies at the University of Sussex. He is co-author (with Chris Reij and Camilla Toulmin) of *Sustaining the Soil* (Earthscan, 1996).

Paperback £16.95   ISBN 1 85383 820 9
Hardback   £45.00   ISBN 1 85383 819 5

Orders to: Earthscan, Freepost 1, 120 Pentonville Road, London N1 9BR
Fax: +44 (0)20 7278 0433

www.earthscan.co.uk

HANDBOOK FOR THE FIELD ASSESSMENT OF LAND DEGRADATION

Michael A Stocking and *Niamh Murnaghan*

With the increasing concern over rural livelihoods and the food security of poor communities in developing countries, it is vital that land quality is maintained. Yet, land degradation is widespread and is lowering the productive capacity of the land in these countries. This practical handbook presents simple, non-technical indicators for assessing land degradation in the field. Based on the perspective of the farmer, the methods selected lend meaning to real farming situations, helping the field professional to understand not only the impact of degradation but also the benefits to be gained from reversing it.

The handbook shows how to calculate indicators such as those of soil loss, explains the interpretation of results and in particular how combinations of different indicators can give conclusive evidence of the severity of land degradation. The focus of the book is firmly on understanding the farmer's interaction with the land, and how environmental protection, food security and the well-being of rural land users may be assured.

With detailed figures, photographs, worked examples and sample forms based on assessment techniques validated by field professionals in Africa, Asia and Latin America, this essential training manual will be invaluable for fieldworkers in NGOs and in governmental and educational institutions. It will also be of interest to researchers and academics in development, environment and agriculture.

Michael Stocking is Professor of Natural Resource Development in the School of Development Studies, University of East Anglia, Norwich.

Niamh Murnaghan is a research associate with the Overseas Development Group at the University of East Anglia, Norwich.

Paperback £25.00 ISBN 1 85383 831 4

Orders to: Earthscan, Freepost 1, 120 Pentonville Road, London N1 9BR
Fax: +44 (0)20 7278 0433

---

www.earthscan.co.uk

ALTERNATIVE IRRIGATION
The Promise of Runoff Agriculture

Christopher J Barrow

'The systematic description and appraisal ... are most praiseworthy... the line drawings are exquisite... This book provides a fascinating ... insight into this most ancient of agricultural practices' Biological Agriculture and Horticulture

Alternative Irrigation is a comprehensive introduction, by one of the world's leading experts, to a neglected and ever more important form of agricultural irrigation. Runoff agriculture uses surface and subsurface water which is often otherwise overlooked and wasted. It enables small farmers as well as commercial agriculturalists to improve yields and the security of harvests, even in harsh and remote environments.

The author introduces the techniques and strategies, as well as the challenges and the potential, of this crucial approach, which can contribute so much to reducing land degradation and improving conservation and sustainability.

This book is an essential tool for teaching and research, as well as for practical application by agricultural and development organizations and workers.

Christopher J Barrow is a senior lecturer at the School of Social Sciences and International Development, University of Wales Swansea. He has been adviser and consultant to organizations such as the World Bank and is author of a number of books, including *Environmental and Social Impact Assessment: An Introduction* and *Developing the Environment: Problems and Management*.

Paperback   £15.95   ISBN 1 85383 496 3
Hardback   £40.00   ISBN 1 85383 495 5

Orders to: Earthscan, Freepost 1, 120 Pentonville Road, London N1 9BR
Fax: +44 (0)20 7278 0433

www.earthscan.co.uk